Community Biodiversity Management

T0252593

The conservation and sustainable use of biodiversity are issues that have been high on the policy agenda since the first Earth Summit in Rio in 1992. As part of efforts to implement *in situ* conservation, a methodology referred to as community biodiversity management (CBM) has been developed by those engaged in this arena. CBM contributes to the empowerment of farming communities to manage their biological resources and make informed decisions on the conservation and use of agrobiodiversity.

This book is the first to set out a clear overview of CBM as a methodology for meeting socio-environmental changes. CBM is shown to be a key strategy that promotes community resilience, and contributes to the conservation of plant genetic resources. The authors present the underlying concepts and theories of CBM as well as its methodology and practices, and introduce case studies primarily from Brazil, Ethiopia, France, India and Nepal. Contributors include farmers, leaders of farmers' organizations, professionals from conservation and development organizations, students and scientists.

The book offers inspiration to all those involved in the conservation and use of agro-biodiversity within livelihood development and presents ideas for the implementation of farmers' rights. The wide collection of experiences illustrates the efforts made by communities throughout the world to cope with change while using diversity and engaging in learning processes. It links these grassroots efforts with debates in policy arenas as a means to respond to the unpredictable changes, such as climate change, that communities face in sustaining their livelihoods.

Walter Simon de Boef is Visiting Professor at the Federal University of Santa Catarina (UFSC), Brazil, and Associate Consultant at the Centre for Development Innovation (CDI) of Wageningen University and Research Centre (Wageningen UR), the Netherlands.

Abishkar Subedi worked at Local Initiatives for Biodiversity, Research and Development (LI-BIRD), Nepal, until 2012 when he assumed the role of Agrobiodiversity and Seed Sector Development Specialist at the Centre for Development Innovation (CDI) of Wageningen UR, the Netherlands.

Nivaldo Peroni is Professor at UFSC, Brazil.

Marja Thijssen is Agrobiodiversity and Seed Sector Development Specialist at the Centre for Development Innovation (CDI) of Wageningen UR, the Netherlands.

Elizabeth O'Keeffe is a freelance copy-editor based in Brazil.

Issues in Agricultural Biodiversity
Series editors: Michael Halewood and Danny Hunter

This series of books is published by Earthscan in association with Bioversity International. The aim of the series is to review the current state of knowledge in topical issues associated with agricultural biodiversity, to identify gaps in our knowledge base, to synthesize lessons learned and to propose future research and development actions. The overall objective is to increase the sustainable use of biodiversity in improving people's well-being and food and nutrition security. The series' scope is all aspects of agricultural biodiversity, ranging from conservation biology of genetic resources through social sciences to policy and legal aspects. It also covers the fields of research, education, communication and coordination, information management and knowledge sharing.

Published titles:

Crop Wild Relatives
A Manual of *in situ* Conservation
Edited by Danny Hunter and Vernon Heywood

The Economics of Managing Crop Diversity On-Farm
Case Studies from the Genetic Resources Policy Initiative
Edited by Edilegnaw Wale, Adam Drucker and Kerstin Zander

Plant Genetic Resources and Food Security
Stakeholder Perspectives on the International Treaty on Plant
Genetic Resources for Food and Agriculture
Edited by Christine Frison, Francisco López and José T. Esquinas

Crop Genetic Resources as a Global Commons
Challenges in International Law and Governance
Edited by Michael Halewood, Isabel López Noriega and Selim Louafi

Community Biodiversity Management
Promoting Resilience and the Conservation of Plant Genetic Resources
*Edited by Walter Simon de Boef, Abishkar Subedi, Nivaldo Peroni,
Marja Thijssen and Elizabeth O'Keeffe*

Forthcoming titles:

Farmers' Crop Varieties and Farmers' Rights
Challenges in Taxonomy and Law
Edited by Michael Halewood

Diversifying Food and Diets
Using Agricultural Biodiversity to Improve Nutrition and Health
Edited by Jessica Fanzo, Danny Hunter, Teresa Borelli and Federico Mattei

Community Biodiversity Management

Promoting resilience and the
conservation of plant genetic resources

Edited by
**Walter Simon de Boef,
Abishkar Subedi, Nivaldo Peroni,
Marja Thijssen and
Elizabeth O'Keeffe**

Routledge
Taylor & Francis Group

LONDON AND NEW YORK

earthscan
from Routledge

First published 2013
by Routledge
2 Park Square, Milton Park, Abingdon, Oxon, OX14 4RN

Simultaneously published in the USA and Canada
by Routledge
711 Third Avenue, New York, NY 10017

Routledge is an imprint of the Taylor & Francis Group, an informa business

British Library Cataloguing in Publication Data
A catalogue record for this book is available from the British Library

Library of Congress Cataloging-in-Publication Data
Community biodiversity management :
promoting resilience and the conservation of plant genetic resources /
edited by Walter Simon de Boef, Nivaldo Peroni, Abishkar Subedi,
Marja Helen Thijssen, and Elizabeth O'Keeffe–First edition.
pages cm.–(Issues in agricultural biodiversity)
Includes bibliographical references and index.
1. Biodiversity conservation–Citizen participation.
2. Sustainable agriculture Citizen participation. 3. Plant genetics.
I. Boef, Walter de.
QH75.C584 2013
334'683–dc23
2012037895

ISBN13: 978–0–415–50219–1 (hbk)
ISBN13: 978–0–415–50220–7 (pbk)
ISBN13: 978–0–203–13059–9 (ebk)

Typeset in Baskerville
by Swales & Willis Ltd, Exeter, Devon

Printed and bound by CPI Group (UK) Ltd, Croydon, CR0 4YY

Contents

Contributors

Fetien Abay Abera is Associate Professor at Mekelle University, Ethiopia, and has been named leading female African scientist. In her research, she addresses farmer innovation and participatory plant breeding. In addition to many other activities, Fetien coordinates the Integrated Seed Sector Development (ISSD) Programme in Tigray region and the 'CBM and Resilience' project in Ethiopia.

Silvio Aguirre Acuña is Sorghum Breeder at the Cooperative Union of Madriz, Nicaragua. He is responsible for the management and conservation of local varieties in seed banks. As breeder he is engaged in generating lines and varieties for distribution in the different agro-ecological conditions of his country.

Ulysses Paulino de Albuquerque is Professor at the Federal Rural University of Pernambuco, Brazil. Ulysses has experience in the management and conservation of local plant resources, with a particular focus on medicinal plants. Ulysses is partner in the 'CBM and Resilience' project.

Sergio Romeo Alonzo Recinos works with the Association of Organizations from the Cuchumatanes, Guatemala, and is a regional coordinator of the Collaborative Programme on Participatory Plant Breeding (PPBMA) in Central America, which is a partner in the 'CBM and Resilience' project. Sergio has been involved in CBM activities in the region for the past ten years, with the aim of improving the livelihoods of communities and enhancing their capacities to adapt to climate change.

Regine Andersen is Senior Research Fellow of the Farmers' Rights Project at the Fridtjof Nansen Institute, Norway; she conducts research in support of implementing farmers' rights under the International Treaty on Plant Genetic Resources for Food and Agriculture and in that context she works with organizations engaged in CBM.

Dao The Anh is Director of the Centre for Agrarian Systems Research and Development, Vietnam. He has been extensively involved in action research in the conservation of local vegetable varieties in Vietnam and currently focuses on intellectual property protection (e.g. geographical indication).

Ana Luiza de Assis is a doctorate student at the Federal University of Santa Catarina (UFSC), Brazil. Ana Luiza focuses her research on traditional knowledge related to cultivated and native forest diversity in Brazil. She conducted her research for completing her MSc in the 'Global Study on CBM and Empowerment.'

Sérgio Guilherme de Azevedo works at the Brazilian Agricultural Research Corporation (Embrapa) Semi-Arid in Brazil. As Rural Development Expert, Sérgio coordinates small-scale farming projects as part of the government's Zero Hunger Programme, promoting the cultivation of local varieties under semi-arid conditions. Sergio participated in the 'Global Study on CBM and Empowerment.'

S.P. Bala Ravi was, until his recent retirement, Project Director at M.S. Swaminathan Research Foundation (MSSRF), India. He has over 40 years of experience in biodiversity management and utilization. He focuses on biodiversity-related legal regimes and promotes the implementation of CBM activities among farming communities for improving their livelihoods.

Rosa Lía Barbieri is Researcher at Embrapa Temperate Agriculture in Brazil, where she is curator of the Cucurbitaceae gene bank. Rosa Lía collects its local varieties; she has organized and participated in various CBM activities. She participated in the 'Global Study on CBM and Empowerment.'

Bharat Bhandari is Senior Agricultural Specialist with the USC Canada, Asia office, in Nepal. Before, he was Programme Coordinator for Local Initiatives for Biodiversity, Research and Development (LI-BIRD), where he coordinated LI-BIRD's CBM programme in Nepal.

Walter Simon de Boef is Visiting Professor at the Federal University of Santa Catarina (UFSC), Brazil, and Associate Consultant at Centre for Development Innovation (CDI) of Wageningen University and Research Centre (Wageningen UR), the Netherlands. Walter works as a scientist and facilitator in local, national, regional and global CBM programmes. He coordinated the 'Global Study on CBM and Empowerment,' and is advisor to the 'CBM and Resilience' project. He further provides advisory services to several seed programmes in Africa.

Marlene Borges is community leader of a group of small-scale farmers in Imbituba, and is President of the Rural Community Association of Imbituba (ACORDI), Brazil. Marlene dedicates her time to the struggle to ensure farmers' access to Areais da Ribanceira, and as such she is engaged in several CBM activities. ACORDI was partner in the 'Global Study on CBM and Empowerment,' and hosts one of the sites to the 'CBM and Resilience' project.

Adriano Canci works with the Association of the Micro-watershed, Guaraciaba, Brazil. Since 2004, Adriano has been coordinating the implementation of a wide diversity of CBM practices including the diversity kit, diversity blocks and participatory crop improvement. The Association was a partner in the 'Global Study on CBM and Empowerment,' and hosts one of the sites of the 'CBM and Resilience' project.

Hugo Fabian Carrera Rueda is Project Coordinator for the Union of Peasant and Indigenous Organizations of Cotacachi, Ecuador. Hugo has 15 years of experience working in the field of rural development, strengthening farmers' organizations, promoting micro-agribusiness and agro-tourism within a larger context of agrobiodiversity management and food security.

Sushanta Sekhar Chaudhury is Senior Scientist and Project Coordinator at MSSRF, India. Sushanta facilitates CBM activities including participatory crop improvement and on-farm management of local rice varieties, documentation of traditional knowledge, registration of farmers' varieties, and strengthening of community-based organizations.

Tadesse Desalegn is Associate Professor of Plant Breeding at Bahir Dar University, Ethiopia. Tadesse is engaged in participatory crop improvement, and coordinates the ISSD Programme in the Amhara region of Ethiopia.

Temesgen Desalegn is Researcher at the Ethiopian Institute of Agricultural Research. Temesgen collaborates with Ethio-Organic Seed Action (EOSA) on the management and use of agrobiodiversity. Temesgen participated in the 'Global Study on CBM and Empowerment.'

Bijaya Raj Devkota is Project Officer at LI-BIRD, Nepal. Bijaya facilitates CBM practices, such as seed banks, participatory varietal selection and grassroots breeding.

Rachana Devkota is Senior Programme Officer at LI-BIRD, Nepal. Rachana has contributed to out-scaling CBM in Bangladesh, India, Nepal and Sri Lanka. Her research work on 'empowerment' has contributed to strengthening the CBM methodology. Rachana participated in the 'Global Study on CBM and Empowerment.'

Emanoel Dias da Silva is associated with Advice and Services for Alternative Agriculture Projects (AS-PTA), Brazil. He contributes to the restoration of agrobiodiversity, agroforestry, access to markets, health and nutrition. Emanoel facilitates and supports the establishment of community networks aimed at sustainable agricultural development.

Terezinha Aparecida Borges Dias is Researcher at Embrapa Genetic Resources and Biotechnology, Brazil. Terezinha promotes interaction between *ex situ* and *in situ* genetic resource conservation, and capacity-building in participatory approaches to CBM.

Joy Angelica P. Santos Doctor is Information and Documentation Officer with South-East Asia Regional Initiatives for Community Empowerment, the Philippines. Joya conducts analysis and documentation of CBM issues and participates in policy and advocacy activities related to CBM.

Joel Donazzolo is Professor at the Federal Technological University of Paraná, Brazil. Joel has been involved in the development of a number of participatory research projects and has recently obtained his doctorate degree with UFSC working with the participatory improvement of feijoa.

Maria Aldete J. da Fonseca Ferreira is Researcher at Embrapa Semi-Arid, Brazil. She collaborates with small-scale farming communities, collecting local Cucurbitaceae varieties and enhancing the capacities of those communities to engage in

CBM through participatory approaches. Maria participated in the 'Global Study on CBM and Empowerment.'

Regassa Feyissa is Director of EOSA, Ethiopia. He is a pioneer of *in situ* conservation and provides overall guidance to CBM activities in Ethiopia. He developed a number of conceptual frameworks and methodologies for practicing CBM.

Mario Roberto Fuentes López works at the Foundation for Agricultural and Forestry Technology Innovation (FUNDIT), Guatemala, and is Regional Coordinator of the PPBMA, partner in the 'CBM and Resilience' project. He focuses his work on the conservation and utilization of maize diversity.

Genene Gezu is Programme Manager at EOSA, Ethiopia. Genene coordinates the organization's agrobiodiversity programmes, and facilitates CBM practices at local, community and national levels. Genene participated in the 'Global Study on CBM and Empowerment.'

Elodie Gras works at AgroBio Périgord, France. Elodie works with the programme 'The Aquitaine Cultivates Biodiversity,' coordinates the seed bank, and facilitates several CBM practices and associated research work.

Clístenes Antônio Guadagnin currently conducts his doctorate research at the Federal University of Pelotas, and is affiliated as extensionist with the Santa Catarina State Enterprise for Agricultural Research and Rural Extension, Brazil. Clístenes works with small-scale farming communities in implementing CBM practices in Guaraciaba.

Raymond P. Guries is Professor at University of Wisconsin-Madison, USA. He has been working with local communities in Costa Rica for ten years, to improve the management of cacao for maintaining diverse shade-grown agro-ecosystems for wildlife and biodiversity conservation.

Sanjaya Gyawali is Visiting Research Fellow at Agriculture and Agri-Food, Canada. Sanjaya previously led the Jethobudho genetic enhancement programme facilitated by LI-BIRD, Nepal.

Natalia Hanazaki is Professor at UFSC, Brazil. Natalia has been focusing on ethnobotany with a local perspective for development, in which local ecological knowledge plays an important part. Natalia participated as researcher in the 'Global Study on CBM and Empowerment,' and is collaborator in the 'CBM and Resilience' project.

Narasimha Hegde is Director of Life Trust, India. He works at grassroots level in documenting and organizing CBM activities in the Sirsi region, focusing on tropical fruit tree genetic resources.

Jair Pedro Henke is Secretary for Agriculture at the municipal government of Guaraciaba, Brazil. Since 2006, Jair has been participating in a programme aimed at managing lands for grazing cattle and supporting several CBM practices.

Rolando Herrera Torres works with the Centre for Research and the Promotion of Rural and Social Development, Nicaragua. Rolando is Regional Coordinator of the PPBMA. He facilitates CBM activities aimed at increasing food security and adapting to climate change, e.g. participatory plant breeding of local varieties of maize and beans, and seed collection.

Danny Hunter is Senior Scientist at Bioversity International, Italy. Danny was team leader of the Taro Genetic Resources Conservation and Utilisation Network, and was involved in the early formation and implementation of the Taro Improvement Programme in the Pacific.

Tolo Iosefa is Assistant Lecturer at the University of the South Pacific, Samoa. He manages the Taro Improvement Programme.

Guy Kastler is farmer and Director of the Farmers' Seed Network (Réseau Semences Paysannes), France. Guy provides support and advice to farmers on legal aspects of establishing seed banks in France. In addition, he has written a number of papers on CBM.

Jennifer Kendall at AgroBio Périgord, France, works with the 'Aquitaine Cultivates Biodiversity' programme, coordinates the seed bank and facilitates several CBM practices and associated research work.

Kamal Khadka is Programme Coordinator at LI-BIRD, Nepal. Kamal leads the 'Agricultural Innovation for Livelihood Security' programme and has experience in implementing CBM practices aimed at improving and conserving local genetic resources.

E.D. Israel Oliver King is Principal Scientist at MSSRF, India. He targets the enhancement of Neglected and Underutilized Species (NUS) through the adoption of CBM practices and approaches. Oliver participated in the 'Global Study on CBM and Empowerment,' and currently coordinates the 'CBM and Resilience' project in South Asia.

Volmir Kist is Post-Doctoral Research Fellow at the University of Viçosa, Brazil. Previous to this, Volmir conducted his doctorate research with participatory maize improvement at UFSC.

Bertrand Lassaigne is farmer and initiator of the 'Aquitaine Cultivates Biodiversity' programme. Bertrand contributes to the Maison de la Semence Paysanne seed bank and contributes to the diversity platform. He has been producing farm-based seed on his farm for many years and is currently engaged in several farmer breeding activities.

Luciane Lazzari works at the Association of the Micro-watershed of Guaraciaba, Brazil. Since 2008, she has been coordinating a participatory research programme on varieties of beans and popcorn.

Bianca Lindner is MSc student at UFSC, Brazil. Bianca focuses in her MSc project on the use and management of a palm species.

Ernani Machado de Freitas Lins Neto is Professor at the Federal University of Piauí, Brazil. Ernani focuses his research on the domestication of fruit trees, in order to better understand the biological and cultural aspects related to their selection and management.

Niels Louwaars is Director of Plantum, the association for the plant reproduction material sector in the Netherlands. Before this, Niels worked as a bio-policy specialist at the Centre for Genetic Resources, the Netherlands (CGN) and the Law and Governance Group of Wageningen University. He has worked on seed systems analysis models, and policy and regulatory issues involved in supporting integrated approaches to seed system development and plant genetic resource management.

Adelar Mantovani is Professor at the Santa Catarina State University, Brazil. Adelar focuses his research on forest ecology, in particular on demography, phenology and the genetic diversity of natural populations of forest species.

Andrea Gabriela Mattos is a doctorate student at UFSC, Brazil. Andrea works with small-scale farming communities documenting associated traditional knowledge on forest resources, and focusing on the management of yerba mate in forest fragments.

Hussein Mohammed is Associate Professor at Hawassa University, Ethiopia. Hussein has been involved in participatory crop improvement, particularly in maize. In addition to other programmes, he coordinates the ISSD Programme in Southern Nations, Nationalities, and Peoples' Regional State.

Anne-Charlotte Moÿ is legal advisor with the Farmers' Seed Network (Réseau Semences Paysannes), France. Anne-Charlotte provides legal support and advice to farmers establishing seed banks in France, sharing and using local varieties in a dynamic manner.

Elaine Mitie Nakamura is an MSc student at UFSC, Brazil. Elaine focuses her research on the use of natural resources and food security.

Rubens Onofre Nodari is Professor at UFSC, Brazil. He has been collaborating with small-scale farming communities on agrobiodiversity management and fruit breeding, facilitating the implementation of CBM practices such as participatory varietal selection.

Juliana Bernardi Ogliari is Professor at UFSC, Brazil. Juliana has coordinated a number of research projects in participatory crop improvement and agrobiodiversity management. She currently focuses her research on assessing the implications of cultivating genetically modified maize on the diversity of local varieties.

Elizabeth O'Keeffe is currently a freelance copy-editor based in Brazil. Before this, she worked as Programme Assistant at Bioversity International in Italy.

Stefano Padulosi is Senior Scientist at Bioversity International, Italy. Stefano focuses on the enhancement of neglected and underutilized species by adopting CBM practices and approaches.

Nivaldo Peroni is Professor at UFSC, Brazil. Nivaldo focuses his research on CBM, biodiversity conservation, landscape domestication, and on the processes that lead to an increase or decrease in the levels of biodiversity. Nivaldo participated as researcher in the 'Global Study on CBM and Empowerment,' and is collaborator in the 'CBM and Resilience' project.

Indra Prasad Poudel is Technical Officer at LI-BIRD, Nepal. Indra has been involved in the Nepal component of the global agrobiodiversity project from its inception, and has contributed to the development of the CBM methodology in Nepal.

Vasudeva Ramesh is Associate Professor at the University of Agricultural Sciences, in Dharwad, India. Vasudeva leads a group involved in promoting CBM activities in the conservation and use of tropical fruit genetic resources in the central Western Ghats. Vasudeva participated as researcher in the 'Global Study on CBM and Empowerment.'

Manuel Ruiz Muller is Programme Director of the Peruvian Society for Environmental Law, Peru. Manuel specializes in environmental policy and law development and works with national and indigenous peoples organizations on *in situ* conservation activities.

Juliana Santilli is an environmental lawyer and public prosecutor in Brazil. She specializes in environmental and cultural heritage law and public policies. Juliana is a co-founder of the Socio-Environmental Institute, and as researcher she addresses the interface between law, agrobiodiversity, CBM and seed systems.

Karine dos Santos is Professor at UFSC, Brazil. Karine works with local communities to improve the management of feijoa, to maintain its genetic diversity and promote participatory research.

Maurício Sedrez dos Reis is Professor at UFSC, Brazil. Maurício works with traditional communities and small-scale farmers, and focuses his research on bratinga management, and on the management of medicinal plants, palm hearts, pinhão, yerba mate and leatherleaf fern.

Pitambar Shrestha is Project Officer at LI-BIRD, Nepal. Pitambar has been contributing through his work to the development of the CBM methodology in Nepal, and has been coaching many organizations in Nepal and elsewhere in the methodology. Pitambar participated as researcher in the 'Global Study on CBM and Empowerment.'

Pratap Shrestha is Regional Representative and Scientific Advisor to the USC Canada in its Asia office in Nepal. Pratap has over 14 years of experience in CBM-related programmes, with both USC Canada and LI-BIRD. He guided the design and implementation of a number of CBM projects, including the CBM South Asia project. Pratap represents USC Canada as associate partner in the 'CBM and Resilience' project.

Sachesh Silwal is Project Officer at LI-BIRD, Nepal. Sachesh has been working with CBM since 2007, in designing, implementing, and monitoring CBM practices, and in enhancing the capacities of farmers and stakeholders, in particular addressing participatory crop improvement and community-based seed production.

Monika Sopov is Senior Advisor at CDI of Wageningen UR, the Netherlands. Her areas of focus are business development, value chain development, cooperative development, food safety and agrobiodiversity.

Juan Carlos Rosas Sotomayor is Professor at the Panamerican Agricultural School, Zamorano, Honduras. He has over 15 years of experience in working with participatory crop improvement, PGR conservation strategies, seed bank management, local seed production and farmer training.

Bhuwon Sthapit is Regional Programme Coordinator at Bioversity International, India. Bhuwon specializes in *in situ* conservation and has been playing a leading role in the development of the CBM methodology in Nepal since 1998. As its coordinator, he has promoted the out-scaling of CBM in a project on tropical fruit tree diversity in Asia. Bhuwon participated as researcher in the 'Global Study on CBM and Empowerment' and represents his organization as associate partner in the 'CBM and Resilience' project.

Sajal Sthapit is Programme Coordinator at LI-BIRD, Nepal. Sajal has been involved in the CBM South Asia project since 2010. His research interest in landscapes and climate change has contributed to the strengthening of the CBM methodology.

Abishkar Subedi is Agrobiodiversity and Seed Sector Development Specialist at the Centre for Development Innovation (CDI) of Wageningen UR, the Netherlands, where he addresses policy and legal aspects of agrobiodiversity management and is engaged in seed sector development in several African countries. Before joining CDI, Abishkar was involved in agrobiodiversity-related projects with LI-BIRD from 2002–2011. He led the LI-BIRD project team in developing the CBM methodology in Nepal and southern Asia. Abishkar participated as researcher in the 'Global Study on CBM and Empowerment.' Currently, he contributes to global CBM out-scaling through the 'CBM and Resilience' project.

Shreeram Subedi is Senior Technical Assistant at LI-BIRD, Nepal. Shreeram has been involved in implementing the Nepal component of the global project on agrobiodiversity, contributing to the development of the CBM methodology.

Saujanendra Swain is Senior Scientist and Coordinator at MSSRF, India. Saujanendra has over 15 years of experience working with agrobiodiversity conservation in Odisha, focusing on the implementation of CBM activities like the community seed banks, value addition, awareness-raising on farmers' rights, and awareness-raising on agrobiodiversity among school children. Saujanendra also contributes to a yearly training programme organized by Wageningen UR/CDI and MSSRF. He participated as researcher in the 'Global Study on CBM and Empowerment,' and participates in the 'CBM and Resilience' project.

César Guillermo Tapia Bastidas is Head of the National Plant Genetic Resources Department of the National Autonomous Institute for Agricultural Research, Ecuador. César has been working with farming communities for over 20 years in implementing CBM activities like seed exchange fairs, community biodiversity registers, community seed banks, agro-tourism and value addition.

Mary Taylor is Plant Genetic Resources Coordinator and Manager of the Centre for Pacific Crops and Trees, Samoa. In association with the Taro Improvement Programme, Mary works with farmers linking them to new genetic diversity to strengthen community adaptation to climate change.

Flavia França Teixeira is Curator of the Embrapa maize gene bank, based at Embrapa Maize and Sorghum, Brazil. She promotes the implementation of a number of CBM practices, including the re-introduction of traditional maize varieties for indigenous peoples and the promotion of value addition activities. Flavia participates in a project aimed at enhancing the resilience of maize to drought.

Kassahun Tesfaye is Assistant Professor at Addis Ababa University, and Senior Researcher in the Environment and Coffee Forest Forum, Ethiopia. Kassahun conducts diversity studies of Ethiopian crops and has contributed to the establishment of the first Biosphere Reserve for Coffee Genetic Resources.

Marja Thijssen is Agrobiodiversity and Seed Sector Development Specialist at the Centre for Development Innovation (CDI) of Wageningen UR, the Netherlands. She coordinated the 'Global Study on CBM and Empowerment' and has been facilitating a number of training programmes concerned with plant genetic resource management with an emphasis on the CBM methodology. Currently, Marja is coordinating seed sector development programmes in Ethiopia, Uganda and at continental level in Africa.

Frédéric Thomas is Researcher at the Institute of Research for Development (IRD), France. Frédéric conducts participatory research with the Réseau Semences Paysannes. He authored and directed a film on the farmers' use of collective trademarks to protect seed production in Vietnam.

Bayush Tsegaye is Programme Advisor for EOSA, Ethiopia. Bayush provides technical advice on CBM-related activities and has been involved in developing and implementing CBM strategies and related training modalities.

Valerie Saena Tuia is Curator of the Centre for Pacific Crops and Trees. Valerie previously worked as Senior Agriculture Officer and Principal Research Officer in Samoa, where she collaborated with farmers on crop diversity management and improvement.

Madhusudan Upadhyay was, until his recent retirement, a Senior Scientist with the Nepal Agricultural Research Council. He led the way towards establishing the national gene bank in Nepal. Madhusudan had a long-standing relationship with LI-BIRD as a key government partner in the implementation of several agrobiodiversity projects.

Karèn Simone Verhoosel is a specialist in multi-stakeholder processes and institutional change at CDI of Wageningen UR, the Netherlands. Karèn has been involved in CBM programmes focusing particularly on participation, power and empowerment. Karèn participated as researcher in the 'Global Study on CBM and Empowerment.' Currently, she is responsible for coordinating the seed sector development programme in Uganda.

Ronnie Vernooy is Genetic Resources Policy Specialist at Bioversity International, Rome. Ronnie has 20 years of research experience in addressing socio-economic, gender and policy aspects of *in situ* conservation, crop improvement, and the sustainable use of agrobiodiversity, from local to national and international levels.

Camila Vieira da Silva is a doctorate student at the Federal University of Rio Grande do Sul, Brazil. Camila focuses her research on bracatinga management in the state of Santa Catarina, and on pinhão and açai collection in particular.

Sofia Zank is a doctorate student at UFSC, Brazil. Sofia focuses her research on ethnobotany addressing the interface between biodiversity management, ecosystem heath and the use of medicinal plants. Sofia conducted her MSc research in the 'Global Study on CBM and Empowerment,' and currently contributes to the 'CBM and Resilience' project.

Alex Anderson Zechini is a doctorate student at UFSC, Brazil. Alex focuses his research on non-timber forest products, on the production and supply of pine nuts in particular, in order to support sustainable harvesting, preserve forest biodiversity and ensure income for family farmers.

Foreword

The area of biodiversity management, which has not received the attention it deserves, involves local communities. *Ex situ* preservation and *in situ* conservation by government departments, research and academic institutions, and scientists have been receiving support and recognition. However, the *in situ* on-farm conservation carried out by rural and tribal families has not been given the recognition it deserves. The M.S. Swaminathan Research Foundation (MSSRF) therefore took up community conservation as its primary area of research and recognition.

The MSSRF's research on biodiversity began in 1989 with a study of genetic diversity in mangrove species and seagrasses. The greatest genetic diversity in mangroves in India was found to occur at Bhitarkanika in Odisha state. From 1990 onwards, the biodiversity programme centred on community biodiversity conservation and enhancement. The study areas were the Kolli Hills in Tamil Nadu, Koraput in Odisha and Wayanad in Kerala. The Kolli Hills and Koraput are in the Eastern Ghats, whereas Wayanad is in the Western Ghats range. In all the areas, the emphasis was on the revitalization of the *in situ* on-farm conservation traditions by tribal and rural families. So far, *in situ* and *ex situ* conservation undertaken by forest departments, universities and research institutions have received attention from biodiversity experts. Unfortunately, community conservation and selection efforts in rural and tribal areas have not received the same attention. As a result, genetic erosion has been severe in such areas. For example, in the Koraput region there were nearly 3500 local varieties of rice 50 years ago. Now this number has declined to fewer than 300. It is estimated that there may be about 150 000 varieties of rice in the world. Such rich intra-specific variability is largely due to human selection, based on genotype × environmental interaction.

It is only in recent years that attention on community conservation, particularly with reference to agrobiodiversity, has started to gain momentum. Some of the steps needed to revitalize community conservation are the following:

- There is need to create an economic stake in conservation through a 4C model of sustainable management of agro-biodiversity, involving concurrent attention on conservation, cultivation, consumption and commerce.
- The community tradition of creating gene, seed, grain and water banks should be revitalized and promoted. This will help to develop sustainable food and water security systems at the local level.

• The concept of farmers' rights promoted by me in the Food and Agriculture Organization of the United Nations (FAO) Council and General Assembly when I was Independent Chairman of the FAO Council during 1981–85 should be converted into legally binding rights.

The speed with which local genetic resources have disappeared through habitat destruction, changes in land use, invasive alien species and changes in culinary habits and climate have made it clear that unless community efforts in *in situ* on-farm conservation and *ex situ* preservation are recognized and rewarded, precious genetic resources will be lost. This led to the concept of farmers' rights in the fora of FAO in the 1980s. MSSRF's early work in Kolli Hills in 1991–95 showed that the rich reservoir of genetic material in nutri-millets, such as the species belonging to *Panicum*, *Setaria*, *Paspalum*, *Pennisetum*, *Eleucine*, etc., will vanish in the near future as a result of competition with cassava and pineapple. MSSRF therefore organized a dialogue on Farmers' Rights and Plant Genetic Resources: Recognition and Reward in January 1994. At this dialogue, the participants approved the first draft of a model Act for Plant Variety Recognition and Rights which I had prepared and presented. It is clear, however, that a mere expression of support for farmers' rights will not yield the desired results. Therefore, a dialogue on Biodiversity and Farmers' Rights was organized at MSSRF in January 1996. At this dialogue, I presented a draft of a Plant Variety Protection and Farmers' Rights Act and this received approval. The suggestion that the rights of breeders and farmers should be mutually reinforcing, and should not be set one against the other, was the basis for an Act integrating farmers' and breeders' rights. This draft provided the basic text for the Government of India's legislation called 'Plant Variety Protection and Farmers' Rights Act'. The International Union for the Protection of New Varieties of Plants (UPOV) convention dealing with plant breeders' rights unfortunately does not accord recognition to the contributions of those who have preserved genetic resources for public good at personal cost. This is why I proposed in my Sir John Crawford Lecture delivered in Washington DC on 1 November 1990, the need to restructure UPOV as the International Union for Breeders' and Farmers' Rights. Breeders and farmers are allies in the struggle for sustainable food security and the eradication of hunger. The Indian legislation is the only one of its kind in the world which deals simultaneously with breeders' and farmers' rights.

Thus, the present book, edited with great care by Walter Simon de Boef, Abishkar Subedi, Nivaldo Peroni, Marja Thijssen and Elizabeth O'Keeffe, is a timely contribution. I hope it will be widely read and used by both professionals and policy-makers. We owe a deep debt of gratitude to the editors and authors for their labour of love for the cause of community-managed conservation programmes.

Professor M.S. Swaminathan
July 2012

Preface and acknowledgements

At the time of writing, world leaders are meeting in Brazil for the Rio+20 Summit. It is almost impossible for me not to return to the impressions I had during the first Earth Summit, 20 years ago. It was at the moment that I, as a graduate from Wageningen University, had joined the Centre for Genetic Resources in the Netherlands, to support its director, Jaap Hardon, in addressing aspects of farmers' management within plant genetic resource studies. At that time, Jaap Hardon was participating in the Keystone Dialogues, a series of off-the-record meetings on plant genetic resources that were held between 1988 and 1991, in Keystone, Chennai and Oslo. Participants from public, private and civil society joined in the dialogues, which finally led to consensus on the urgent need to save the world's resources for food and agriculture. In light of the Keystone dialogues and the Convention on Biological Diversity (CBD), we joined together with a wide platform of interested parties from public, private and civil society to discuss a global agenda that included *in situ* conservation, farmers' rights, benefit-sharing, participatory plant breeding and other elements on the global plant genetic resources agenda.

Looking back, I realize that CBD – with its three objectives of conservation, sustainable use and benefit-sharing – made a tremendous impact on my professional life, and on the lives of many others working in the field of plant genetic resources. The CBD created a momentum for us to explore the various ways of achieving *in situ* conservation of plant genetic resources, or what we today call 'on-farm management' or 'community biodiversity management'. We have always emphasized the key role that farmers have played, and continue to play, in the development of crops that feed the world; however, with agricultural development in the hands of scientists, government and the corporate world, against a backdrop of the green revolution, recognition of this role has dwindled. The CBD, and later the Global Plan of Action for the Conservation and Sustainable Utilization of Plant Genetic Resources for Food and Agriculture, inspired us to engage in the Community Biodiversity Development and Conservation (CBCD) Programme. At the same time, the International Plant Genetic Resources Institute (IPGRI, now Bioversity International) took on a leading role with a project aimed at strengthening the scientific basis of *in situ* conservation on-farm of plant genetic resources. What we, and several others, did over the course of those years was try to put into practice many of the objectives outlined in the CBD.

Later, in the first decade of the twenty-first century, two non-governmental organizations (NGOs) in Asia – Local Initiatives on Biodiversity, Research and

Development (LI-BIRD, Nepal) and the M.S. Swaminathan Research Foundation (MSSRF, India) – began to be recognized in their work on implementing *in situ* conservation of plant genetic resources using a common methodology that we now refer to as 'community biodiversity management'. I have been associated with Centre for Development Innovation (CDI) (Wageningen UR), of Wageningen University and Research Centre, since 1998; and with the Federal University of Santa Catarina (UFSC, Brazil) Post Graduate Programme on Plant Genetic Resources, since 2003. At both the CDI and UFSC, we used these Asian experiences as references. Together with Bhuwon Sthapit of Bioversity International; Marja Thijssen of CDI; Resham Gautam, Bimal Regmi, Abishkar Subedi and Pitambar Shrestha of LI-BIRD; Saujanendra Swain and Oliver King of MSSRF; Genene Gezu of Ethio-Organic Seed Action (EOSA, Ethiopia); Aldete Fonseca of the Brazilian Agricultural Research Corporation (Embrapa); Adriano Canci of the Association of Micro-Watersheds in Guaraciaba; and Nivaldo Peroni and Natalia Hanazaki of UFSC, we were able to build upon the experiences of LI-BIRD and MSSRF, and begin to translate CBM into diverse contexts around the world. At CDI, we incorporated CBM in the annual training programmes that are organized in Wageningen. We also included CBM as a major topic for addressing *in situ* conservation in the yearly training programme that we organize with our colleagues from MSSRF in Jeypore and Chennai, India. Bhuwon Sthapit of Bioversity, and staff from MSSRF, LI-BIRD, UFSC and Embrapa became vital resource persons in our training programmes in different corners of the world, thereby contributing to those programmes, and becoming platforms for innovation and learning with the CBM methodology.

It was only when Marja Thijssen and I found an opportunity to develop a proposal in response to a call addressing agrobiodiversity within the partnership programme 'Globalization and Sustainable Rural Development', of the Directorate General for International Cooperation of the Netherlands and Wageningen University Research Centre, that we pulled our global CBM network together into one programme. This was the moment when CDI, LI-BIRD, MSSRF, UFSC, Embrapa, EOSA and Bioversity joined hands to study the degree to which the CBM work being carried out in various corners of the world contributes to the empowerment of farming communities in the conservation and use of agrobiodiversity. While implementing our study, we gradually developed an informal partnership with the Réseau Semence Paysanne (RSP) in France, because of the commonalities in our work with CBM with small-scale farmers and agrobiodiversity in the South and the North. We focused our study on empowerment, as we considered this a precondition for CBM as an effective methodology for implementing the *in situ* conservation of agrobiodiversity, and in this way we were able to review our work and methodology within the context of the CBD's objectives of conservation, sustainable use and benefit-sharing.

Work in the study coincided with the second call for proposals of the Benefit-Sharing Fund of the International Treaty on Plant Genetic Resources for Food and Agriculture (ITPGRFA) in 2010. Most members of our group joined forces, while other key players such as the Collaborative Programme on Participatory Plant Breeding in Central America (PPBMA), Community Technology Development Trust (CTDT, Zimbabwe), as well as various others, became partners. We submitted a proposal on approaching CBM as a methodology for contributing to on-farm management and enhancing community resilience in 12 countries, through what we call nodes of community-based organizations, non-governmental organizations and universities. Our proposal was approved, and partners began implementing the project in 2012, in a consortium that is being coordinated by LI-BIRD in Nepal and supported through Bioversity and CDI.

It was in this context in which the idea to compile a book on community biodiversity management evolved. This historic background creates a perspective of learning and reflection with regards to the achievements we have made over the last 20 years of work, following the first Rio Earth Summit. We are not in a position to develop a State of the World report, but are in the fortunate position of operating in five important countries in the world of plant genetic resources – Brazil, Ethiopia, France, India and Nepal. We compiled this book with experiences from those countries, and complement these with many others. The broad network of organizations involved is reflected in the diversity of experiences. We share experiences and lessons learned, and relate these to several practices and policy dialogues. We explore the role of plant genetic resource professionals and programmes in implementing the CBM methodology, and thereby contributing to the fulfilment of the objectives of the CBD and the ITPGRFA.

As such, this book is part of a learning process, which seeks ways to document and share what we consider to be achievements in reaching practical outputs in our domain of plant genetic resources, while matching these to more contemporary topics such as food security, local livelihood development and community resilience. Many achievements have been made across the world in the implementation of what was discussed and agreed upon in Rio twenty years ago, and we hope that by sharing those experiences, and transforming them into a coherent and logical methodology for what we refer to as community biodiversity management, we can contribute to mainstreaming this work in practices, programmes and policies. Another challenge is to create a space for CBM in professionalism and in the institutional frameworks that guide us in the conservation and use of our plant genetic resources. If we continue to engage in such processes, we can maintain our commitment to what was agreed upon, and seek ways to capitalize upon those lessons learned to address the challenges that lie ahead of us today (i.e. the current agenda of Rio+20).

This book is the result of the collective efforts of farmers, community leaders, field workers, students, researchers, professors and plant genetic resource professionals and many others, i.e. people who in one or another way are engaged in community biodiversity management. It is through the dedication of those people to strengthen communities in the conservation and use of plant genetic resources, and document and share their experiences and learning processes, that we have been able to compile this book. It is a privilege to work with them and facilitate the sharing of their experiences with a global platform of practitioners, and other interested and concerned people. We appreciate and recognize their vital contribution to this book.

I would like to thank Marja Thijssen, Abishkar Subedi and Nivaldo Peroni for their ideas and contributions in compiling this volume; moreover, their enthusiasm in contributing to the compilation and editing the chapters of the book. I realize that since they are already fully engaged in their day-to-day duties in programme management, training, research and education, it has been hard to create time and space, and find the necessary peace to contribute to such a complex editorial process.

I want to share my appreciation for Elizabeth O'Keeffe, who has played a vital role in the long road for compiling a book with so many chapters, communicating with authors of dissimilar backgrounds, copy-editing the texts to a common style, and ensuring that the book is accessible to our target audience, in both its logic and language. Every sentence and word has passed through her careful thoughts, thereby making a precious contribution.

A special thanks to Tim Hardwick and Ashley Wright of Earthscan/Routledge, who supported us throughout the editorial process. We would also like to recognize the contributions of Carlos Augusto Ribeiro Machado, Mavi Peña and Mike Wilkinson, in translating texts from Portuguese, French and Spanish into English.

Many of the authors who contributed to this book took part in the 'Global Study on CBM and Empowerment' and are currently participating in the 'CBM and Resilience' project. These authors are affiliated with a multitude of organizations located in diverse locations across the globe, allowing for a wealth and diversity of experiences to be shared in this book. In acknowledgement of these organizations, which we list in alphabetical order, and on behalf of Wageningen CDI and the co-editors of the book, I would like to recognize the contributions made by colleagues and staff members associated with the following organizations[1]: ACORDI, ASA/PB, AS-PTA, ADMs in Guaraciaba, Epagri, UFSC, UnB, and UFRPE (Brazil); IDRC (Canada); PPBMA (Central America), including its partners ASOCUCH and FUNDIT (Guatemala), EAP/Zamorano (Honduras) and CIPRES and UNICOM (Nicaragua); DENAREF and UNORCAC (Ecuador); BDU, EIAR, EOSA, HU and MU (Ethiopia); Agro-Bio Périgord, Bio d'Aquitaine, IRD, Reverse and RSP (France); College of Forestry, Sirsi and MSSRF (India); LI-BIRD, NARC, and USC Canada Asia office (Nepal); CDI and CGN at Wageningen UR (the Netherlands); FNI (Norway); SPDA (Peru); SEARICE (the Philippines); USP and SPC (Samoa); CASRAD (Vietnam) and Bioversity International.

We would like to thank the Directorate General for International Cooperation (DGIS) of the Netherlands for their generous financial support of the Global Study on Community Biodiversity Management and Empowerment under the DGIS-Wageningen UR Partnership Programme 'Globalization and Sustainable Rural Development.' The management of Wageningen CDI has supported us throughout the development of the book. In addition, we appreciate the support of the Secretariat of the International Treaty on Plant Genetic Resources for Food and Agriculture, and their recognition of the project Community Biodiversity Management and Resilience, of which the current book is an integral component.

Walter Simon de Boef
21 June 2012
Florianópolis, Brazil

Note

1 The full names of these organizations are provided in the list of abbreviations.

Abbreviations

ABP	Agro Bio Périgord (France)
Abrasem	Associação Brasileira de Sementes e Mudas (Brazilian Seed Association)
ABS	access and benefit-sharing
ACAP	Annapurna Conservation Area Project (Nepal)
ACORDI	Associação Comunitária Rural de Imbituba (Rural Community Association of Imbituba, Brazil)
ADCS	Agriculture, Development and Conservation Society (Nepal)
ADM	Associação de Desenvolvimento da Microbacia (Association of Micro-watershed Development, Brazil)
AgroBio Périgord	L'Association de Développement de l'Agriculture Biologique en Dordogne (Association for the Development of Organic Agriculture in the Dordogne, France)
ANDES	Asociación para la Naturaleza e Desarrollo Sostenible (Association for Nature and Sustainable Development, Peru)
ASA/PB	Articulação no Semiárido Paraibano (Semiarid Network in Paraíba, Brazil)
AS-PTA	Assessoria e Serviços a Projetos em Agricultura Alternativa (Advice and Services for Alternative Agriculture Projects, Brazil)
ASOCUCH	Asociación de Organizaciones de los Cuchumatanes (Association of Organizations from the Cuchumatanes, Guatemala)
ASSO	Associação dos Pequenos Agricultores Plantadores de Milho Crioulo, Organico e Derivados (Association of Local Maize Variety Farmers and Processors, Brazil)
ATK	associated traditional knowledge
BDU	Bahar Dar University (Ethiopia)
BUCAP	Biodiversity Use and Conservation in Asia Programme
CAPA	Centro de Apoio ao Pequeno Agricultor (Support Centre for Small-Scale Farmers, Brazil)
CASRAD	Centre for Agrarian Systems Research and Development (Vietnam)
CBD	Convention on Biological Diversity

CBDC	Community Biodiversity Development and Conservation
CBM	community biodiversity management
CBO	community-based organization
CBR	community biodiversity register
CCP	commercial coffee plantation
CDI	Centre for Development Innovation (the Netherlands)
CE	Centro Ecológico (Centre of Ecology, Brazil)
CENARGEN	Embrapa Recursos Genéticos e Biotecnologia (Embrapa Genetic Resources and Biotechnology, Brazil)
CGIAR	Consultative Group on International Agricultural Research
CGN	Centre for Genetic Resources, the Netherlands
CGRFA	Commission on Genetic Resources for Food and Agriculture
CIAL	comité de investigación agrícola local (local agricultural research committee)
CIAT	Centro Internacional de Agricultura Tropical (International Centre for Tropical Agriculture, Colombia)
CIPRES	Centro para Investigación y la Promoción del Desarrollo Rural y Social (Centre for Research and the Promotion of Rural and Social Development, Nicaragua)
CIRAD	Centre de Coopération Internationale en Recherche Agronomique pour le Développement (Agricultural Research Centre for International Development, France)
CLIP	social analysis of collaboration, conflict, legitimacy, interests and power
CNPT	Centro Nacional da Populações Tradicionais (the National Centre for Traditional Populations, Brazil)
CORPEI	Corporation for Export and Investment Promotion (Ecuador)
CRI	Crop Research Institute (China)
CSB	community seed bank
CSO	civil society organization
CSR	community seed reserve
CST	Centre for Sustainable Technologies (India)
CSWC	Community Seed Wealth Centres (Nepal)
DADO	District Agricultural Development Office (Nepal)
DGIS	Directorate General for International Cooperation of the Netherlands
DR-CAFTA	Dominican Republic-Central America Free Trade Agreement
DUS	distinctness, uniformity and stability
EAP/Zamorano	Escuela Agrícola Panamericana, Zamorano (Panamerican Agricultural School, Zamorano; Honduras)
EIAR	Ethiopian Institute of Agricultural Research (Ethiopia)
Embrapa	Empresa Brasileira de Pesquisa Agropecuária (Brazilian Agricultural Research Corporation)
EOSA	Ethio-Organic Seed Action (Ethiopia)
Epagri	Empresa de Pesquisa Agropecuária e Extensão Rural de

	Santa Catarina (Santa Caterina State Enterprise for Agricultural Research and Rural Extension, Brazil)
EU	European Union
FAO	Food and Agriculture Organization of the United Nations
FC	forest coffee
FEPAGRO	Fundação Estadual de Pesquisa Agropecuária (the State Foundation of Agricultural Research, Brazil)
FFS	farmers' field school
FGC	forest garden coffee
FIPAH	Fundación para la Investigación Participativa con Agricultores de Honduras (Foundation for Participatory Research with Honduran Farmers)
FNI	Fridtjof Nansen Institute (Norway)
FOMRENA	Fondo Regional de Tecnologías Apropiadas en Manejo Sostenible de Recursos Naturales (Andean Foundation for the Promotion of Sustainable Technologies for Natural Resource Management)
FRB	Fondation pour la Recherche sur la Biodiversité (Foundation for Biodiversity Research, France)
FSO	Farm Seed Opportunities (Europe)
FUNAI	Fundação Nacional de Índio (National Foundation for Amerindian People, Brazil)
FUNDIT	Fundación de Innovación Tecnológica Agropecuaria y Forestal (Foundation for Agricultural and Forestry Technology Innovation, Guatemala)
GEF	Global Environment Facility
GM	genetically modified
GMO	genetically modified organism
GMRI	Guangxi Maize Research Institute (China)
GPA	Global Plan of Action
HH	household
HU	Hawassa University (Ethiopia)
IBAMA	Instituto Brasileiro de Meio Ambiente e Recursos Naturais Renováveis (Brazilian Institute of the Environment and Renewable Natural Resources)
IBC	Institute for Biodiversity Conservation (Ethiopia)
ICAR	Indian Council of Agricultural Research
ICDS	Integrated Child Development Services Scheme (India)
ICMBio	Instituto Chico Mendes de Conservação da Biodiversidade (Chico Mendes Institute for Biodiversity Conservation, Brazil)
ICRISAT	International Crops Research Institute for the Semi-Arid Tropics (India)
IFAD	International Fund for Agricultural Development
IGC	Intergovernmental Committee on Intellectual Property and Genetic Resources, Traditional Knowledge and Folklore

INCA	Instituto Nacional de Ciencias Agrícolas (National Institute of Agricultural Sciences, Cuba)
INCRA	Instituto Nacional de Colonização e Reforma Agrária (National Institute for Colonization and Agrarian Reform, Brazil)
INIAP-DENAREF	Instituto Nacional Autónomo de Investigaciones Agropecuarias – Departamento Nacional de Recursos Filogenéticos (National Plant Genetic Resources Department of the National Autonomous Institute for Agricultural Research, Ecuador)
INRA	Institut National de la Recherche Agronomique (National Agricultural Research Institute, France)
IPR	intellectual property rights
IRABS	International Regime on Access and Benefit-Sharing
IRD	Institut de recherche pour le développement (Institute of Research for Development, France)
ISE	International Society of Ethnobiology
ISSD	integrated seed sector development
ISTA	International Seed Testing Association
ITPGRFA	International Treaty on Plant Genetic Resources for Food and Agriculture
IUCN	International Union for Conservation of Nature
IUPGRFA	International Undertaking on Plant Genetic Resources for Food and Agriculture
JFPM	Joint Forest Planning and Management (India)
KARC	Kulumsa Agricultural Research Centre (Ethiopia)
KHABCoFED	Kolli Hills Agrobiodiversity Conservers' Federation (India)
KIDEKI	Kishna Dekhi Kishan Samma (Nepal)
KKRGCS	Kalinga Kalajeera Rice Growers Cooperative Society (India)
LI-BIRD	Local Initiatives for Biodiversity, Research and Development (Nepal)
MAPA	Ministério da Agricultura, Pecuária e Abastecimento (Ministry of Agriculture, Livestock and Food Supply, Brazil)
MARD	Ministry of Agriculture and Rural Development (Vietnam)
MDA	Ministério de Desenvolvimento Agrário (Ministry of Agrarian Development, Brazil)
MFSC	Ministry of Forestry and Soil Conservation (Nepal)
MMA	Ministério do Meio Ambiente (Ministry of the Environment, Brazil)
MMC	Movimento das Mulheres Camponesas (Women Farmers' Movement, Brazil)
MOAC	Ministry of Agriculture and Cooperatives (Nepal)
MOF	mixed ombrophilous forest
MPA	Movimento dos Pequenos Agricultores (Small-Scale Farmers' Movement, Brazil)
MSSRF	M.S. Swaminathan Research Foundation (India)

MST	Movimento dos Trabalhadores Sem Terra (Landless Workers' Movement, Brazil)
MU	Mekelle University (Ethiopia)
NAFED	National Agricultural Cooperative Marketing Federation of India Ltd
NARC	Nepal Agricultural Research Council
NCVESC	National Centre for Variety Evaluation and Seed Certification (Vietnam)
NEABio	Núcleo e Estudos em Agrobiodiversidade (Nucleus for Agrobiodiversity Studies, Brazil)
NEG	new environmental governance
NGO	non-governmental organization
NPGRP	national plant genetic resource programmes
NPSC	northern plateau of Santa Catarina (Brazil)
NSCU	Sistema Nacional de Unidades de Conservação da natureza (National System of Conservation Units, Brazil)
NSN	Nayakrishi Seed Network (Bangladesh)
NTFP	non-timber forest product
NUS	neglected and underutilized species
ORMAS	Orissa Rural Development and Marketing Society (India)
PBR	plant breeders' right
PCI	participatory crop improvement
PGE	participatory genetic enhancement
PGR	plant genetic resources
PGRC/E	Plant Genetic Resources Centre/Ethiopia
PGRFA	plant genetic resources for food and agriculture
PLAR	participatory learning and action research
PNPCT	Política Nacional de Desenvolvimento Sustentável dos Povos e Comunidades Tradicionais (National Policy for the Development of Traditional People and Communities, Brazil)
PPB	participatory plant breeding
PPBMA	Programa Colaborativo de Fitomejoramiento Participativo en Mesoamérica (Collaborative Programme on Participatory Plant Breeding in Central America)
PPVFR Act	Protection of Plant Varieties and Farmers' Rights Act (India)
PRA	participatory rural appraisal
PRR	Programa de Reconstrucción Rural (Rural Reconstruction Programme, Honduras)
PVP Law	Plant Variety Protection Law (Vietnam)
PVS	participatory varietal selection
Resex	reserva extrativista (extractive reserve, Brazil)
RSP	Réseau Semences Paysannes (Farmers' Seed Network, France)
SDC	Swiss Agency for Development and Cooperation
SDR	reserva de desenvolvimento sustentável (sustainable development reserve, Brazil)

SEARICE	South-East Asia Regional Initiatives for Community Empowerment (the Philippines)
SFB	scientist-supported farmer breeding
SFC	semi-forest coffee
SHGs	Self-help groups
SINTRAF	Sindicato dos Trabalhadores da Agricultura Familiar (Small-Scale Farmers' Trade Union, Brazil)
SNNPR	Southern Nations, Nationalities and Peoples' Region (Ethiopia)
SoD	Seeds of Diversity (Canada)
SOS/E	Seeds of Survival/Ethiopia
SPC	seed production cooperative
SPDA	Sociedad Peruana de Derecho Ambiental (Peruvian Society for Environmental Law)
SWOT	analysis of strengths, weaknesses, opportunities and threats
TaroGen	Taro Genetic Resources Conservation and Utilisation Network (South Pacific)
TIP	Taro Improvement Programme (South Pacific)
TK	traditional knowledge
TRIPs	Agreement on Trade-Related Aspects of Intellectual Property Rights
UBINIG	Unnayan Bikalper Nitinirdharoni Gobeshona (Policy Research for Development Alternatives, Bangladesh)
UCODEP	Unidad y Cooperación para el Desarrollo de los Pueblos (Union for Economic Cooperation and Development of the People, Ecuador)
UFPE	Federal University of Pernambuco, Brazil
UFSC	Universidad Federal Santa Catarina (Federal University of Santa Catarina, Brazil)
UnB	University of Brasilia, Brazil
UNEP	United National Environment Programme
UNICOM	Unión de Cooperativas de Madriz (Cooperative Union of Madriz, Nicaragua)
UNORCAC	Unión de Organizaciones Campesinas e Indígenas de Cotacachi (Union of Peasant and Indigenous Organizations of Cotacachi, Ecuador)
UPOV Convention	International Convention for the Protection of New Varieties of Plants
USC	Unitarian Service Committee (Canada)
USDA	United States Department of Agriculture
USP	University of the South Pacific
VBCDC	Village level Biodiversity Conservation and Development Committee (Nepal)
VCU	value for cultivation and use
VDC	Village Development Committee (administrative unit in Nepal)

VKC	village knowledge centre (India)
VRC	Variety Release Committee
Wageningen UR	Wageningen University and Research Centre (the Netherlands)
WIPO	World Intellectual Property Organization
WTO	World Trade Organization

Part I

Community biodiversity management and *in situ* conservation

1.1 General introduction

Marja Thijssen, Walter Simon de Boef, Abishkar Subedi,
Nivaldo Peroni and Elizabeth O'Keeffe

The debate on how to implement *in situ* conservation

The Convention on Biological Diversity (CBD) defined *in situ* conservation of agrobiodiversity as being the conservation of domesticated and cultivated species in the
surroundings where they have developed their distinctive properties (CBD, 1992a).
Defining *in situ* conservation of agrobiodiversity was the first step, but the major
challenge over the past two decades has been its translation into conservation practices that fit in the context of the livelihoods of small-scale, and often poor, farmers
(Jarvis *et al.*, 2011). How to implement *in situ* conservation of agrobiodiversity on-
farm has been much debated. It became gradually clear that the flexible nature of
farmers' management and utilization of crops and local varieties did not match the
perspective of conservationists in their design of *in situ* conservation on-farm (Hardon and De Boef, 1993). Part of the problem was that the various organizations
involved with agrobiodiversity conservation had contrasting objectives with regards
to the implementation of this conservation strategy (De Boef, 2000). Those organizations dedicated to the conservation of plant genetic resources (PGR) focused on
supporting farming communities to continue to use local varieties. Activist non-
governmental organizations (NGOs) aimed to contribute to the empowerment of
farming communities to exercise their (farmers') rights, for example, by addressing
their access to and control over local PGR. Development organizations (NGOs)
incorporated conservation strategies in their efforts to contribute to sustainable
livelihoods.

Over the years, it became clear that the dynamic nature of the management and
utilization of local crops and varieties by farmers did not match the methodologies
used by PGR conservationists, who basically control and freeze all dynamics for the
sake of ensuring the conservation of the PGR concerned. For implementing *in situ*
conservation on-farm, they now needed to support dynamic processes and team
up with farming communities that have their own ideas and priorities in relation
to agricultural development and the use of their genetic resources. The effect these
new aspects and uncertainties had on the field of expertise and professionalism of
PGR should not be underestimated, particularly with regards to limiting the degree
to which formal PGR programmes could advance in the implementation of *in situ*
conservation of agrobiodiversity on-farm (De Boef, 2000).

Introduction to the term 'on-farm management'

The First Report on the State of the World's Plant Genetic Resources for Food and Agriculture replaced the term '*in situ* conservation on-farm' with 'on-farm management', while maintaining '*in situ* conservation' as the overarching term (FAO, 1996). 'On-farm management' better accommodates the dynamism and facilitates the creation of links between PGR conservation and other crop development activities, such as plant breeding and seed production (Almekinders and De Boef, 2000).

The Second Report on the State of the World's Plant Genetic Resources for Food and Agriculture (FAO, 2011) provides ample insights into the advances being made concerning the on-farm management of PGR, for example:

> The last decade has seen an increase in the use of participatory approaches and multi-stakeholder teams implementing on-farm conservation projects (p. 45).

> Scientific understanding of the on-farm management of genetic diversity has increased. While this approach to the conservation and use of plant genetic resources for food and agriculture is becoming increasingly mainstreamed within national programmes, further efforts are needed in this regard (p. 22).

> In many countries NGOs play a very important role at the farm and community level in promoting and supporting the conservation and management of plant genetic resources (p. 126).

Based on those observations, development and conservation organizations now approach PGR in the context of sustainable and livelihood development, rather than as a means to achieve conservation. Jarvis *et al.* (2011) indicate that most initiatives are small, with modest aims and a limited area for implementation and application. They have been developed to solve problems associated with the use and maintenance of PGR by farmers and their communities. The initiatives target a multitude of crops, in dissimilar locations and situations. Each individual experience, while alone modest, contributes to a comprehensive body of knowledge and experiences on how to associate conservation-oriented practices with farmers' livelihood development. With this book, we bring together experiences of different organizations that support farmers in what is today referred to as the methodology of community biodiversity management (CBM), which aims to contribute to implementing *in situ* conservation and the on-farm management of agrobiodiversity.

Community biodiversity management

Strengthening community institutions

CBM is a methodology with its own set of practices, which aims, through a participatory process, to build community institutions and strengthen their capabilities to achieve the conservation and sustainable use of PGR. As part of the process, a local organization, either a government agency for rural development and extension services, or an NGO, engages itself in strengthening capabilities, supporting and mentoring decision-making processes, and reinforcing local institutions, until they are fully autonomous in decision-making at community level, while considering agrobiodiversity as one of their important livelihood assets. Such autonomous decision-making contributes to the empowerment of farming communities, which ultimately results in a situation where the community purposely and collectively manages its biodiversity in a sustainable manner. Once this situation is achieved, the organizations supporting the community have achieved their goal (i.e. *in situ* conservation or on-farm management). The authors of a number of chapters in this book were actively involved in a global study on CBM and empowerment conducted in 2009–2010, and coordinated by the Centre for Development Innovation (CDI) of the Wageningen University and Research Centre, the Netherlands. The study aimed to build a scientific foundation of CBM, based on an assessment of CBM practices in different sites across the globe, in support of the assumption that *in situ* conservation can only be achieved when farming communities are empowered in their agrobiodiversity management.

Conservation and development organizations that aim to support community-based organizations in the implementation of what we refer to as the CBM methodology, require knowledge, skills and expertise for building social institutions at community levels (i.e. for enhancing social capital in agrobiodiversity conservation and management). The CBM methodology follows a number of strategies for self-organization and self-governance that are similar to those of common property management and community-based natural resource and biodiversity management.

Emergence of a common methodology

Despite the fact that many development and conservation organizations practise CBM as a means to contributing to sustainable livelihood development and *in situ* conservation, its concept and theory, and the practical sharing of experiences concerning CBM, is limited. This gap is identified in the Second Report on the State of the World's Plant Genetic Resources for Food and Agriculture (FAO, 2011). Development organizations, NGOs in particular, focus their work on practice, rather than on sharing the concepts and experiences that guide and inspire them. Often, concepts emerge by practice and learning, rather than by pre-designing a process based on empirical reasoning and scientific scrutiny. With this book, we aim to overcome this gap and provide a foundation to what we refer to as the CBM methodology, supported by structured documentation and the presentation of experiences. We want to foster an understanding of the CBM methodology by public research and conservation institutions. Despite the fact that professionals from these institutions usually dominate global and national conservation debates, they quite often have little practical experience in, or knowledge of, this area.

CBM evolved gradually as a methodology. Many CBM experiences take place in isolation; some claim that the methodology only works under specific conditions. With this book, we aim to link and compare experiences from five countries – Brazil, Ethiopia, France, India and Nepal – that play an important and strategic role in the international debate on the conservation and sustainable use of PGR. Despite a diverse array of specific historical, political, cultural, biological and agro-ecological contexts, it is possible to identify common conditions in which the CBM methodology contributes effectively to the empowerment of farming communities, and leads to *in situ* conservation or on-farm management of PGR. Those commonalities cross barriers between centres of origin or diversity and beyond, as well as the old North and South divisions. Organizations involved in CBM may use dissimilar terminologies but they often share the same goals. In the process of evaluating an agrobiodiversity project in Nepal, a number of reflections were made concerning the collaboration between scientists, development workers and farming communities. The CBM methodology evolved in response to these reflections, as described by Subedi *et al.* (Chapter 1.2). We consider the transformation of these experiences into a structured process with a set of practices to be the foundation of the CBM methodology.

We chose to focus on five major countries, complemented with interesting experiences from other locations, covering all regions of the world, as illustrated in Figure 1.1.1, to compile the comprehensive body of knowledge that makes up this book. By focusing on just a few countries, we could gain more in-depth insights into the experiences and contextualization of CBM. With the variation of experiences, we aim to contribute to a global and common knowledge on CBM.

CBM and socio-environmental change

In the context of climate change, CBM and its practices can be converted into a basket of options that enhance the capabilities of farming communities to respond to unpredictable variations in their ecological, social and economic environment and thereby enhance community resilience. CBM has the potential to contribute to resilience as a way of promoting ecosystem health, which is required for agriculture to meet food security and sustainability. Diversity itself, and the capabilities of farming communities to use diversity in a conscious and collective manner, are fundamental

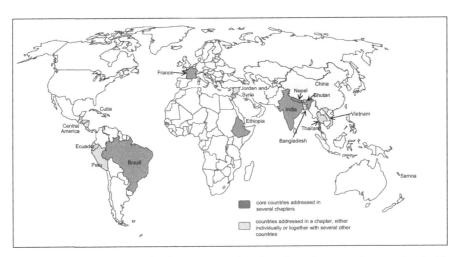

Figure 1.1.1 Map illustrating the five core countries, plus a few other countries, associated with the community biodiversity management experiences described in this book.

attributes of resilience. CBM incorporates aspects of agrobiodiversity conservation and use, and community resilience and empowerment, into one single methodology or process.

The authors of several chapters of this book are currently participating in a CBM and resilience project, which is being implemented by community organizations, development organizations and universities in over 12 countries in Sub-Saharan Africa, Central and South America, and South Asia. Through an interactive, participatory process, they are developing strategic action plans that aim to mainstream CBM and its practices for achieving on-farm management of agrobiodiversity, and for strengthening community resilience at local, national and global levels. This is described in more detail in Chapter 1.2.

CBM: inspiring a new PGR professionalism

Higher education and professional development institutions that include the conservation and use of PGR and biodiversity in their curricula primarily depend on the existing body of knowledge, which emphasizes *ex situ* conservation and scarcely incorporates approaches that address conservation and use within a livelihood and sustainable development setting. This bias needs to be addressed in the formation of future conservation and development professionals. At present, no specific methodology or practices have been put in place for preparing future conservation and development professionals with the right attitude, skills and capabilities. Universities have a role to play in ensuring that future generations of these professionals are better equipped to take on the facilitation roles required to make CBM work, while also contributing through research to the design of community-based strategies. The universities need to address complex aspects of development studies, such as empowerment, sovereignty and resilience, in their curricula. New PGR professionals need to have the capabilities to work through participation and learning, and to approach the science of PGR and biodiversity with paradigms other than that of positivism, which dominates the field of agronomy and biology, and the science of PGR conservation in general.

With this book, our intention is to present a body of knowledge and experiences in such a way that it can be used in those educational programmes, thus filling an important gap. Furthermore, we aim to ensure that what we offer, in terms of inputs to professional and academic development, matches the interest of current and future professionals, and adequately prepares them with the appropriate capabilities and skills for involvement in the increasingly complex world of PGR conservation and use. This book will be an important resource, supporting the universities involved in the CBM and resilience project, and others who work with agrobiodiversity and PGR in their research and educational activities, thereby contributing to this new professionalism, as further described by De Boef *et al.* (Chapter 7.1).

CBM and farmers' rights: new governance in PGR

The topic of farmers' rights over genetic resources is controversial. The debate is complex, conceptual and political, and it often loses contact with those who really matter (i.e. the farming communities who manage biodiversity). Both the Convention on Biological Diversity (CBD) and the International Treaty on Plant Genetic Resources for Food and Agriculture (ITPGRFA) have tried to elaborate mechanisms that recognize and facilitate farmers' rights. However, as with the *in situ* conservation of local crop and varieties, farmers' rights may just remain a concept or a discussion on paper. By using the CBM methodology and set of practices, organizations can support and enhance the capabilities of farming communities to exercise their farmers' rights in distinct ways. CBM encourages farmers to exercise custodianship over genetic, crop and landscape biodiversity. Its collective nature motivates communities to document, monitor, and make decisions concerning their PGR. Through CBM, farmers use their genetic resources in crop improvement, and exchange and market the seed of local or newly developed varieties. Farming communities increasingly transform their biological resources, through value addition, into unique products that contribute to their income. It can be concluded that farmers' rights are being increasingly implemented through the application of the CBM methodology, and as such they have moved beyond concepts and debates. The link between CBM and farmers' rights, coupled with the recognition and strengthening of farming communities as custodians of agrobiodiversity, contributes to discussions on a new governance in PGR, as further described by de Boef *et al.* (Chapter 7.1).

Objectives and outline of the book

The major goal of this book is to contribute to the global understanding of CBM as a methodology with the potential to contribute to the *in situ* conservation of PGR, the implementation of farmers' rights, and an enhancement of the capabilities of farming communities to strengthen their resilience in the face of changing socio-environmental conditions. In order to reach this goal, we address some basic concepts and theories underlying CBM, and share and demonstrate the lessons learned through working with CBM practices in a diversity of countries. With this compilation of insights and case studies, we hope that the book can act as a source and reference for relevant global, regional and national fora and their follow-up actions; and will encourage professionals and farming communities, to assume their responsibilities in implementing the process and its practices.

The book is divided into seven sections (Parts I–VII). Part I introduces the CBM methodology and its practices, outlining the historical background of the methodology, which evolved over the course of implementing an agrobiodiversity project in Nepal. The foundation of the CBM methodology in Nepal is complemented with similar experiences from Brazil, Bhutan, Ethiopia, France, Thailand and Vietnam. Part I concludes by pulling several examples of CBM practices together, to examine the relationship between CBM and *in situ* conservation.

Part II addresses the variation that exists among CBM practices. It presents a number of practices that contribute to raising awareness, understanding and monitoring diversity, and building the associated community institutions; and highlights experiences with diversity fairs, biodiversity registers and seed banks in Brazil, Ecuador, Nepal, and Central America. Community seed banks, which can be considered drivers for CBM processes owing to their strong emphasis on building community institutions, are compared in Bangladesh, Ethiopia, France, India and Nepal. The establishment of a community trust fund in Nepal, with the specific aim of sustaining the CBM processes, concludes the section.

In addition to crops and varieties, community management can be applied to the conservation and use of agrobiodiversity at landscape level. Part III opens by highlighting the importance of integrating a landscape management approach in sustainable livelihood development. The specific relationship between traditional communities and biodiversity in Brazil is then described, followed by a discussion on how the field of ethnobotany can contribute to CBM. Several case studies from Brazil are presented in this section, including one that details the Sustainable Development Reserve, a conservation unit that is unique to Brazil, and which aims to safeguard landscapes through human management. Other case studies described include those that focus on the management and use of coffee forests in Ethiopia; yerba mate/ Araucaria forests and *umbuzeiro* landscapes in Brazil; and forest landscapes and the *garcinia* species in the Western Ghats of India.

In order to sustain communities' engagement in biodiversity management, it is essential to add value to biodiversity and its management. Part IV explores CBM from this angle and shares some examples. Key concepts for promoting value addition and value chain development, and for creating a market environment to promote the use of biodiversity, are introduced and discussed. A number of case studies deal with value addition for the sustainable community management of biodiversity, involving various citrus species in Nepal; traditional millet species and a local rice variety in India; local wheat varieties in France; and rice varieties with a geographic indication in Vietnam. The last two case studies clearly show how the CBM methodology and objectives may conflict with common or normal agricultural development policies and regulations.

Since the 1990s, the field of participatory crop improvement has been recognized as an instrument for enhancing the effectiveness of breeding programmes. Part V approaches participatory crop improvement (PCI) within a CBM context. An overview of the variation of methods is provided, assessing the extent of community participation, the degree to which PCI has been embedded in the CBM processes, and its contribution to conservation and community resilience. Case studies of grassroots breeding in Nepal, a methodology for working with neglected crops, are presented. The participatory genetic enhancement of local varieties, with the aim of ensuring their continued cultivation, is discussed, with case studies focusing on a particular group of local rice varieties in Nepal, and composite maize varieties in Brazil. Part V also shares a number of other experiences, including the domestication of a fruit species in Brazil, for which farmers and a university group have joined forces; the conservation and use of local varieties in Central America, achieved through the application of several methods of participatory crop improvement; the promotion of farmer breeding in India, by sharing scientific principles with farmers; the recovery of local maize varieties, through exchange and participatory evaluation in France; and the broadening of the genetic base of taro in the Samoa, South Pacific, by obtaining access to new germplasm.

Part VI addresses issues concerning policies and rights of relevance to CBM and, in particular, farmers' rights. A group of authors who actively participate in global debates on genetic resource policies sets the scene. Farmers' rights are discussed from a more global perspective, but in the context of their association with CBM practices. Special attention is given to how farmers' rights have been incorporated in legal frameworks in India. Examples are shared on how access and benefit-sharing has moved beyond a concept and is being implemented in various ways in China, Jordan, Syria, Peru and Nepal. The implementation of the Brazilian access and benefit-sharing law is discussed, with the focus on how, through its protection of genetic resources, it may indirectly be contributing to the appropriation of the commons. Variety and seed laws, which are tailored to serve agro-industrial modes of production in France and Brazil, are examined within the context of how small-scale farmers in both countries are using, sharing and conserving local crops and varieties. The European Commission developed two regulations with the aim of supporting the conservation of varieties, but which in reality have the opposite effect; this issue is looked at from the perspective of French farmers. A concluding chapter illustrates the difficulties created by seed and variety laws and their implementation, as experienced by farmers in France, in maintaining and using genetic diversity within the informal system.

The final section of the book (Part VII) places this comprehensive body of experiences at a more conceptual level, associating CBM with community resilience. The recognition of CBM as a methodology for contributing to *in situ* conservation and enhancing community resilience has unambiguous implications on the required professionalism of those working with communities in CBM, which in this book is referred to as a new professionalism in PGR. This is followed by a discussion on how to take the next step for contributing to new forms of governance that better enable and incorporate vital aspects of CBM and farmers' rights, with professionals contributing to, and governments creating, an enabling environment for the recognition and empowerment of communities in managing PGR. Synthesizing the experiences of several cases throughout the book, this section untangles the relationship between CBM and empowerment. Similarly to empowerment, community resilience is a buzzword that is gaining force in the world of PGR. Within the context of current global debates on PGR and resilience, this section looks at community resilience from an ecological and social perspective, also exploring the challenges faced by the further evolution of CBM and its practices. In addition, resilience is approached in a more political context, associating it with autonomy, farmers' rights, sovereignty and empowerment. The section concludes by exploring opportunities and challenges for using the CBM methodology and its practices, and advocates that the methodology is not only relevant in the context of PGR conservation, but also that it should be positioned for designing strategies and frameworks to strengthen community resilience.

1.2 The evolution of community biodiversity management as a methodology for implementing *in situ* conservation of agrobiodiversity in Nepal

Abishkar Subedi, Pitambar Shrestha,
Madhusudan Upadhyay and Bhuwon Sthapit

Participation in a global programme: initial steps

The recognition of *in situ* conservation as a strategy by the Convention on Biological Diversity inspired a number of global programmes. In 1995, Bioversity International (then known as the International Plant Genetic Resources Institute), together with partners in nine countries – Burkina Faso, Ethiopia, Hungary, Mexico, Morocco, Nepal, Peru, Turkey and Vietnam – began to implement a global programme entitled 'Strengthening the scientific basis of *in situ* conservation of agrobiodiversity'. The programme aimed to contribute to achieving a better understanding on where, when and how *in situ* conservation of agrobiodiversity can be successful; what factors influence *in situ* conservation; and how this information both directly and indirectly adds value to diversity in terms of social, economic, environmental and genetic benefits (Sthapit and Jarvis, 2005).

The Nepal component of the global programme was jointly implemented between 1997 and 2006 by the Nepal Agriculture Research Council (NARC), Local Initiatives for Biodiversity, Research and Development (LI-BIRD) and Bioversity International, and was funded by the Directorate General for International Cooperation in the Netherlands, the Swiss Development Cooperation, and the International Development Research Centre in Canada. In the current chapter, we share how this agrobiodiversity project gradually evolved through several phases, from a single component within a global scientific programme, to one that can be considered individually as a global reference (FAO, 2011) for the development of the methodology referred to as community biodiversity management (CBM). The agrobiodiversity project started in 1997, in three ecosites representing high-hill (Jumla district), mid-hill (Kaski district) and terai or lowland ecosystems (Bara district); these agro-ecosystems were selected in order to cover the variation required to study *in situ* conservation. Figure 1.2.1 indicates the locations of these sites in Nepal.

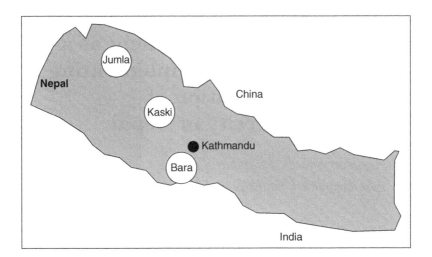

Figure 1.2.1 Location of the agrobiodiversity ecosites in Nepal.

Phase 1: Strengthening the scientific basis of *in situ* conservation on-farm (1997–2002)

In order to create institutional innovation, we managed the Nepal project in such a manner that governmental, non-governmental and international organizations worked together with local communities. We established multi-stakeholder platforms in which we used participatory methodologies to study the *in situ* conservation of agrobiodiversity on-farm. We consider this phase to have been important for the development and implementation of a number of practices, which were later consolidated in the CBM methodology. These practices include: (i) sensitizing local communities, and strengthening their capacity, with regards to agrobiodiversity conservation and use; (ii) locating and documenting genetic diversity and associated traditional knowledge; (iii) characterizing local varieties; (iv) understanding the value of plant genetic resources (PGR); and (v) designing options for enhancing the benefits of PGR, creating conditions to sustain *in situ* conservation on-farm.

Phase 2: Redirecting the project to meet the actual demands of the community (2002–2003)

The agrobiodiversity project was successful in creating awareness, influencing policies and achieving a better understanding on the scientific basis of where, when and how *in situ* conservation of agrobiodiversity can be managed successfully. However, in discussions with community representatives, we gradually realized that the success stories had not been adequately translated at community levels. We further noted that farmers' participation in the project activities over the years had begun to decline. We used a mode of participation that in hindsight we consider to have been overly consultative; community members had provided services to researchers without seeing the benefits from those interactions themselves. Our actions in communities had served our interest in generating knowledge and understanding, furthering our academic career and publications, but they also responded to globally set research objectives. Community members challenged us on this, indicating that they had gained little from all their work. This reflection came as a shock to us, as we realized that we had been implementing participatory methodologies that directly served our and not the community's interest. We realized that in order to truly contribute to the implementation of *in situ* conservation on-farm, we needed to change the type of collaborative or participatory interactions, with the ultimate aim of reaching a mode of participation that is referred to as self-mobilization (Pretty, 1995).

This response of the community to scientist-driven research should be placed within the context of the political transformation in which Nepal has been engaged over the past decade. In practical terms, the lead researchers engaged in the project, particularly from the government sector, had to devolve decision-making to farming communities and community-based organizations. It was not safe for any government-employed researcher to travel to villages and this diminished their capacity to be directly engaged in local research and development activities. A more direct implication that motivated the response of the local community to research-driven interventions in their community, as described above, was that local leaders questioned the relevance of any external development agency project to the livelihoods of small-scale farmers.

In response to this political but also professional context, we recognized that, through our original scientific- and conservation-guided approach, we had created institutional modalities that only functioned with project support. The actions of non-governmental organizations (NGOs) and national partners, but also those of the communities, were dependent on the availability of project resources. The modalities we had used had created limited ownership of the community institutions over the conservation actions. Based on these reflections, we decided to conduct a joint project review. In order to respond adequately to the critique, the design of the review was interactive, meaning that all participating farmers' groups, nodal farmers, project staff members and local stakeholders joined in the process (Subedi *et al.*, 2005a). Through the review, the following series of major gaps were identified in the modalities we had up until that time been using:

- We did not adequately integrate livelihood priorities, or the interests of the communities and their institutions, into the design and implementation of the project.
- We followed a common design for achieving externally formulated objectives, which had serious implications, leading to restricted local ownership.
- We used and designed several new participatory methodologies that primarily served our own scientific agenda; farmers and their communities were consulted in our research process, but they did not have the benefit of setting their own agenda and developing their own institutions.
- We organized the capacity-building of local institutions on an ad hoc basis, or on the basis of project needs, without considering the agenda and needs of the community.

- We applied mechanisms for monitoring and evaluation that complied with a scientific agenda, but which did not facilitate learning among local institutions and partners, as a result of which they pulled out of the network or reduced their involvement.
- We developed complex practices such as for participatory crop improvement and for the design of community biodiversity registers without building a proper foundation in the community, among farmers and their community institutions; as a consequence, they did not see the immediate benefits of those complex and long-term processes.

The joint review of the agrobiodiversity project resulted in the significant reorganization and restructuring of the project. The lessons learned through this review were of vital importance to LI-BIRD, for transforming the capabilities gained through scientific research into a community-based methodology.

Phase 3: Learning with community-based conservation actions (2003–2006)

The critical feedback obtained during the review exercise led to a rethinking of the whole modality of the agrobiodiversity project. In addition, we carried out an analysis of strengths, weaknesses, opportunities and threats (SWOT), and conducted sustainable livelihood analysis. We translated these lessons from tools into actions and the agrobiodiversity project continued in two ecosites, in Kaski and Bara. Ongoing conflict in Nepal disrupted travel to the project sites, and as such researchers were unable to carry out all the monitoring visits, particularly in the Jumla ecosite. Mainly because of this conflict, which caused fieldwork restrictions, LI-BIRD had to discontinue its activities in the Jumla ecosite in 2002.

In the Kaski ecosite, supported by a grant from the Small Grant Programme of the United Nations Development Programme/Global Environment Facility in Nepal, we integrated different participatory methods into livelihood-based agrobiodiversity management. We moved beyond our original focus, which was solely on crops, integrating wetland and forest resources in our approach. This wider approach, coupled with the reflection of the communities, challenged us to seek methods and tools to address biodiversity in a broader sense. We aimed towards social organization and autonomy at local level, fostering the interest of community members and their institutions in biological resources. To ensure follow-up and sustainability we teamed up with local government agencies. Our focus on strengthening community-based organizations (CBOs) in Kaski resulted in the establishment of Pratigya Cooperative, Rupa Fisheries Cooperative and the Kishna Dekhi Kishan Samma (KIDEKI). Each of these has taken on a leadership role in managing various resources: agriculture and horticulture (Pratigya Cooperative); wetlands (Rupa Fisheries Cooperative); and forest diversity (KIDEKI). The establishment of these CBOs led to enhancing the capacities of local farmers and their groups (Subedi *et al.*, 2005b).

Many of the practices that we had previously been implementing as single activities, such as the diversity fair, diversity blocks and value addition activities, have now become an integral part of livelihood-based community action plans, and as such are part of a larger process. An example of this can be seen in the participatory genetic enhancement of the Jethobudho local rice variety, a process that contributed to its conservation but also created opportunities for income generation by farmers and their local institutions. This experience is further described by Silwal *et al.* (Chapter 5.5).

In close collaboration with these local organizations in Kaski, we were successful in introducing, experimenting with, and refining the modalities of the CBM fund, which are described in more detail by Shrestha *et al.* (Chapter 2.9). The CBM fund, as a mechanism for local financing, emerged as a mandatory step for sustaining conservation-oriented community action plans. It stimulated immediate interest among the farmers' groups and a number of local biodiversity-based farmer entrepreneurs. These groups were quick to take out loans from the CBM fund, from which they were able to reap economic benefits, while providing services to the community and society through the conservation of agrobiodiversity.

Even though the agrobiodiversity project was active in the Bara ecosite, a study carried out there following its first phase showed that farmers in Bara continued to lose local crops and varieties at an alarming rate (Chaudhary *et al.*, 2004). Based on the awareness raised by the outcomes of the study, several farmers' groups reorganized themselves into a registered local institution, which they called the Agriculture, Development and Conservation Society (ADCS). Its first intervention, the establishment of a community seed bank, responded to the most immediate challenge (i.e. the loss of local varieties, which is associated with aspects of seed security) (Shrestha *et al.*, 2005). The community seed bank activities of the ADCS were instrumental in bringing together whole farming communities on topics related to the conservation and use of agrobiodiversity. The ADCS linked other practices to the community seed bank. The diversity fair was used as a mechanism to collect and share local varieties and associated knowledge. The diversity kit was effective for distributing the seed of rare but appreciated varieties; while the diversity block was crucial for recycling, multiplying and measuring the varietal performance of local varieties. Shrestha *et al.* (Chapter 2.2) elaborate on how these practices contribute to raising awareness and enhancing the understanding of communities on agrobiodiversity. The ADCS used the community biodiversity register, which is further described by Subedi *et al.* (Chapter 2.4), to locate custodians and to document the traditional knowledge associated with local crops and varieties. Plant breeders associated with LI-BIRD teamed up with the ADCS to engage in participatory plant breeding.

A CBM fund was first set up by the ADCS in Bara to sustain the community seed bank operations, but it gradually evolved, like the CBM fund in Kaski, to promote livelihood- and agrobiodiversity-based enterprises, while also meeting conservation goals. Shrestha *et al.* (Chapter 2.9) describe how the maintenance of local varieties was included as a precondition to allow farmers to access the fund. Conservation goals have in such a way become embedded in the livelihood development strategies of local institutions.

Phase 4: Emerging CBM methodology (2005–2006)

After our continuous efforts and experiences with pilot projects in two ecosites, we synthesized the whole process of CBM evolution. We began to understand which key

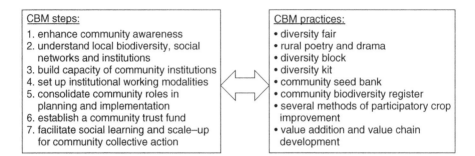

Figure 1.2.2 Steps and practices of the methodology for community biodiversity management (CBM), as developed in Nepal. Participatory crop improvement in the context of CBM in Nepal includes the following methods: grassroots breeding, participatory enhancement of local varieties, participatory genetic enhancement of local varieties, participatory varietal selection and participatory plant breeding.

steps the farming communities followed in the two ecosites; and which practices they employed to obtain social, economic and environment benefits. We at LI-BIRD, and among partners such as NARC, as well as our development partners, also began to reflect on lessons learned over the process of the evolution towards establishing CBM, and contributing to community empowerment. During this process of reflection and learning, we identified the seven steps of the process that we had developed, supported by a set of practices. Figure 1.2.2 shares the steps and accompanying practices. The learning process and reflection laid the foundation to developing an overarching methodology that is today recognized as CBM (Subedi *et al.*, 2005a).

From these experiences, we defined CBM as a community-based and participatory methodology for strengthening the capacity of farmers and farming communities to manage their biodiversity for the benefit of the household and the community. The CBM methodology facilitates the identification, conservation, management, exchange and value addition of agrobiodiversity, through community actions. It aims to build the capacity of local communities to make their own decisions on the conservation and use of agrobiodiversity. It results in autonomy in decision-making on agrobiodiversity conservation and use at the community level; stimulates ownership; and supports community-based conservation and sustainable livelihood options, with minimal external inputs and risks. The processes of consolidation resulted in the publication of a manual on the CBM methodology with its practices (Sthapit *et al.*, 2006).

Phase 5: Scaling-up CBM at national, regional and global levels (2008–2012)

Once CBM had matured as a methodology, LI-BIRD was able link up with the Nepalese Ministry of Agriculture and Cooperatives (MOAC) in out- and upscaling CBM in locations beyond the two original ecosites. Several ministries took on individual CBM practices: the community biodiversity register was adopted by the Ministry of

Forestry and Soil Conservation; and the community seed bank was incorporated by the MOAC into their development programmes. Various stakeholders use the CBM methodology for contributing to conservation and livelihood development in 16 districts of Nepal. The MOAC uses the methodology in ten districts, supported by LI-BIRD. LI-BIRD began to encourage other NGOs in South Asia (Bangladesh, India and Sri Lanka) to use the methodology, through the CBM South Asia programme, with financial support from the Development Fund, Norway.

A number of international organizations became interested in CBM. Bioversity International and partners embraced the methodology in a regional project on cultivated and wild tropical fruit tree species. The project is being implemented in 36 communities in India, Indonesia, Malaysia and Thailand, with financial support from the United Nations Development Programme/Global Environment Facility. Ramesh *et al.* (Chapter 3.7) describe how the CBM methodology and its practices faced new challenges in the interface between the natural forest ecosystem, home gardens and semi-commercial orchards in the Western Ghats of India.

The Centre for Development Innovation (CDI), at Wageningen University and Research Centre in the Netherlands, incorporated the CBM methodology as a key strategy for implementing *in situ* conservation in its annual global and regional training programmes, and in national training programmes that it conducted in Brazil, Ethiopia, Ecuador, India and Thailand.

In 2009, CDI, in collaboration with LI-BIRD, Bioversity International and partners in Brazil, Ethiopia, India and Nepal, launched a global study on CBM and empowerment to compare different practices and realities at 15 selected sites. The global study addressed in detail the assumption that CBM is a methodology for achieving *in situ* conservation; it must also contribute to the empowerment of communities in managing their genetic resources. The study further shaped conditions for the establishment of a global network of organizations engaged in CBM. This informal platform includes CBM nodes made up of CBOs, NGOs and local universities. CBM nodes operate in 12 countries in Africa, Asia and Latin America. LI-BIRD, with support from Bioversity International and CDI, developed a joint project for exploring mechanisms through which CBM as a methodology can support the on-farm management of agrobiodiversity and enhance community resilience in the context of adaptation to climate change. The project, which is currently being implemented with financial support from the Benefit-Sharing Fund of the International Treaty on Plant Genetic Resources for Food and Agriculture (ITP-GRFA), aims to develop a number of regional and national strategic action plans to further mainstream CBM in the context of on-farm management and community resilience.

The future of CBM: building resilience in dynamic agrobiodiversity hot spots

More than a decade of continuous involvement and experiences has enriched our knowledge and insights on how communities can effectively assess, monitor, add value to and manage agrobiodiversity. Furthermore, it shows that agrobiodiversity is a major source of livelihood for rural communities in developing countries like Nepal, but also in developed countries like France. Our experiences in Nepal, but also those of our partners in Brazil, Ethiopia, India and many other countries, show that the conservation and use of agrobiodiversity are components of the livelihood development of farming communities in parts of the world that are also considered to be the centre of origin or diversity of our crops.

Global and national priorities concerning long-term *in situ* conservation are still largely focused on wild or natural biodiversity, which covers 10–15% of the land of this planet on a permanent basis (Soutullo, 2010). Such priorities have not yet been translated into ways to manage agrobiodiversity in a sustainable way, and to reach a credible number of farming communities, in areas that are consistent in size with the potential value of the resources maintained. Therefore, we consider the logical next step to be that of consolidating the CBM methodology to identify and recognize agrobiodiversity-rich landscapes as permanent sites for the *in situ* conservation of local crops and varieties. With the CBM methodology we have the instruments in our hands to enhance the capacities of communities responsible for the conservation and use of agrobiodiversity. The development of conservation frameworks that recognize those communities as custodians of those landscapes is now required. We know that there are key differences in approaches and methods to managing 'rhinos' and 'rice'. However, through the evolution of CBM, we have gained an increasing body of empirical evidence on how we can better manage and sustain these dynamic and evolving productive landscapes. The dynamic management of agrobiodiversity in those hot spots that link local processes to global priorities is an investment towards sustaining agrobiodiversity for use by future generations.

1.3 Community biodiversity management

Defined and contextualized

Pratap Shrestha, Pitambar Shrestha, Abishkar Subedi,
Nivaldo Peroni and Walter Simon de Boef

Community biodiversity management

Community biodiversity management (CBM) is a methodology for promoting conservation and the sustainable utilization of biodiversity at local level, with an emphasis on agrobiodiversity or plant genetic resources. CBM distinguishes itself from other strategies that target *in situ* conservation, or on-farm management, by its focus on increasing the decision-making power of communities and securing access to and control over their biological and genetic resources for sustainable livelihood management. The CBM methodology integrates local knowledge and practices and is driven by local social systems, local rules and institutions (Sthapit *et al.*, 2006). In the current chapter, we provide definitions for terms commonly associated with CBM, to establish a common vocabulary. We then look at CBM and the components or steps that form the basis of its implementation, based on experiences in Nepal. However, our main intention, in this general introductory chapter, is to focus on the methodology, rather than describe the processes that result in community biodiversity management.

Definitions

Plant genetic resources and agrobiodiversity

Plant genetic resources (PGR) can be defined as all plant materials that have an actual or potential value (IBPGR, 1991, p. 74). The reference to resources suggests an economic and political connotation. *Agrobiodiversity* is often approached as an ecological service rather than as a resource, owing to its broader and ecological association, and to its delineation within the three levels of biodiversity (i.e. system, species and genetic diversity). Based on Professor Harold Brookfield's definition of agrodiversity (Brookfield, 2001), De Boef and Thijssen (Chapter 1.8) describe agrobiodiversity as 'a dynamic and constantly changing patchwork of relations between people, plants, animals, other organisms and the environment, always coping with new problems, always finding new ways'. This description has major implications for defining the relationship between CBM and the on-farm management of agrobiodiversity as a conservation strategy.

Communities

In the context of the dynamics of agrobiodiversity, *communities* are approached as open systems that have inflows and outflows of people, goods, knowledge, information, skills and resources (Cumming *et al.*, 2006). This means that communities are also open systems with regards to PGR and associated traditional knowledge (ATK). According to Wilson (2012), interactions between communities and society can be identified at regional, national and global levels. At the regional level, the migration of people into and out of the community (e.g. for work, education or marriage) may result in both an inflow and outflow of PGR and ATK. At national level, there are many ways to activate such flows, including the implementation of policies related to the dissemination of seed of improved varieties among farmers, and mechanisms that promote agricultural research and extension; the implications of agricultural subsidies on the use of particular crops and varieties; and the management of protected areas of agrobiodiversity at the landscape level. At the global level, flows may be connected to global trends concerning intellectual property rights and the introduction of hybrid and genetically modified varieties; or to food trade regulations and fluctuations in food prices, which affect food and seed security. Within CBM, communities are approached as open systems simply because of the multitude of scalar interactions; we know only too well that with PGR there is no escape from the forces of globalization. To approach communities as closed systems, with the aim of conservation through community biodiversity management, does not fit in with the dynamics and constantly changing patchwork of relations that define agrobiodiversity. Rather, community codes of conduct (norms and regulations) could be used to regulate the open systems in favour of biodiversity conservation and sustainable utilization.

Local knowledge and local management

Local knowledge is unique to a given culture or society (Warren *et al.*, 1995). The term is interchangeable with *traditional knowledge*, which, however, has stronger links to a cumulative body of knowledge and beliefs that were handed down over generations by cultural transmission, and which concern the relationship of living beings (including humans) with one another and with their environment (Berkes *et al.*, 1995). Thus, traditional agrobiodiversity management refers to such historical continuity. Because of the dynamic nature of agrobiodiversity, and the openness of communities that manage agrobiodiversity, the term *traditional* needs constant redefinition. As such, we prefer to use the term *local management* in this context.

Local crops and varieties, and informal seed systems

As with knowledge and management, we apply the term *local crops and varieties* to those PGR that are maintained and used by farmers in their production system. Communities, as open systems, have continuous inflows and outflows of crops, and local or improved varieties, as well as less tangible elements like ATK and technologies. We avoid using the term *landraces*, because of the conflicting views that exist concerning whether the term should only be applied to those varieties that are found in centres of origin or crop genetic diversity; or whether it can also be applied to those that are found beyond those centres. The term landrace has a strong territorial and static connotation, which seems incompatible with the more dynamic perception of agro-biodiversity and communities as open systems. Consequently, we use the term *local variety* throughout the book. Local varieties are managed by farmers, and are thus maintained either through *farm-saved seed* or through *informal seed systems*. These varieties are not necessarily considered traditional, as many of them originate as modern varieties maintained by farmers, and as such may have gone through a process of dynamic management (i.e. local maintenance, selection and exchange).

CBM as an endogenous or exogenous process

We understand the term *community biodiversity management* to have two distinct meanings: (a) a methodology with a set of practices, used by both communities and external actors supporting those communities, for the maintenance and sustainable use of agrobiodiversity; or (b) a situation in which community organizations achieve the maintenance and sustainable use of agrobiodiversity, in a conscious and autonomous manner.

Thus, according to the first meaning, CBM can be associated with actors (though not necessarily exclusively) outside the community, which trigger the process of management. With regards to the second meaning, the CBM process is initiated by the community itself, without the involvement of any external actors. In addition, CBM can also be a combination of both meanings, for example, where the CBM process is initiated internally within a community, but requires the involvement of external actors, such as conservation and development organizations, in order to fully implement CBM as a methodology, and result in CBM as a sustained situation.

When we refer to CBM as an externally initiated process that gradually becomes embedded in community structures, we consider it an *exogenous process*. But when CBM is an existing situation, resulting from a process that is initiated within the community, we consider it an *endogenous process*. External actors can act as catalysts for advancing such endogenous CBM processes. In both exogenous and endogenous CBM processes, the objectives and modalities of the external agents, for introducing or strengthening CBM (as methodology), and contributing to the empowerment of the community in conservation and livelihood development, are quite different.

CBM practices

CBM as a methodology incorporates many *practices*, as used both by community-based organizations (CBOs) concerned with conservation and use, and by other external actors. Practices can be single actions that are geared towards raising awareness of conservation and diversity, enhancing understanding, building capabilities within CBOs, and encouraging communities to make informed decisions over their agrobiodiversity. Such single-action practices include rural diversity drama and poetry, diversity fairs, diversity kits and diversity blocks, as described in more in detail by Shrestha *et al.* (Chapter 2.2). The practices may also result in the development of community institutions such as the community seed bank, which is detailed in a global overview by Shrestha *et al.* (Chapter 2.8); and the community biodiversity register, as described by Subedi *et al.* (Chapter 2.4). CBM practices are also approached as multi-year processes for value addition and value chain development, as presented in various chapters in Part IV. Several methods of participatory crop improvement that are embedded within the general process of CBM, and which can as such be considered CBM practices, are described in Part V.

CBM components

Experience in Nepal has shown that a series of *steps* lead to the establishment and promotion of the CBM methodology to manage agrobiodiversity by collective community action. However, a diversity of realities and variations exist around the world today in using the CBM methodology, and this diversity is partially linked to whether the CBM processes are endogenous or exogenous. Consequently, when considering the CBM methodology in a global context, we refer to those steps as *components* (i.e. the CBM methodology does not require all the components to be implemented in all the situations, and they do not need to follow the same sequence of order as would be implied by using the term steps). These components are summarized here below (Sthapit *et al.*, 2006).

Enhancing community awareness of conservation and diversity

This component is essential for initiating the CBM process. It is important that community members are engaged in the organization of activities, in order to raise awareness. They realize the value of conservation, discuss conservation needs, allocate the roles and responsibilities for community members and other actors, and identify and understand their own working modalities for the organization of practices such as diversity and/or seed fairs, rural drama and poetry, and the implementation of diversity blocks and diversity kits. It is important that, in the end, the CBOs that are involved in the process are autonomously capable, and recognize the value, of continuously organizing such awareness-raising events.

Increasing the community's understanding of diversity, social networks and organizations

In order to enhance the community's understanding of local biodiversity, social networks and organizations, and to develop appropriate conservation and utilization strategies, a number of participatory rural appraisal (PRA) tools are used. Four-cell analysis is an effective PRA tool for identifying common, unique and rare plant genetic resources. In comparison with other PRA tools, such as variety listing, mapping and matrix ranking, four-cell analysis enables the participants to develop diversified livelihood options and community-based conservation strategies (De Boef and Thijssen, 2007). Social seed network analysis is useful for understanding the structure of seed networks and for identifying farmers with specific roles as nodal, access providers or connectors within those networks. The tool is also relevant for enhancing the understanding of flows of related knowledge, both within and beyond farming communities (Poudel *et al.*, 2009; Abay *et al.*, 2011).

Establishing institutional working modalities

This component is essential for ensuring CBM can function and that it results in autonomy. Key organizations are identified for the coordination of CBM within the community, and their roles and responsibilities, and institutional norms are defined. A work plan is prepared and community-level indicators for monitoring performance are identified. A CBM committee is established within those local organizations in which community members and the CBOs are represented. Specific CBM practices, such as the community seed bank and the community biodiversity register, which involve complex organizational processes and institutional set up, can be instrumental in the formation of leadership roles in the community, thereby creating a foundation in human resources for the establishment of a CBM committee. In addition to planning, implementation and management of CBM activities, the key responsibilities of the committee include coordinating and overseeing the implementation of CBM strategies and plans. Furthermore, the committee also enforces the codes of conduct established for the management of genetic resources, including community-based mechanisms for access and benefit-sharing. Its capabilities are further strengthened by establishing institutional linkages with, and seeking resources from, service providers outside the community.

Strengthening the capabilities of community organizations

This component focuses on strengthening the capabilities of CBOs in carrying out analyses for increasing efficiency and self-confidence, but also for enhancing their capacity in social mobilization (communication, mobilization of members and participatory decision-making), organizational management (administrative and financial management) and institutional development (institutional policy, strategies and governance). In addition, the CBOs need to develop their capacity to establish locally driven guiding principles for use in their operations, and to design, implement and monitor the effectiveness of codes of conduct. They may also become engaged in complex processes and procedures in order to organize biodiversity access and benefit-sharing, and a community biodiversity fund at community level. The application of this component should be based on an assessment of the existing situation. The Venn diagram can be used to identify and explore existing relationships between CBOs and other actors operating in a community. A SWOT (strengths, weaknesses, opportunities and threats) analysis of CBOs can help to identify what is needed to strengthen the capabilities of those organizations, and explore which local organizations may take up coordination and/or facilitation roles (De Boef and Thijssen, 2007). The need to set up a new CBO may emerge during this component, around a specific user group.

Consolidating the community's role in planning and implementation

This component encourages the community to assume its responsibilities in CBM. The CBM committee is a community institution or legal unit that has the authority to act on behalf of local communities. It coordinates the activities of various farmers' groups or partner CBOs, each of which develops annual action plans through village meetings. The organization of activities for strengthening capabilities, by the committee members and CBOs, supports this community-level planning process.

Establishing a CBM trust fund

The CBM trust fund is an effective way of organizing community members, developing community ownership and motivating the community and their organizations to implement CBM action plans. It may also provide a mechanism to transfer to the community any funds generated from the use of genetic resources through access and benefit-sharing mechanisms. It is important to strengthen the capabilities of the CBM committee members for managing the fund, as it can be a motivating factor for collective actions in the community.

Monitoring and evaluating activities

Procedures for monitoring and evaluating activities within the community must also be incorporated in the CBM process. The CBM committee establishes an annual CBM calendar, which presents action plans that address the roles and responsibilities of participating organizations and community members. This CBM calendar is helpful for reviewing and monitoring progress, but it also promotes transparency within the community and among the organizations involved. Review meetings are regularly held for monitoring and evaluating community actions.

Social learning and scaling-up for collective community action

This component supports the involvement of an increasing number of households from within the community, and from other farming communities, for adopting and adapting similar CBM processes and practices. Progress concerning the scaling-up of CBM activities to other households and communities is reviewed during community meetings, where successes and failures are shared, and innovations and new practices are identified. Ways of facilitating collective learning include diversity fairs, educational tours and community visits, as well as travelling and learning workshops for farmers. The synthesis of social learning and good practices can be used to inform policy-makers and influence the creation of policy environments that support and enable CBM as an approach towards integrating conservation and development.

Diversity of situations sharing common processes

CBM, as initially developed in Nepal and other locations in Asia, acquired various nuances and interpretations in order to be accepted as a global and common methodology. The very notion of community, which is commonly defined as an open system, within the context of agrobiodiversity, still varies from sites with almost a thousand households, to those with fewer than 40 farming households. The concept of agrobiodiversity, which can refer to wildlife or landscapes, but which also refers to local crops and varieties used by agricultural settler societies in Brazil, is quite different in countries with an ancient agricultural background, such as India and Nepal, or in those that are prominent centres of origin, such as Ethiopia. In France, agrobiodiversity is associated with particular regions and the livelihood strategies of small-scale farmers who oppose and divert away from mainstream agro-industrial production systems.

The non-governmental organization Local Initiatives for Biodiversity Research and Development (LI-BIRD) in Nepal, in partnership with local farming communities, and national and international partners, elaborated CBM as a methodology, which is implemented through a process consisting of a number of steps that use different participatory tools and practices, as presented by Subedi *et al.* (Chapter 1.2). Similar methodologies, not necessarily referred to as CBM, were developed simultaneously in South and South-East Asia (FAO, 2011). The M.S. Swaminathan Research Foundation (MSSRF) is an important reference point for CBM in India. The South-East Asia Regional Initiatives for Community Empowerment (SEARICE), which is active in several Asian countries, supports CBM processes that are embedded within local government structures, and influences the development of genetic resource policy frameworks favourable to CBM processes. In Ethiopia, the Seeds of Survival (SoS) programme initiated by the national gene bank, with support from the Unitarian Service Committee, Canada, and continued by the NGO Ethio-Organic Seed Action (EOSA), is also an important reference for CBM. In Latin America, CBM-type activities are being implemented by public, non-governmental and community-based organizations in several Andean countries, while in Central America a participatory crop improvement programme incorporates many CBM practices. In Brazil, the Seeds of Passion network, in the state of Paraíba, which involves over 200 community seed banks, has been functioning for over 20 years and is both a national and global reference point for CBM. Endogenous CBM processes are being strengthened and institutionalized at landscape level in Brazil, in the form of a conservation regime referred to as the sustainable development reserve. The CBM methodology and its practices are also of great value to organizations of small-scale farmers in France. These farmers are engaged in endogenous CBM processes that are identical to those in Nepal, India, Ethiopia and Brazil; they support those farmers and their organizations in their search for more sustainable and regionally oriented production systems, and for regaining sovereignty in their use of seed and varieties. This shows that CBM is relevant beyond the old North–South divisions, for those farmers and farming communities who value agrobiodiversity as an asset in their livelihood strategy.

Thus, a common CBM methodology purports to be applicable in dissimilar situations, linking the empowerment of farming communities to the achievement of on-farm agrobiodiversity management. The rather complex local realities have their own specific historical, political, cultural, and even biological and ecological contexts, which link CBM to interactions between people, their biological resources and their environment. Each reality may appear unique, in terms of the type of agrobiodiversity managed; the structure and organization of the community and its organization; the endogenous and/or exogenous processes through which community management is initiated; and the relationship, type of interactions and duration the community has with external actors. However, the current book aims to show that despite the diversity of realities it is possible to identify common properties of how the CBM methodology contributes to achieving on-farm management, together with conservation and livelihood development.

1.4 On-farm management of plant genetic resources through community seed banks in Ethiopia

Regassa Feyissa, Genene Gezu, Bayush Tsegaye and Temesgen Desalegn

Conservation efforts targeting threatened local crops and varieties

Ethiopia is the centre of origin for a number of crops, such as coffee (*Coffea arabica*) and safflower (*Carthamus tinctorius*); it is also the cradle of domestication of major agricultural crops, including tetraploid durum wheat (*Triticum turgidum*), barley (*Hordeum vulgare*), sorghum (*Sorghum bicolor*), and a number of legume and oilseed crops (Engels and Hawkes, 1991). Over 90% of Ethiopia's agricultural production still depends on local varieties that are genetically diverse, have evolved in specific environments and contain a multitude of varied and desirable traits. Strategies for agricultural development aim to increase production and productivity by promoting just a few crops and modern varieties. Plant genetic resources (PGR) that were previously maintained and used by farmers have become at risk of genetic erosion (Worede *et al.*, 1999). Since the 1970s, the Ethiopian government, in collaboration with international donors, and conservation and development agencies, has been attempting to counter this threat. In 1976, the Plant Genetic Resources Centre/Ethiopia (PGRC/E), now the Institute of Biodiversity Conservation (IBC), was established as the national gene bank of Ethiopia. Since its foundation, IBC has aimed to balance both *ex situ* and *in situ* conservation. Such an integrated approach was, and continues to be, an example for PGR programmes in Africa, and in the world as a whole.

The founder of the institute, Dr Melaku Worede, has always emphasized the need to ensure that activities concerning the conservation of PGR reach farmers indirectly, through efforts in breeding and research, and that these activities should be structured in such a way that the farmers themselves can both contribute to conservation and directly benefit from this conservation through use (Worede and Mekbib, 1993). Consequently, the institute, under Dr Melaku's leadership, has been promoting *in situ* conservation, through on-farm management, which has distinguished Ethiopia from many other public programmes (FAO, 1996, 2001). Over the years, the institute has collaborated with a number of international organizations or programmes, including the Seeds of Survival Programme of the Unitarian Service Committee of Canada (USC, Canada); the Global Environment Facility (GEF); Bioversity International; and the Community Biodiversity Development and Conservation (CBDC) Programme. Dr Regassa Feyissa, leading author of this chapter, was responsible for implementing several of these programmes in the 1990s. IBC has played a critical role in initiating on-farm management activities in Ethiopia, as well as in other countries in Africa.

The integration of both off-farm and on-farm conservation efforts is essential. The situation in Ethiopia following the tragic drought of 1984 is in itself a good argument for promoting such integration. When the gene bank attempted to restore the PGR that were lost as a consequence of the drought, it was unable to reach those farming communities hardest hit with the amount of seed needed to support a quick recovery. Since gene banks normally hold samples that are small in size, the multiplication of seed for redistribution to farmers after the crisis took years. No matter how efficient conventional gene banks are in multiplying and disseminating seed, it will always take a long time for farmers to benefit from the back-up service that the gene banks can provide them with.

In 2003, Dr Regassa Feyissa established the non-governmental organization (NGO) Ethio-Organic Seed Action (EOSA). This NGO entered a vital niche (i.e. organic seed of local varieties), and as such was able to support many of the country's on-farm management activities through development projects. EOSA provides technical support and oversight for practices contributing to on-farm management, and promotes the use of local crops and varieties in agricultural development. Genene Gezu and Bayush Tsegaye, co-authors of this chapter, have been working on EOSA programmes for a number of years. The organization works closely with local and national government extension agencies, including national agricultural research institutes and universities. Co-author Temesgen Dessalegn provides support to several farmers' groups associated with EOSA, through his research work at the Holeta Agricultural Research Centre. Professionals like these have been working with farmers' groups engaged in on-farm management in Ethiopia for more than 20 years, thereby laying a foundation for motivating farmers to assume their responsibilities in processes that can be described as community biodiversity management (CBM).

Vital role of farmers' dynamic management of local diversity

Following a long-standing tradition, farmers retain the seed stock of local varieties unless disruptive circumstances prevent them from doing so. In this way, these farmers have been conserving varieties for generations, maintaining seed or planting material in their gardens, backyards, farm fields and traditional storage facilities (Worede *et al.*, 1999). Farmers cultivate local varieties, and develop new ones that will most likely be adapted to their specific environment, but that will above all satisfy farmers' specific needs and preferences. They maintain a high degree of genetic heterogeneity in their varieties, thus ensuring continued adaptability under changing conditions (Teshome *et al.*, 1999). This adaptability is the main reason why farmers usually prefer genetically diverse varieties instead of uniform, stable varieties. This use of diversity within varieties, and across locations, serves both conservation and crop improvement purposes. It offers the maximum opportunity for the continuation of co-evolutionary processes, without compromising either diversity or productivity. The informal seed system can therefore be considered the basis for the dynamic management of local varieties by farmers (Balcha and Tanto, 2008). The fact that the informal seed system is the major source of seed

(Bishaw *et al.*, 2008) reflects a need to integrate on-farm management strategies into agricultural development in Ethiopia. Such integration has the potential to result in win–win situations, in linking the conservation and use of local PGR with agricultural development, under changing climatic conditions. In this way, conservation and diversity go hand in hand with agricultural development, and are no longer contradictory forces (De Boef *et al.*, 2010).

Community seed banks: a central pillar of on-farm management

The need to integrate the formal gene bank with the informal seed system led to the establishment of community seed banks (CSBs). IBC began supporting the establishment of CSBs in the 1990s, in collaboration with its national and international partners. On its foundation, EOSA took over responsibility for supporting the CSBs. Today, we at EOSA continue to be engaged in a process of training, equipping and motivating scientists, extensionists and farmers in the conservation and development of PGR, with the CSB playing a central role in this conservation effort. Shrestha *et al.* (Chapter 2.8) provide a global overview of the CSB as a CBM practice, including some highlights of our experiences in Ethiopia.

Farmers need be able to access the right quantity of affordable seed, of their preferred varieties, at the right time. We noticed that the CSBs did not initially respond adequately to this need, since their original focus was on maintaining seed security in more general terms. IBC, at that time known as the Plant Genetic Resources Centre/Ethiopia, began to reintroduce local varieties in 1989, in collaboration with the Seeds of Survival/Ethiopia (SoS/E) programme of the USC Canada. The local varieties were multiplied by certain CSB members and disseminated among all the other members, with some small samples of seed being maintained for conservation. In this way, seed security became linked to PGR conservation, through the CSBs. In 1993–1994, IBC, with support from, and in collaboration with, the Global Environment Facility (GEF) and the Community Biodiversity Development and Conservation (CBDC) Programme, integrated the establishment of community seed banks and on-farm conservation into the SoS/E programme. Over the years, the USC, Canada has continued to provide support for the genetic enhancement of local varieties, seed multiplication and distribution (Feyissa, 2000).

Currently, the CSBs are organized as legal entities, under the umbrella of so-called 'conservation cooperatives', and have a well-established structure for management and governance. After almost two decades of investment in human resources and infrastructure, and support through institutional and technical assistance, the conservation cooperatives are now well organized and sustainable. In Table 1.4.1, we share the current status of CSBs associated with EOSA's conservation and development programmes. Figure 1.4.1 indicates their distribution in Ethiopia. Today, the CSBs serve as community centres for organizing local support for farmers in PGR conservation and climate change adaptation. They act as community-level seed and germplasm reserves, where farmers store samples for retrieval during planting season, and where representative samples of local varieties are set aside as germplasm reserves. As a CBM practice, they reduce the threat of genetic erosion, while increasing food

Table 1.4.1 Characterization of community seed banks that are currently supported by Ethio-Organic Seed Action (EOSA) in Ethiopia, 2011

Region	Location	District	Year of establishment	Type of organization	No. of members		No. of crops in conservation and seed production		Amount of seed in '000 kg		Other activity (2011)
					2003[a]	2011	2003[a]	2011	2003[a]	2011	
Amhara	South Wello	Kallu, Harbu	1998	Cooperative	223	377	3	11	8.7	10.5	Number of women members increased
	South Wello	Wore Ilu	1998	Cooperative	362	672	8	14	9.8	13.7	Number of women members increased
	South Wello	Kallu, Fontenina	2009	Farmer group	—	233	—	6	—	3.5	Cooperative legalization applied for by the CSB
Oromia	East Shewa	Lume	1998	Cooperative	1000	234	7	6	36.3	72.5	Expansion of services, foundation laid for new CSB
	East Shewa	Gimbichu	1998	Cooperative	300	512	4	6	36.3	81.4	Women's groups organized separately for economic empowerment
	West Shewa	Dendi	2011	Cooperative	—	73	—	8	—	6.1	New CSB established, based on lessons learned from East Shewa

a Source: Tanto and Balcha (2003).

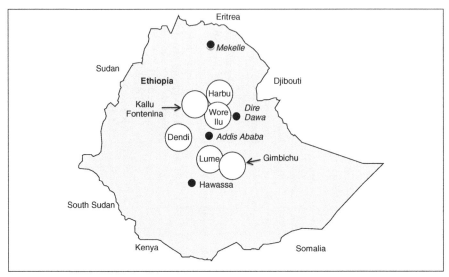

Figure 1.4.1 Location of the community seed banks associated with Ethio-Organic Seed Action (EOSA) in Ethiopia, 2011.

security; moreover, they contribute to social organization at community level, thereby enhancing autonomy of the community in livelihood development.

We also use practices like diversity fairs, seminars and workshops, disseminating information through publications, to raise awareness of the value of local crops and varieties at research and policy levels. Our aim is to contribute to a better integration of formal and informal seed systems, in a manner that contributes to PGR conservation and use. A direct output of this advocacy and awareness work is that some regional governments have started to support and finance the establishment of CSBs.

Lessons learned for sustaining farmers' contribution to on-farm management of PGR

CSBs are crucial for enhancing farmers' awareness of the value and potential of local crops and varieties. Over time, through their involvement in the CSBs, the pride of communities concerning their heritage of crops and varieties has been growing, motivating them towards the restoration, conservation and use of PGR. The function of the CSBs was initially to create such awareness and act as a point of access to PGR, and, subsequently, to serve as a local backup facility, where local crops and varieties could regain their original or even increased value. In this way, should these communities be faced with droughts like those in the 1980s, they will be better prepared, and will no longer be dependent on IBC for restoring their local varieties.

An important lesson learned is that on-farm management cannot rely on providing direct production subsidies to farmers (i.e. paying farmers to maintain varieties). Instead, the focus should be on enhancing the use and values of local crops and varieties, and promoting non-market and market incentives for farmers to continue to cultivate them (Tsegaye, 2005). This approach is important for making the practice of conservation economically viable and self-supporting, and for embedding it in farmers' livelihood strategies.

The controversy over whether *in situ* and *ex situ* conservation are cost effective makes little sense as both are important and each has its particular advantages and shortfalls. While the relationship between farmers and their varieties remains limited and evolutionary processes are interrupted under *ex situ* conservation, the strategy is nevertheless an important security measure in case of disasters.

The experiences gained over the course of our work in using CSBs to achieve on-farm management have resulted in the maintenance and use of a higher degree of crop diversity by participating farmers (Bezabih, 2008). CSBs benefit communities with their organizational capacity and by raising awareness of genetic resources in livelihood development (Engels *et al.*, 2008), thus enhancing the social capital needed to sustain on-farm management as a conservation strategy.

1.5   The diversity kit

Restoring farmers' sovereignty over
food, seed and genetic resources
in Guaraciaba, Brazil

*Adriano Canci, Clístenes Antônio Guadagnin,
Jair Pedro Henke and Luciane Lazzari*

The decline in use of farm-saved seed and varieties

Access to quality seed has become one of the underlying developmental needs of farming communities for achieving food security and sustainable development. The advent of the green revolution in Brazil heralded a massive loss of farmers' varieties, which were replaced by a few high-yielding and hybrid varieties. Small-scale farmers in Guaraciaba, in the western part of the state of Santa Catarina (Figure 1.5.1), recall that up until as recently as the 1970s they were only growing their own local varieties of major staple crops. However, by the 1990s these had already vanished from most of their farms. The resulting dependency of farmers on the seed of external sources has been a burden for them, in terms of the high costs involved with such cultivation.

The Micro-watershed Development Programme was implemented by Santa Catarina State Enterprise for Rural Development and Extension (Epagri), from 2004 to 2009, with funding from the World Bank. In the municipality of Guaraciaba, communities associated with micro-watershed areas decided to dedicate several activities

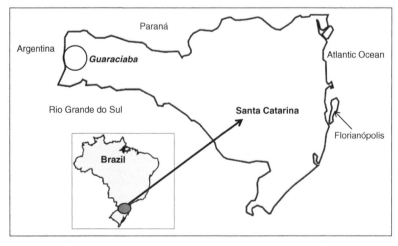

Figure 1.5.1 Map illustrating the location of Guaraciaba, Santa Catarina State, Brazil.

within the programme to community biodiversity management (CBM). One of the most successful of these activities, the diversity kit, focused on addressing the aforementioned decline in use of farm-saved seed and varieties.

We developed the concept of the diversity kit with the aim of regaining seed security and enhancing our subsistence in food production, thereby restoring our self-esteem and food sovereignty. The diversity kit itself is an assemblage of seed that is selected by farmers, containing both local and improved varieties but without either hybrid or transgenic seed. We became involved in the diversity kit process in Guaraciaba through our various roles – as micro-watershed development project facilitators (Adriano Canci and Luciane Lazzari); municipal extensionist (Clístenes Antônio Guadagnin); and leader of a community-based organization (CBO), the Rio Flores Micro-watershed Development Association, which was set up and supported as part of the Micro-watershed Development Programme (Jair Pedro Henke).

The origin of the diversity kit as a concept

In 2004, during several meetings held with the communities in Guaraciaba over the course of implementing the micro-watershed programme, one of the farmers, Iracema do Carmo Weimann, asked us to revitalize the farmers' habit of saving and exchanging seed to produce, rather than purchase, vegetables for home consumption. She inspired us to seek strategies that would support and facilitate farming communities to reduce household costs. This topic was identified as a priority, especially with regard to basic staple foods like rice, beans and several vegetables. In 2005, several farmers, joined by co-authors Adriano and Clístenes, participated in a training course at Chapecó, where success stories related to CBM in Nepal were shared by Bhuwon Sthapit of Bioversity International, who highlighted the CBM practice of the diversity kit. Soon after this course, we adopted the process of the diversity kit as a strategy for farmers to regain access to their own food, seed and varieties.

The diversity kit and its multi-step process

The diversity kit in Guaraciaba evolved as a collective approach to share seed, ideas and practices that promote the debate concerning the restoration, conservation, use and management of plant genetic resources (PGR). The process was innovative because it attempted to facilitate a participatory and multi-stakeholder approach to connect different segments of society, ranging from government, farmers' associations, Epagri, universities, local and regional institutions and the donor community. This process was made up of the following five steps, which are illustrated in Figure 1.5.2.

Step 1: Understanding the local context

We carried out participatory rural appraisals to understand the food production status at household level. The diagnosis included an inventory of crop varieties to document the availability of local diversity and the associated knowledge. We used

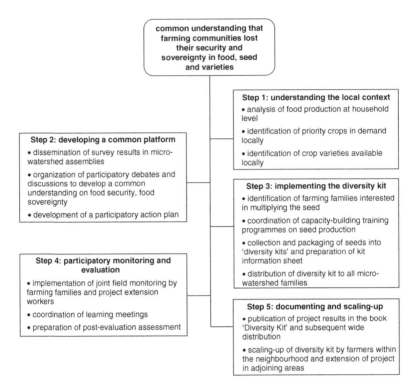

Figure 1.5.2 The steps of the diversity kit process in Guaraciaba, Brazil.

preference-ranking during focus group discussions with farming households to iden-tify the crops most in demand.

Step 2: Developing a common platform

The CBO provided farming households, Epagri and project staff with a common platform for sharing the results of the diagnosis, selecting crops for seed production, and identifying farmers who were interested in, and committed to, seed production. Bhuwon Sthapit, together with academics from the Federal University of Santa Cata-rina, shared experiences concerning participatory methods on diversity assessment, participatory varietal selection and seed conservation. This contributed to achieving a common understanding on project goals and to developing a participatory imple-mentation plan.

Step 3: Implementing the diversity kit

We identified a number of farming households for producing quality seed, based on their social status. The assumption was that by providing services to the CBO, they would earn additional income, and this would also enhance their capacity to continue to generate such income in the future. We strengthened the seed production capacities of those farming households identified. Epagri purchased the 'basic' seed that farmers needed for producing seed for the diversity kits.

Each kit consisted of several small packets, containing at least four to eight different crop species, and up to five varieties of each crop that had been identified during the first two steps. We provided detailed information on all crop varieties included, covering name, use-values and cultivation practices. Each kit box was labelled with pictures of local rice and bean crops, illustrating local food culture. We distributed the kits to all farming households in the community.

Step 4: Participatory monitoring and evaluation

The diversity kit project facilitated the organization of joint field monitoring activities on a regular basis, involving seed-producing farmers and project staff. We organized learning and sharing meetings with the CBOs responsible for monitoring the progress in the micro-watersheds, and for making necessary plans for the next cropping season.

Step 5: Documenting and scaling-up the diversity kit

We integrated field monitoring and a post-evaluation process as part of a learning process, and subsequently disseminated the information extensively within the CBOs. We documented the experiences in several chapters of the book, *Diversity Kit* (Canci *et al.*, 2010), which was written by farmers, project staff, and PGR professionals. The book was widely distributed, locally, regionally and nationally, and led to a scaling-up of the practice in the adjoining villages and municipalities.

The diversity kit: more than just a box of seed

Increase in access to seed of choice

Over the course of the project, a total of 300 farming households were able to benefit from the diversity kit, obtaining access to more than 16 food crops and 53 varieties. As a result of the implementation of the diversity kit, farmers obtained better access to quality seed of varieties and crops of their choice. One major change resulting from the diversity kit process was that an increasing number of households began to grow more crops and a wider range of varieties. Members of 16 farming households produced seed for the diversity kits, and their capacities in quality seed production and multiplication were enhanced during the process. Now, these farmers produce and exchange farm-saved seed, thus revitalizing the informal seed system.

Restoration of knowledge and socio-culture

Knowledge on traditional use-values and even local crop cultivation practices had almost completely disappeared in western Santa Catarina. This accompanied the loss of local crops and varieties as a result of the industrialization of agriculture. The seed of local varieties, and the detailed information sheets contained in the diversity kits, contributed to the restoration of local knowledge and practices. In addition, the diversity kits incorporated other, more subjective elements, such as the pleasure experienced by farming households in producing their own food, and an overall increase in their well-being and happiness. The image depicted on the kit boxes, showing a father and a mother harvesting food, reinforces cultural values shared by communities over time.

Self-reliance on household food production and food sovereignty

At the beginning of the project, it became clear to us that the dependency of farmers on markets to get staple foods, such as rice and beans, was a major challenge. Our target was to rebuild self-reliance on farm-grown food. The importance of this goal was further highlighted by the results of a survey we carried out in Guaraciaba in 2007. The survey showed that before the kits were distributed only 25% of farming households were growing rice and 50% beans. Upon distribution of the kit in some of the communities these figures increased to 58% for rice, and to approximately 90% for beans. In addition to this, more than 35% of households interviewed considered the fact that they were able produce healthy food again at household level an important benefit (Vicente *et al.*, 2010).

We used the kit as a tool for debating food security and sovereignty, for enriching the area and promoting sustainable development. Furthermore, the kit facilitated the raising of awareness of the need to reduce, or even abandon, the use of pesticides in food production, for improving the dietary habits, health and general well-being of the farming households. The right of households to protect their food production and cultivation, with the assurance that they are consuming healthy products, promotes true food security and sovereignty.

Lessons learned

We observed that the farmers' interest in experimenting with new varieties was one of the main incentives for planting the seed included in the diversity kits. This illustrates the fact that the farmers' innovative spirit and curiosity, despite having been repressed by the green revolution and industrialization of agriculture, is still alive and can be reinvigorated. It also shows the profound relationship of farming households with their environment, and their continuous observation of the results of their work. The economic impact of the diversity kit is also an important factor. When a household produces its own food, it ensures an increase in its income by reducing the costs of household food purchases. This factor, coupled with the fact that the farmers are now cultivating varieties of a known origin and quality, thus avoiding heavily processed and industrialized food, enhances the quality of food and livelihoods. It is clear that the diversity kit project was successful in motivating farming households to produce their own food. Moreover, they now realize that their participation in this CBM practice has been an important tool for regaining food sovereignty.

1.6 Practices that contribute to the empowerment of farming communities for managing their agrobiodiversity in Asia

Joy Angelica P. Santos Doctor

SEARICE: catalysing community biodiversity management

The South-East Asia Regional Initiatives for Community Empowerment (SEARICE) has been working with farmers since 1996 to enhance their skills in implementing community biodiversity management (CBM) in five countries in Asia: Bhutan, the People's Democratic Republic of Laos, the Philippines, Thailand and Vietnam. Our work is firmly rooted in processes that encourage awareness-raising and contribute to the empowerment of farming communities in the management of their agrobiodiversity and seed systems. Through these processes, farmers are made aware of their own strengths and weaknesses, allowing them to (re-)gain access to, and control over, their livelihood assets. In addition, these processes strengthen the farmers' capacity to make decisions autonomously, in a manner that conserves and sustains their agrobiodiversity. It is our goal to enhance the empowerment of farming communities by building their capabilities to engage in local and innovative research, and by promoting the recognition of, and respect for, men and women farmers, as co-equals of scientists and plant breeders.

SEARICE plays a catalytic role within communities in the five target countries for achieving CBM. It creates a space for small-scale farmers to assess their situations, and to address socio-economic and environmental issues, so that they can become autonomous, propelling forces for influencing the development of policies, and for defining their political agenda. In this way, the farmers can identify what they want and need, and can address, both in capabilities and through advocacy, the structural constraints that have been preventing them from achieving their human potential.

Farmer breeding: ensuring conservation and control over local varieties

Through the work of our partners in Asia, farmers have become conscious of the growing corporate control over seed, and of the need to assume responsibilities concerning their own access to and control over plant genetic resources (PGR). Through participation in farmer field schools organized by SEARICE partners, farmers realized that PGR were gradually being lost. To address this loss, our partners launched efforts to conserve local varieties, building bridges between plant breeders and farmers. One of the ways to achieve this was to carry out farmer and participatory breeding. Breeding institutions provided pre-breeding materials to farmers, who then began to produce new materials according to their own preferred traits, and adapted to farmer management. These farmer-bred varieties are resistant to pests and diseases, do not depend on chemical inputs and have shorter maturity periods. Such characteristics lower production costs and increase productivity (Almekinders *et al.*, 2006). The experiences of several organizations that have been collaborating with SEARICE on the implementation of CBM are summarized below:

- In Bhutan, the Biodiversity Use and Conservation in Asia Programme (BUCAP) was initiated by the National Biodiversity Centre, in close partnership with the *dzongkhag* (district) agriculture offices in various dzonghkhags, and with the Renewable Natural Resources and Research Centres in four regions of the country. Farmers selected 32 rice and two maize varieties, which were recognized by the Bhutan Varietal Release Committee. Previously, farmers relied only on just a handful of varieties, selected at the International Rice Research Institute in the Philippines.
- In Laos, a national project team, consisting of the Plant Protection Centre of the Department of Agriculture, the Ministry of Agriculture and Forestry and the Rice and Commercial Crop Research Centre, coordinated a project that produced 67 rice varieties, over a period of nine years. The Provincial and District Agriculture and Forestry Offices implemented the project at provincial and district levels.
- The Philippine Rice Research Institute, a public breeding programme, released 55 inbred lines between 1994 and 2004, whereas SEARICE partners in the Philippines released 209 farmer-bred varieties during the same period.
- Since 1995, SEARICE partners in Vietnam have produced 250 farmers' varieties, from crosses and selections; 45 traditional varieties; and over 400 local varieties, which are being maintained by farmers through rehabilitation. This is in sharp contrast to the previous situation, where the farmer's role in rice breeding had become virtually non-existent, following the introduction and cultivation of the green revolution rice variety, *Than Nong* 8 (IR8).

Farmers' rights and advocacy

Farmers associated with our projects expressed the need to influence policies and the political arena, in order to advance farmers' rights within the context of CBM. This motivated us to work together to make changes in the policy arena. Those changes we were successful in achieving include, among others:

- the development of local mechanisms to recognize and protect farmers' rights (i.e. to save, use, exchange and sell seed, and improve their own varieties);
- a ban on the introduction of genetically modified organisms in certain areas;
- the incorporation of farmers' seed in local government seed policies, including those related to organic agriculture programmes;
- the participation of farmers in policy discussions with decision-makers at national levels, and in international agreements and treaties.

CBM case studies

Here I share three case studies describing the experiences of groups of small-scale farmers who work with our partner organizations, involving strengthening their capacity to influence decision-makers to adopt CBM to help the *in situ* conservation of PGR.

Thailand: reviving the Bun Kao Mai ceremony

Bun Kao Mai is a traditional Buddhist ceremony dedicated to 'Mother Rice' or to the rice spirits, which celebrates and refers to diversity in rice varieties, bountiful harvests and excellent seed. Following the green revolution, the famers had little to offer to the rice spirits and so *Bun Kao Mai*, along with its blessings of diversity and excellent seed, faded.

In 1996, the Joko Learning Centre, through the local civil society network Hug Muang Nan (Love Nan Group), in Nan Province, partnered with SEARICE to implement the Community Biodiversity Development and Conservation (CBDC) programme, similar to the CBDC programme that was implemented in Ethiopia, as described by Feyissa (Chapter 1.4). CBDC's work in Nan Province focused on the *in situ* conservation and development of PGR, and on rice in particular. CBDC in Nan Province included a series of practices, as part of a CBM process, aimed at repatriating local rice varieties; supporting household seed storage capacities; establishing community seed banks; distributing seed to farmers; facilitating local seed exchange; implementing participatory varietal selection and plant breeding; and conducting on-farm trials with local varieties through farmers' field schools. In addition, PGR conservation and development was included in the curricula of primary education and monk schools. These efforts led to the revival of local rice varieties; the development of a multitude of farmers' varieties; and a wide availability of seed, which in turn resulted in better harvests.

Consequently, farmers had enough reasons to express their gratitude to the rice spirits, and as such CBDC was instrumental in the revival of the *Bun Kao Mai* ceremony. Farmers believed that *Bun Kao Mai* inspired them to work towards CBM. The project evolved into an event that exhibited CBM principles and practices, and raised awareness was on the need to conserve PGR, among policy-makers, politicians, monks, teachers, students, farmers, tourists and other visitors. Consequently, the *Bun Kao Mai* event became a platform for motivating the community to learn about local, regional and global issues relating to agricultural biodiversity.

Local government was so inspired by the *Bun Kao Mai* event that they allocated funds to ensure its regular organization. In addition, the work of teachers and students, in encouraging young people to appreciate the importance of the local culture associated with rice, was recognized and they obtained numerous awards for their efforts in this regard. The local government also formulated a policy supporting the work of farmers in seed production and variety development. Through networking in the *Bun Kao Mai* platform, the Joko Learning Centre has become a member of the National Health Committee, which formulates national policies on food safety. Furthermore, Nan Province has been designated a model for the creation of provincial community rice production centres.

Bhutan: contributing to the development of enabling policies

A number of practices that contribute to the on-farm management of PGR were initiated through the Biodiversity Use and Conservation in Asia Programme (BUCAP) in 2001. Its overall goal was to contribute to poverty alleviation through the conservation and sustainable use of PGR. The programme included practices aimed at achieving on-farm conservation, and the development and use of rice and maize genetic resources in 11 provinces, directly benefiting over 289 households, in 18 project sites in Bhutan.

In the initial stages of BUCAP, the science of PGR conservation was a fairly new subject for our Bhutanese partners. One of our immediate priorities was to enhance the technical capacity of those partners. We also wanted to foster close ties with the national gene bank in Bhutan, which secures PGR for holistic and sustainable agricultural development, and provides farmers with free and easy access to the seed of its accessions. Over 1200 germplasm accessions are being conserved in the national gene bank and about 300 accessions are being held in three community seed banks. In addition, new crops and crop varieties have been made available to the farmers through participatory varietal selection (PVS) in high yielding and locally adapted crops and varieties. In Bhutan, BUCAP has contributed to enhancing household food security and broadening the use of local crops and varieties. Some examples of the results of BUCAP include:

- Two new maize varieties have been added to the five existing varieties.
- 30 new rice varieties have been evaluated and introduced through PVS.
- Upland rice has been introduced into 13 communities as a new crop, to diversify the maize-based cropping system.
- New varieties of vegetables, legumes and cereals have been introduced into five communities, promoting food and nutrition security at household level.
- Local rice and maize varieties are being maintained by farmers through improved seed purification and storage management, thereby contributing to on-farm management.

As part of our advocacy work at SEARICE, we provided space within BUCAP for policy-makers to interact with farmers, and become acquainted with the results of the project. This exposure resulted in the development of provisions on farmers' rights to save and exchange seed for inclusion in the Biodiversity Act of Bhutan. Furthermore, the Variety Release Committee (VRC) has acknowledged farmers' initiatives on PVS, and recognized farmers' varieties. PGR activities have been mainstreamed in the development plan of the Royal Government of Bhutan. The draft National PGR policy has recognized the informal seed system and promotes sustainable agricultural development through the conservation and use of PGR.

Vietnam: recognizing farmers as breeders

In Vietnam's Mekong Delta, the main rice-producing region of the country, demand for rice seed, based on the total area available for plantation, is around 0.55 million tons. The formal seed supply system, which includes research institutes, seed centres, extension services and seed companies (government, non-government and foreign), cannot meet such a demand. The formal system deals primarily with formally released varieties and markets the seed through labelled units of volume, officially sanctioned by law.

The need for quality seed that meets the demand of farmers prompted SEARICE partners to establish seed clubs. In order to do this, farmers were first trained through the farmers' field school (FFS) method. The aim of the seed clubs is to engage these farmers in the seed production and distribution of quality seed for community use. Since 2010, 358 seed clubs have been set up, involving 5857 farmers in seed production in the Mekong Delta. Our partners provide the seed clubs and individual farmers with segregating materials from early generations (F_2–F_5), in order to facilitate breeding and selection.

The role of the seed clubs in the Mekong Delta has become significant, as they have been able to respond adequately to the high demand for rice seed, and provide more seed than the formal seed supply system can. In 2008, for instance, the seed clubs produced and supplied more than 16% of the total seed demand in the Mekong Delta, while the formal seed sector only produced about 3.5%. Also in 2008, in An Giang, one of the 13 provinces where the seed clubs have been established, the seed clubs were able to supply more than 90% of the provincial seed demand. In recognition of this major achievement, the Vietnamese government acknowledged the SEARICE-assisted programme in An Giang province as a breakthrough programme.

Farmers have benefitted from the seed clubs through lower seed prices. The price of seed marketed by the seed clubs was 26% lower than that of seed sold through the formal system. The formal system has to cover costs of transport, processing, storage and packing. Seed club members do not incur these costs as they sell directly to customers at harvest time. In terms of quality, the formal seed system is controlled by field inspections and seed testing. Seed quality assurance in the seed clubs is based on direct contact between buyers and sellers, including observation of seed production demonstration plots.

The seed clubs also produced a diverse number of varieties. These varieties are in high demand for a number of reasons, including for use in diverse cultivation conditions; for increasing resistance to ever-changing pests and diseases; and for meeting household production and taste requirements, and specific demands for quality in local markets. Two varieties produced by farmers gained recognition in the Mekong Delta. The HD1 and TM3 varieties have good characteristics and are suitable in acid sulfate and saline soils. A farmer in Kien Giang province, Mr Nguyen Van Tinh, developed the HD1 variety, which was officially certified on 23 December 2010, at national level, by the Ministry of Agriculture and Rural Development. The whole process from selection and tests, to documentation and official release, took nine years. Results of the national rice variety testing, which was carried out on the winter–spring crop of 2006–2007, showed that the HD1 variety had a good yield and was of good quality. It was further reported to be resistant to brown plant hopper and yellow draft disease. This achievement confirms the fact that farmers, through breeding, are capable of releasing good varieties that can compete with those produced by the formal sector. The case of HD1 is the first time that a farmers' variety has been officially released by the Vietnamese government. Results from the national tests show that almost all of the promising farmers' varieties pass the varietal trials and can be released at national level, in the same way that varieties produced by the formal sector are.

Although the seed clubs made the informal seed supply system flourish, their activities ran afoul of Vietnam's Plant Variety Protection (PVP) Law. The Law prevents farmers from breeding rice varieties and sharing them, and consequently the Tan Binh seed club (Dong Thap) and Ngo Khue enterprise have been fined for supplying the seed of non-released farmers' varieties within their community. However, in 2008, recognizing the impediments of the PVP Law, and acknowledging the work of seed clubs, the Ministry of Agriculture and Rural Development issued Decision 35/2008/BNN, which officially recognizes the existence of the informal seed system, and acknowledges farm-saved seed. This policy has opened up opportunities for farmers to develop new varieties and produce seed of those local varieties with support from local authorities (Tin *et al.*, 2011).

Lessons learned

The experiences of SEARICE and its partners in Asia, as presented through the case studies described above from Thailand, Bhutan and Vietnam, show that in order to contribute to the empowerment of farmers and their communities in agrobiodiversity management, we need to address aspects related to seed supply, breeding and PGR conservation. Only through such an integrated approach, can we contribute to the empowerment required to achieve *in situ* conservation, or the on-farm management of PGR. As illustrated by the case studies from Thailand and Vietnam, SEARICE and its partners focus on seed supply and breeding, by using methods such as farmers' field schools to strengthen farmer organizations in agrobiodiversity management.

Policy and legal frameworks can create impediments, but because of the significant results of our work with partners at grassroots level in the five target countries, we have been able to influence the adaptation of genetic resource policies to address farmers' rights in the context of seed production and farmer breeding, and recognize farmers as custodians of agrobiodiversity. This recognition at policy level and, above all, the self-recognition of farmers, contributes to their empowerment and motivates their continued participation in community biodiversity management.

Note

The current chapter is based on reports prepared by Ditdit Pelegrina, Normita Ignacio, Golda Hilario and Singay Dorji; interviews with Samruoy Phadphon and Pimonphan Sakitram from the Joko Learning Censtre and the Hug Muang Nan Foundation in Thailand; and on a research study published in Tin *et al.* (2011).

1.7 The Maison de la Semence Paysanne and diversity platform

Promoting agrobiodiversity in France

Jennifer Kendall and Elodie Gras

Bertrand Lassaigne: a pioneer in recognizing the value of farmers' varieties

After receiving his agricultural diploma, Bertrand Lassaigne worked for several technical institutes and seed companies, evaluating and registering varieties, and assessing the effectiveness of several phytosanitary products. In the late 1980s, he decided to return to the land of his grandparents in Périgord and become a farmer. He attended a seminar on organic farming and resolved to adopt the practice of organic farming. He began selling his products locally, often directly to users who appreciated their local origin. Since most seed in the market is not organic, he also started to produce his own seed. Maize seed and varieties remained a problem because only hybrid varieties were available.

In 2000, he discovered that a particular company was selling genetically modified (GM) maize in south-west France, which was a shock to him. The production of organic maize was still in its infant stage in the region where his farm was situated. He considered the possibility that one of neighbours would start to cultivate GM maize as catastrophic, as he wanted to maintain his production absolutely free of genetically modified organisms (GMOs). Because of the absence of local or farmers' maize varieties in France, Bertrand decided to return to the origin of maize, Guatemala, where he observed that small-scale farmers were still growing traditional and open-pollinated varieties. The voyage further motivated him to promote the use of local diversity in his region, and also more widely in France (Zaharia, 2009).

Initial steps: restoring the conservation and use of local maize varieties

An experimental project was set up in 2001 at AgroBio Périgord, the Association for the Development of Organic Agriculture in Périgord, in south-western France (Figure 1.7.1), with the aim of finding alternatives to hybrid maize seed. We evaluated 11 maize varieties that Bertrand had collected from farmers in Guatemala. As it turned out, the varieties were unfortunately not adapted to the French climate and production system. During this same period, small-scale farmers across France, as well as Italy and Spain, began to abandon the cultivation of hybrid maize and look for local or farmers' varieties (*variétés paysannes*), following the introduction of

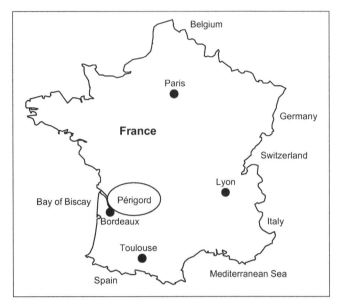

Figure 1.7.1 Map illustrating the location of Périgord in France.

Source: AgroBio Périgord, unpublished data.

GM maize. In the second year of our project, we evaluated local maize varieties from several regions in France and neighbouring countries. But our first experience with the material from Guatemala, although unsuccessful, was crucial as it led us to the decision to carry out tests on local maize varieties on an annual basis.

In this chapter, we share the experiences of Agrobio Périgord with the diversity platform and Maison de la Semence Paysanne (or small-scale farmers' seed bank), which have gradually become drivers for promoting farmers' use, exchange and experimentation of local varieties in France. The chapter is complemented by that of Lassaigne and Kendall (Chapter 5.4), which details experiences working with farmer and participatory maize breeding in France.

We refer to small-scale farmers, throughout this chapter, using the French term *paysans*. These farmers, or paysans, use a diversity of crops and farm-saved seed, and primarily sell their produce at local, regional and, increasingly, organic markets. They operate parallel to the agro-industrial chain, which is dominated by large-scale farmers who use hardly any diversity in their production. In comparison with farmers in many developing countries, the scale of operation of paysans in France, in terms of area and economy, is by no means small. However, their mode of production and use of diversity is quite distinct from the mainstream agricultural sector.

Diversity platform

Following our decision to search for open-pollinated maize varieties, we focused on gathering varieties from farmers who maintain local varieties for their own personal use. We searched in regions across France and beyond, where specific culinary traditions ensured the conservation of local varieties. The National Agricultural Research Institute of France (INRA) provided us with some accessions from its gene bank and we began to test their capacity to adapt to the climatic and growing conditions in our region. Little by little, we expanded the number of maize varieties, maintaining those that had interesting characteristics. This work evolved into a 'diversity platform', which is a space where farmers can experiment with, promote and learn about local varieties. Every year, the platform establishes five types of diversity blocks, each with a specific function (Bio d'Aquitaine/AgroBio Périgord, 2011):

1   *Introductory blocks* are used to evaluate varieties that have been newly introduced under common production and management conditions; the varieties are compared with those varieties that are well adapted, are included in demonstration blocks and are further multiplied.
2   *Multiplication blocks* are used to maintain varieties; the maintenance is conducted under controlled pollination.
3   *Breeding blocks* involve farmer and participatory breeding, through a number of steps, which are described by Lassaigne and Kendall (Chapter 5.4).
4   *Variety demonstration blocks* are used to compare the selected local varieties. During regular field days, farmers evaluate the varieties, thereby contributing to awareness raising and enhancing farmers' skills.
5   *Crop diversity demonstration blocks* aim to raise awareness among visitors on new crops that may be interesting to include in their production system.

The diversity of varieties and crops included in the diversity platform in 2010 are illustrated in Figure 1.7.2.

Maison de la Semence Paysanne

The idea to establish a seed bank began to take shape in the mind of Bertrand during a trip to Brazil in 2003. In southern Brazil, the concept of community seed banks has been adopted as a means to enhance the autonomy and seed security of farmers. Maison de la Semence Paysanne literally means 'house of small-scale farmers' seed'; and it has become an integral part of our dynamic management of agrobiodiversity. Its members are from the Périgord region and beyond, and as such we cannot really call it a community seed bank, unless the term 'community' can be understood in this context as a network of farmers or group of people that share a common interest.

Objectives, collection and members

The objectives of the Maison de la Semence Paysanne are to contribute to the revitalization of local varieties, and ensure the maintenance of farm-saved seed that is free from GMO contamination. Furthermore, we aim to enhance farmers' autonomy; strengthen their capacity to develop and select varieties that are adapted to organic farming and for low-input agriculture; and maintain agrobiodiversity and associated

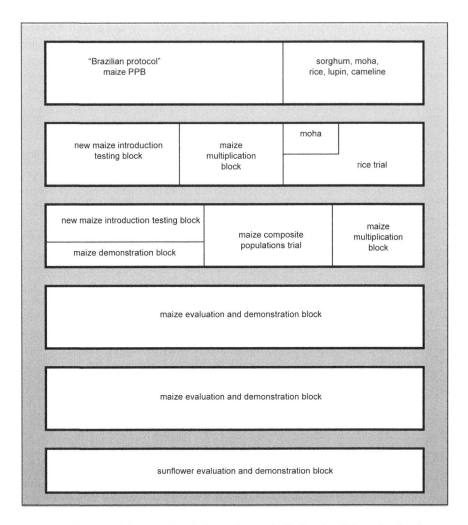

Figure 1.7.2 Layout of the diversity platform of AgroBio Périgord at Ribeyrolle, Le Change, France, 2010.

Source: AgroBio Périgord (2010).

knowledge for future generations. We began with a collection of composite maize varieties that had almost disappeared from farmers' fields in France. Today, we have over 100 maize varieties, more than ten sunflower varieties, and several varieties of soybean, buckwheat, moha (Hungarian grass), lupine, and a number of vegetable and fodder crops. The Maison de la Semence Paysanne has evolved from regional to national coverage. In 2010, more than 250 farmers from all over France were associated with this seed bank. By 2010, it was active in more than 12 regions of France, with more than five farmers from each region participating in the maintenance, experimentation, selection and demonstration of local varieties. Figure 1.7.3 shows its gradual expansion in members and its coverage of farmers across regions of France. In 2011, although the Maison de la Semence Paysanne had about 100 maize varieties in stock, farmers were only cultivating about 25 varieties. Although it can be assumed that, in addition to the 25 varieties formally registered as being cultivated, some farmers continued to cultivate another dozen or so, on a more informal basis. We disseminated seed to more than 60 new farmers as a way of motivating the farmers to cultivate a greater diversity of varieties.

Farmers formally join the Maison de la Semence Paysanne for only a couple of years, but many of them continue to cultivate and exchange local varieties after their

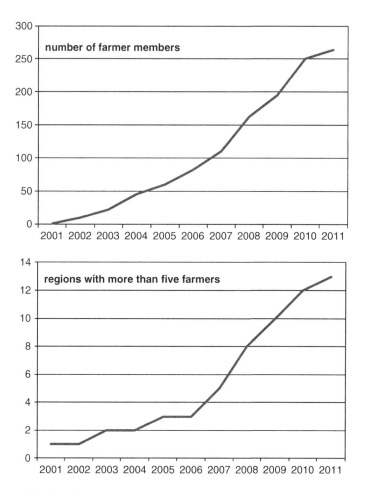

Figure 1.7.3 Number of member farmers and number of regions with more than five member farmers associated with the Maison de la Semence Paysanne of AgroBio Périgord, 2001–2010.

Source: AgroBio Périgord (2010).

membership ends. In addition, more than 30 farmers and amateur gardeners contribute to the conservation, experimentation and exchange of seed of several vegetable varieties in the seed bank.

We do not have large quantities of seed for many local varieties, because formal gene banks supply much of the seed and they only provide us with small user samples. Another reason why we do not maintain large quantities of seed of local varieties is because farmers are put off from using them due to their low productivity. However, such local varieties may have other interesting traits like taste, flavour, colour, historical or cultural values, or else they may be considered farmers' patrimony. Moreover, they are still a source for the future creation of composite varieties. Maize and sunflower are, however, allogamous species, and cannot be conserved in a multiplication platform. They need to be isolated to avoid hybridization; with a minimal distance of isolation of 300 m for maize, and 700 m for sunflower. Amateur gardeners multiply the seed of rare maize and/or sunflower varieties in their gardens and return all the seed to the Maison de la Semence Paysanne. This shows the importance of this kind of partnership between farmers and gardeners for conserving and using agrobiodiversity.

The collection: accessing, evaluating and including new varieties

Our first sources of varieties are the farmers themselves, who send us materials with the aim of contributing to conservation. In addition, we are sometimes able to identify varieties that we do not yet have in our collection during diversity fairs in France or other European countries. In these cases, we approach the farmers and ask them to share some seed. We also look for specific materials, approaching the national gene bank at INRA, or research organizations, to complement our collection. Before including materials in our collection, we conduct observation trials. We look for local varieties in a manner that balances agronomic, economic, environmental and nutritional qualities, including adapted varieties and/or those that have interesting traits. Once we decide to include a variety in our collection, we insert the relevant passport data into our database, providing the ethnobotanical description, including name, origin, place, name of the farmer or institution providing the materials, and the number of years the variety has been cultivated by the farmer. Before any seed is included in the Maison de la Semence Paysanne, we test it to detect the presence of any GMOs using leaf analysis. In this manner, we keep our collection GMO-free. Figure 1.7.4 illustrates the flows of varieties within our network, the diversity platform and Maison de la Semence Paysanne.

Membership: facilitating the exchange of seed

The farmers and gardeners who wish to participate in the Maison de la Semence Paysanne, by contributing seed and multiplying varieties, sign a contract for experimentation and multiplication. Above all, they adopt a moral stance in contributing through use to the maintenance and improvement of the local variety. Members return part of the seed they produce to restock our collection, in a process that echoes that of the community seed banks in many developing countries, or of the seed net-

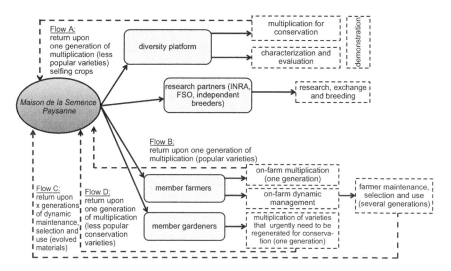

Figure 1.7.4 Overview of the variety flows between the Maison de la Semence Paysanne, diversity platform, research partners and member farmers.

Notes: INRA, National Agricultural Research Institute of France; FSO, Farm Seed Opportunities project.

Source: AgroBio Périgord, unpublished data.

works in northern America and other European countries, as described by Shrestha *et al.* (Chapter 2.8). Through the contract, the Maison de la Semence Paysanne and its members engage themselves in a type of experimentation that enables them to bypass the restrictive variety and seed legislation in France, which considers the exchange and sale of non-certified seed and non-registered varieties illegal, as described in more detail by Kastler (Chapter 6.8). The varieties that are cultivated by each member are registered, as well as their production data. A database records this information and allows us to closely monitor the use of local varieties, acting as a kind of variety register similar to the community biodiversity register, as described by Subedi *et al.* (Chapter 2.4).

A local group with national dynamics

As AgroBio Périgord is a local farmer organization, the first goal was at local and not national level. However, we were the first and only organization to work on local maize varieties and, consequently, many farmers from different regions of France approached us to participate in the Maison de la Semence Paysanne. So, now the organization is structured in two levels, one with the local group of farmers and amateur gardeners in Périgord, and the other with a national network of farmers' groups (Bio d'Aquitaine/AgroBio Périgord, 2011). In 2010, one of these groups, which is based in the Poitou-Charentes region, established the organization Cultivons la Biodiversité en Poitou-Charentes ('we cultivate biodiversity in Poitou-Charentes').

Organizational structure

The Maison de la Semence Paysanne of AgroBio Périgord is an institutionally, politically and morally independent organization. It depends on public funding, primarily in the form of subsidies from the regional government, for 70% of its budget. We therefore consider it vital that a structure for the collective and dynamic transfer of knowledge and material is developed, to ensure that the social relations that make the Maison de la Semence Paysanne work do not disappear when funding is no longer be available. Our aim is to embed what we do in existing farmers' networks and practices.

Practices contributing to CBM in France

The diversity platform, the Maison de la Semence Paysanne, and farmer and participatory breeding of local varieties are intrinsically linked. The Maison de la Semence Paysanne has a role in conceptualizing and institutionalizing the collective management of local varieties. It provides a structure for the legal exchange and distribution of seed of local varieties. The diversity platform motivates, capacitates and unites farmers for using diversity. We realize that in our context, a community is not the same as those in developing countries. Our activities in AgroBio Périgord are organized at a regional and national level, through our linkage with the Réseau Semences Paysannes (National Seed Network). Our community has become a national network of farmers who approach diversity in a common and collective manner. In our efforts towards achieving the conservation and cultivation of agrobiodiversity, we use similar approaches to those used by the farmers who originally inspired us in Guatemala, Brazil and other countries. We hope that by sharing our experiences, we can demonstrate that cultivating agrobiodiversity is a matter that concerns us all, whether we are small-scale farmers in the North or the South. We are all farmers who need and use the diversity of our local varieties in a dynamic manner to sustain our livelihoods, thereby providing our services to society in producing good and healthy food. Moreover, we work to guarantee food sovereignty and conserve biodiversity for future generations.

1.8 Community biodiversity management and *in situ* conservation of plant genetic resources

Walter Simon de Boef and Marja Thijssen

The dilemma of implementing *in situ* conservation and on-farm management

Despite the fact that both the Convention on Biological Diversity (CBD) and the International Treaty on Plant Genetic Resources for Food and Agriculture (ITPGRFA) recognize the importance of the *in situ* conservation strategy, formal plant genetic resources (PGR) programmes have been slow to implement it. As indicated by Thijssen *et al.* (Chapter 1.1), those programmes face the dilemma of how to incorporate *in situ* conservation activities into their day-to-day work, or rather, following the terminology of the Food and Agriculture Organization of the United Nations (FAO, 1996), how to contribute to on-farm management. Few are purposely engaged in on-farm management. Dias *et al.* (Chapter 2.5) describe, for example, how public gene banks associated with the Brazilian Agricultural Research Corporation (Embrapa) promote the reintroduction of accessions to farming communities. Feyissa *et al.* (Chapter 1.4) outline how, since 1989, the Institute of Biodiversity Conservation in Ethiopia (then known as the Plant Genetic Resources Centre/Ethiopia) has been implementing activities for contributing to on-farm management by supporting the establishment of community seed banks. Other national and international PGR programmes have a strong association with non-governmental organizations (NGOs). The agrobiodiversity project in Nepal, described by Subedi *et al.* (Chapter 1.2), was successful because of the partnerships it fostered between the NGO Local Initiatives for Biodiversity, Research and Development (LI-BIRD) and national (National Agricultural Research Council) and international (Bioversity International) PGR programmes. The PGR programme in Ecuador has developed a strong linkage with a community-based organization (CBO), as illustrated by Tapia and Carrera (Chapter 2.3). Following the reintroduction of gene bank accessions, the national programme supported the CBO through practices such as diversity fairs, diversity blocks and awareness-raising on agrobiodiversity in tourism and education. Kendall and Gras (Chapter 1.7 of) describe how the Maison de la Semence Paysanne of AgroBio Périgord accesses germplasm from the public gene bank in France, though the farmer management of these materials is further sustained through informal structures independent from the public conservation programme. Moreover, most public PGR programmes that contribute to on-farm management are either experimental (e.g. the agrobiodiversity project in Nepal), or depend on specific funding (most other examples). They are neither implemented as part of mainstream PGR programmes, nor funded through government resources destined for PGR conservation.

The Second Report on the State of the World's Plant Genetic Resources for Food and Agriculture confirms this observation, indicating that NGOs and associated CBOs have a strong capacity for, and the most experience in, implementing on-farm management (FAO, 2011). The current chapter analyses experiences of public, non-governmental and community-based organizations engaged in conservation and development, as case studies. Our aim is to provide insights into, and experiences on, moving beyond the dilemma of how to implement on-farm management of PGR.

In situ conservation of agrobiodiversity as an oxymoron

While 'conservation' implies the halting of change, or the maintenance of PGR in their particular state, when aiming to achieve *in situ* conservation by farmers, what are targeted are the active dynamics of farmer or community management of local crops and varieties. Accordingly, the term '*in situ* conservation of agrobiodiversity' emerges as an oxymoron (i.e. it seems contradictory but, surprisingly, expresses a reality). The Global Plan of Action (GPA; FAO, 1996) found a way of dealing with this dilemma, by adopting the term 'on-farm management'. However, the limited implementation of on-farm management since the GPA was formally adopted remains a drawback to the commitments undertaken by national governments, as set out in both the CBD and the ITPGRFA.

Analytical framework that emphasizes dynamism and interactions in agrobiodiversity

To move beyond the dilemma, we assess the efforts of conservation and development organizations by using an analytical framework inspired by the work of Harold Brookfield. We have adapted his description of 'agrodiversity' (Brookfield, 2001) to define agrobiodiversity as follows.

Agrobiodiversity is a dynamic and constantly changing patchwork of relations between people, plants, animals, other organisms and the environment, which always copes with new problems, and always finds new ways.

We formulated three research questions:

1   If agrobiodiversity constitutes the dynamic and constantly changing patchwork of relations between people, plants, animals, other organisms and the environment, how can this dynamism and these relations be maintained?
2   Should '*in situ* conservation of agrobiodiversity' be treated as an oxymoron?
3   With what approach and practices can conservation and development organizations contribute to the *in situ* conservation of agrobiodiversity?

In order to characterize the practices aimed at contributing to on-farm management, we analyse the practices, examine their impact on the dynamics, and assess the approach towards institutionalizing relationships for the maintenance of agrobiodiversity by rural communities. We aim to respond to the questions, thereby gradually contributing to the social construction of 'community biodiversity management' (CBM) as a methodology for achieving *in situ* conservation.

Matrices for assessing practices and responding to the questions

For case studies, we primarily use those practices that are shared in several chapters in this volume. The cases are characterized and assessed using three matrices. We characterize the practices in the first matrix for dividing the responsibilities among conservationists, development agents and farming communities using the following criteria: (a) location of the practice; (b) type of material maintained; (c) person responsible for maintenance or management; and (d) objective of the practice.

In the second matrix, we disentangle the effect of the practices on dynamics for three domains: (a) genetic and ecological dynamics, assessing the linkage between the material (seed and varieties) and the biotic, abiotic and human environment; (b) social and economic dynamics, assessing the linkage with dynamics, in social and economic structures and values; and (c) cultural dynamics, assessing the linkages with issues such as ethnicity and identity, cosmovision and spirituality. The terms used to describe the dynamics are elaborated in Table 1.8.1.

In the third matrix, we address the effects of the practices on institutional aspects at the farming community level, looking at: (a) whether awareness is raised among farmers and rural communities on the conservation and use of agrobiodiversity; (b) the type of (community) organization established or required to implement the practice and contribute to a degree of collective action in PGR management; (c) the degree of autonomy of the community in conservation and associated activities; and (d) the degree of recognition that these community efforts gain within wider PGR frameworks.

Table 1.8.1 Characterization of terms used to describe the effect of practices of on-farm management on the dynamics of agrobiodiversity

Typology	Characterization within domain
Disconnected	The dynamics no longer function; the practice results in 'freezing' the dynamics
Managed for	The dynamics are sustained; the practice aims to contribute to conservation
Managed for use	The dynamics are sustained, motivated by the farming communities' use of agrobiodiversity
Revitalized	The dynamics are restored, following their loss or degradation
Strengthened	The dynamics are strengthened, becoming more vital than before the practice
Market-supported management	The dynamics are strengthened; the market supports this process

Source: De Boef *et al.* (2012).

Case studies

Wheat gene management zones in Turkey (TU)

A gene management zone was established in the early 1990s in Turkey for the *in situ* conservation of wild relatives of wheat (*Aegilops* spp., *Tricicum* spp.). The zone is managed as a habitat at the Ceylanpinar State Farm, which is run by the national PGR programme. Neighbouring rural communities lost their access to the area with the establishment of the gene management zone. The basic purpose of the zone is to sustain the ecosystems through evolutionary changes and ensure the continuity of the crop's wild relatives (Ertug Firat and Tan, 1997).

Coffee forest conservation in Ethiopia (ET)

The national PGR programme in Ethiopia uses *in situ* strategies to conserve wild and semi-wild coffee (*Coffea arabica* L.) within so-called coffee forests. Feyissa *et al.* (Chapter 3.6) describe how this *in situ* effort ensures the conservation of the Afromontane forests with their natural coffee populations. Farmers collect forest coffee, but are not allowed to encroach upon the natural forest and damage the coffee stands and shade trees. They sell an increasing amount of their produce in special markets, where forest coffee gets a premium price.

Community seed banks in India (IN)

The M.S. Swaminathan Research Foundation (MSSRF) supports community seed banks for maintaining food security in the tribal villages of the Jeypore Tract, in Orissa State. The seed banks conserve local varieties and ensure the provision of quality seed of local varieties in times of emergency. If required, they provide the farmers with grain so that they can maintain their seed for cultivation. The community seed banks are gradually being transformed into CBOs to ensure their sustainability. When the seed banks reach this point of sustainability and autonomy, MSSRF slowly withdraws its support, while maintaining, for security reasons, a repository of the local varieties in a gene bank that is managed under a black box arrangement (see Chapter 2.8).

Diversity platform and Maison de la Semence Paysanne in France (FR)

AgroBio Perigórd is a CBO in France that aims to support farmers in their use of agrobiodiversity, while also contributing to their autonomy. As described by Kendall and Gras (Chapter 1.7), each year the CBO cultivates a wide diversity of local varieties of maize, sunflower and several other crops in a diversity platform that functions for awareness-raising and training, experimentation and demonstration. 'Maison de la Semence Paysanne' means, loosely translated, small-scale farmer seed bank. More than 250 member farmers and amateur gardeners, from several regions of the country, benefit from, and contribute to, the restoration and use of local varieties.

Diversity theatre in Nepal (NP1)

Diversity theatre has become an important practice for creating awareness of biodiversity in Nepal. Local actors put on plays based on traditional stories and myths concerning local crops and varieties. The practice creates awareness among communities and their members on the importance of these varieties, and motivates CBOs to take control of decision-making and the management of their biodiversity assets. Shrestha *et al.* (Chapter 2.2) share more details on these practices.

Community biodiversity registers in Nepal (NP2)

Another practice that was further developed by LI-BIRD and partners in this project is that of the community biodiversity register. This practice is vital for actively engaging CBOs in biodiversity management. As described by Subedi *et al.* (Chapter 2.4), the community biodiversity register contains records of PGR. It prevents knowledge erosion, protects against biopiracy and strengthens community decision-making in conservation and development.

(Re-)introduction of local varieties in Brazil (BR1)

The Krahô tribe is an indigenous group that lives in the Brazilian savannah in the state of Tocantins. Following the advice of local extension agencies in the 1970s, the Krahô farmers began to purchase seed instead of maintaining and using their own. When Embrapa started supporting the Krahô in the late 1990s, leaders of the tribe demonstrated an interest in agrobiodiversity. They visited a gene bank to identify maize accessions that were comparable to those they had lost in the 1970s. As detailed by Dias *et al.* (Chapter 2.5), Embrapa continued to provide support to the Krahô for more than ten years. This support led to the joint organization of an annual seed fair, for celebrating diversity and promoting seed exchange among several indigenous peoples.

Diversity kits for promoting food security in Brazil (BR2)

Farming communities in the municipality of Guaraciaba, in the state of Santa Catarina, decided to use diversity kits to restore food security and revive the habit of growing their own food. Community members received a kit containing the seed of several maize, rice and pulse varieties. The diversity kit became an instrument for restoring farmers' autonomy in food, seed and varietal security (see Chapter 1.5).

Community management of faxinais in southern Brazil (BR3)

Farmers in the southern part of the state of Paraná manage, in a collective manner, domesticated landscapes that are referred to as *faxinais*. They extract *pinhão* seeds from the araucaria trees (*Araucaria angustifolia*) and leaves from yerba mate plants (*Ilex paraguariensis*); both products are primarily commercialized. Farming households associated with the faxinais also rear local breeds of pigs in the communal areas. All products obtain premium prices in urban areas. These communities, through their close relationship with the forest, are considered traditional people (see Chapter 3.2). Peroni *et al.* (Chapter 3.4) describe the efforts made to recognize the faxinais as a sustainable development reserve within the Brazilian system for nature conservation, as a step towards guaranteeing the protection of the landscape and supporting the associated livelihood of the *faxinalense* peoples.

Results: assessment of the case studies

Case characterization

We provide a characterization of the nine case studies in Table 1.8.2. A clear distinction emerges concerning the responsible organizations. The first two case studies (TU and ET) are examples of situations where conservation organizations and their professionals are in control. This is in contrast to the other case studies, where community committees exist or are established within CBOs with an emphasis on agrobiodiversity management. IN and FR are comparable, since both cases address community seed banks. In India, MSSRF takes a livelihood approach, supporting CBOs in reaching seed and varietal security. AgroBio Périgord in France is a CBO in which a national network of farmers form a community of farmers with a shared interest in the conservation and use of local varieties. In Nepal, LI-BIRD focuses more on the knowledge and information components related to agrobiodiversity, through diversity theatre and the biodiversity register (NP1 and NP2). The three case studies from Brazil illustrate a wide variation in situations in which conservation activities are embedded within existing CBOs, thus on-farm management strengthens or revitalizes structures and capacities associated with the conservation and use of local crops and varieties (BR1 and BR2) and agricultural landscapes (BR3).

Assessment of dynamics

The assessment of agrobiodiversity dynamics distinguishes the case studies through their contribution to conservation and use (Table 1.8.3). A distinction can be made between the three case studies that focus mainly on habitat conservation (TU, ET and BR3). Forest coffee conservation (ET) and the collective management of the faxinais (BR3) have a strong community or human component, which is strengthened by the niche market for speciality coffee or non-timber forest products. However, for the case study on wheat gene management zones (TU), both the socio-economic and cultural dynamics are frozen; human dynamics are considered a threat. The community seed banks in IN and FR lead to revitalizing and strengthening the dynamics of the three domains. NP1 and NP2 strengthen the social and cultural dynamics associated with local crops and varieties in particular. BR1 enhances cultural dynamics through the reintroduction and revitalization of ethnic, spiritual and also social and economic (in terms of autonomy) domains of diversity. BR2, the diversity kit, focuses on access to quality seed and on the use of local crops and varieties (genetic and ecological dynamics), but also strengthens the socio-economic and cultural dynamics, through supporting awareness-raising and collective actions targeting the sustained use of agrobiodiversity.

Table 1.8.2 General characterization of case studies and practices aimed at contributing to on-farm management of agrobiodiversity

Case study[a] /practice		General characterization			
		Location	Material	Responsible	Objective
		Where?	What?	Who?	Why?
TU	Wheat gene management reserve	Fenced reserve	Habitats with wild relatives of wheat	State farms manager/ curator	Continued co-evolution, research
ET	Coffee forest conservation	Denominated area	Coffee forests	Curator/area manager	Conservation and use
IN	Community seed banks	Community seed bank	Local varieties	Community committee	Livelihood security and conservation
FR	Diversity platform and Maison de la Semence Paysanne	Community seed bank	Local varieties	Community-based organization	Conservation and use
NP1	Diversity theatre	Community members	Agrobiodiversity	Non-governmental organization	Awareness-raising
NP2	Community biodiversity register	Community members	Information on agrobiodiversity	Community committee	Documentation, protection and monitoring
BR1	(Re-)introduction	Community members	Local varieties; gene bank accessions	Curator and community structures	Awareness raising and reintroduction
BR2	Diversity kit	Community members	Local crops and varieties	Community committee	Household food security and conservation
BR3	Collectively managed forests	Community forest	Landscape with NTFPs[b] and pigs	Community committee	Conservation and use

Source: De Boef et al. (2012).

a Country codes: TU, Turkey; ET, Ethiopia; IN, India; FR, France; NP, Nepal; BR, Brazil.

b NTFPs, non-timber forest products.

Table 1.8.3 Case study analysis on the effect of practices aimed at contributing to the on-farm management of agrobiodiversity, on three domains of agrobiodiversity dynamics

Case study[a] / practice	Effect of practice on dynamics of agrobiodiversity[b]		
	Genetic and ecological domain	Social and economic domain	Cultural domain
TU Wheat gene management reserve	Managed for conservation	Disconnected	Disconnected
ET Coffee forest conservation	Managed for conservation	Market-supported management	Strengthened
IN Community seed banks	Managed for use	Strengthened	Strengthened
FR Diversity platform and Maison de la Semence Paysanne	Managed for use, revitalized	Strengthened	Revitalized and strengthened
NP1 Diversity theatre	—	Strengthened	Strengthened
NP2 Community biodiversity register	—	Strengthened	Strengthened
BR1 (Re-)introduction of varieties	Revitalized	Revitalized	Revitalized
BR2 Diversity kit	Revitalized, strengthened	Revitalized, strengthened	Revitalized, strengthened
BR3 Collectively managed forests	Managed for sustainable use	Market-supported management	Strengthened

Source: De Boef et al. (2012).

a Country codes: TU, Turkey; ET, Ethiopia; IN, India; FR, France; NP, Nepal; BR, Brazil.

b Effects of the practices on the specific agrobiodiversity dynamics are characterized in Table 1.8.1.

Assessment of institutional aspects

The institutional assessment demonstrates the embedding of conservation practices within local or community structures (Table 1.8.4). TU, for example, by focusing on wheat gene management, involves no such structures; its conservation intervention is carried out excluding local communities from the conservation effort. In contrast, ET and BR3 address habitat and landscape conservation as a strategy, with strong human influences. In ET, farmers are enabled to market forest coffee. BR3, collectively managed forests, forms a basis for the establishment of a conservation unit, with the aim of maintaining all the dynamics, in all domains. The other case studies either support the establishment of conservation-oriented CBOs (IN and FR), or strengthen existing organizations (NP2, BR1, BR2). In both structure and function, these CBOs are or become autonomous, and in the end should operate independently from formal or non-governmental conservation and/or development organizations. IN, FR, NP2, BR1, BR2 and BR3 illustrate how CBOs are increasingly able to make conscious decisions on the conservation and use of agrobiodiversity, by recording and maintaining diversity. BR1 and BR2 enhance the access to genetic diversity for CBOs and farmers. FR facilitates farmers' access to local varieties at both local and national level; however, in order to obtain access and bypass seed laws farmers have to become members. NP1, FR, BR1, BR2 and BR3 increase the understanding of, and strengthen local structures for, community (agro-)biodiversity management. In particular, NP2 contributes to the establishment of community institutions (like the community biodiversity register). BR3 explores ways to establish a formal conservation unit associated with the traditional livelihood and landscape management system.

Agrobiodiversity conservation strategies, dynamism and relations

The assessment of the case studies shows that when defining what constitutes '*in situ* conservation', we tend to refer to a situation where farming communities utilize and maintain agrobiodiversity. Consequently, on-farm management of agrobiodiversity should be described as a situation consisting of 'relations between people, plants, animals, other organisms and the environment'. The practices serve within a process in which relations and dynamism are continued or strengthened. As such, the conclusion can be drawn that, as a conservation strategy, '*in situ* conservation' is indeed an oxymoron. It is a process, or an emerging property, to which conservation and development organizations contribute (De Boef, 2000). The set of practices, or the CBM methodology, becomes the means by which such dynamic processes are continued, strengthened or revitalized.

As described by Shrestha *et al.* in Chapter 1.3, CBM can only be achieved by recognizing and strengthening communities and their institutions from the onset of any intervention by a conservation or development organization. Although several of the case studies share isolated practices, CBM is a methodology that requires integration; it links conservation to sustainable livelihood development. The key point is to institutionalize local level decision-making on agrobiodiversity conservation and use, within a livelihood context, but also for creating sustainable structures. The case

Table 1.8.4 Case study analysis on the institutional aspects of practices aimed at contributing to the on-farm management of agrobiodiversity

Case study[a]/practice		Institutional analysis[b]				
		Awareness	Local organization	Autonomy	Recognition	
TU	Wheat gene management reserve	None	State farm	Not relevant	PGRFA programme
ET	Coffee forest conservation	Unique resource with market value	Forest reserve and marketing cooperatives	Cooperatives linked to value chain	Recognized through premium price
IN	Community seed banks	Diversity and livelihood security	CBO	CBO, NGO facilitated	CBO is recognized custodian
FR	Diversity Platform and Maison de la Semence Paysanne	Diversity and access to local varieties	CBO	CBO organized at local and national level	CBO functions through membership
NP1	Diversity theatre	Focus of practice	NGO facilitated	NGO facilitated	Custodians themselves become aware
NP2	Community biodiversity register	Supported through documentation	CBO managed	CBO monitoring and decision-making; NGO facilitation	Legal recognition of farmers and CBO as custodians
BR1	(Re-)introduction of local varieties	Culture of diversity is revitalized	Indigenous CBO	Indigenous CBO, public gene bank supported	Indigenous people are recognized for their custodianship
BR2	Diversity kit	Household food security, food and variety sovereignty	CBO	CBO managed, public project supported	Farmers are custodians as a consequence of becoming more aware
BR3	Collectively managed forests	Faxinais associated with culture of conservation and livelihood	CBO	CBO managed, formally recognized as formal conservation unit	CBO and collective system foundation for formal conservation unit

Source: De Boef *et al.* (2012).

a Country codes: TU, Turkey; ET, Ethiopia; IN, India; FR, France; NP, Nepal; BR, Brazil.

b CBO, community-based organization; NGO, non-governmental organization; PGRFA, plant genetic resources for food and agriculture.

study of domesticated landscapes in Brazil illustrates how the community management of biodiversity is recognized and can be formalized as a conservation unit in a sustainable development reserve. CBM reinforces the capacity of farming or user communities and their institutions, as highlighted in the third matrix. What is critical is that the CBM methodology facilitates and enhances decision-making, for securing community access to and collective control over their resources.

In this way, CBM is socially constructed as a participatory method that aims to recognize and contribute to the empowerment of community institutions to capitalize on their assets in terms of agrobiodiversity and associated knowledge. It also supports the collective organization and management of resources, while enhancing the notion of (bio)culture that is associated with local varieties, crops and landscapes. Embedded within a livelihood approach, CBM bridges conservation and development objectives that were for a long time considered conflicting. It ensures that communities have the capacity to manage the biological and genetic resources that they depend upon, now and in the future.

CBM aims for communities to take control of the maintenance and use of their genetic resources, and as such it is clearly distinguished from *ex situ* conservation strategies. For on-farm management to be effective, farming communities must be in control and able to make decisions (i.e. govern their agrobiodiversity). Conservation practices are no longer solely embedded in formal research and development structures, disconnected from farming communities; they are deeply rooted within sustainable livelihood development strategies that aim to contribute to the empowerment of farming communities to manage agrobiodiversity.

Note

The current chapter has been adapted from the following article: De Boef, W.S., Thijssen M.H., Shrestha, P., Subedi, A., Feyissa, R., Gezu, G., Canci, C., De Fonseca Ferreira, M.A.J., Dias, T., Swain, S. and Sthapit, B.R. (2012) 'Moving beyond the dilemma: practices that contribute to the on-farm management of agrobiodiversity', *Journal of Sustainable Agriculture*, vol 36, pp. 788–809.

Part II

Practices contributing to community biodiversity management

2.1 Practices contributing to community biodiversity management

Introduction

Walter Simon de Boef and Abishkar Subedi

Synopsis

Community biodiversity management (CBM) is a process or methodology made up of a set of practices. It is the strength of the sequence and combination of practices in which community members engage themselves, either autonomously or supported by conservation and development organizations, that results in the collective and conscious management of agrobiodiversity. Only when this situation emerges can the *in situ* conservation or on-farm management of agrobiodiversity be achieved (De Boef and Thijssen, Chapter 1.8). Part II focuses on sharing a range of practices that are commonly used by different types of organizations, complementing the series of more introductory, conceptual descriptions of CBM processes presented in Part I.

Experiences of the global agrobiodiversity project implemented by Local Initiatives for Biodiversity, Research and Development (LI-BIRD), together with partners in Nepal, are presented in this section. Shrestha *et al.* (Chapter 2.2) introduce the diversity fair, diversity block, diversity kit and various practices that relate to biodiversity and culture. Subedi *et al.* (Chapter 2.4) explore the practice of the community biodiversity register (CBR), and Shrestha *et al.* (Chapter 2.9) elaborate on how the CBM trust fund sustains CBM processes.

Tapia and Carrera (Chapter 2.3) share the experiences of the national PGR programme in Ecuador, which uses a wide variety of practices to support farming communities in Cotacachi in assuming their responsibility in CBM. Dias *et al.* (Chapter 2.5) show how public gene banks in Brazil are engaged in the reintroduction of accessions into indigenous, traditional and small-scale farming communities, as a means of contributing to on-farm management.

Dias (Chapter 2.7) describes the modalities of the Seeds of Passion network of community seed banks (CSBs) in Paraíba, Brazil, which is one the oldest and largest networks of CSBs in the world, with more than 200 functioning CSBs. Fuentes and Alonzo (Chapter 2.6) describe how in Guatemala and Honduras, community seed reserves, a variation on CSBs, have been designed and are being implemented as a strategy for strengthening community resilience. Based on our observation that CSBs are common and are, in many cases, drivers for the CBM processes, Shrestha, together with other CBM colleagues (Chapter 2.8), compares CSB modalities in nine countries and tries to identify some common lessons learned.

Community biodiversity management and practices

CBM as a methodology incorporates many practices that are used both by community-based organizations (CBOs) concerned with conservation and use, and by external stakeholders that aim to strengthen communities in conservation and livelihood development. Practices can be single actions that are geared towards enhancing awareness of conservation and diversity; increasing an understanding of diversity, associated social structure and institutions; strengthening capabilities within CBOs; and encouraging communities to make informed decisions over their agrobiodiversity. Such single action practices include diversity fairs, diversity kits and diversity blocks. Practices can also aim towards the development of community institutions, such as the CSB and the CBR, requiring a process of capacity development, and the establishment and consolidation of institutional and working modalities. These are not single-action practices; they constitute complex, multi-year learning processes. Other, similar practices address more entrepreneurial aspects, such as the value addition of products of local crops and varieties; and the seed production and marketing of local varieties. In the current chapter, we aim to look at the implementation of those practices shared in Parts I and II, within the wider context of the construction of the CBM methodology.

Overview of practices

In Table 2.1.1 we give an overview of the practices described in the first two parts. The diversity fair, diversity block and diversity kit, as well as a number of practices that address the more specific cultural, religious and spiritual aspects of biodiversity, are very common. As single-action practices, they contribute to a larger process, independent from the original objective of the organization facilitating CBM. It is logical that when the public gene banks are engaged in CBM, the reintroduction of gene bank accessions is common. In Ecuador, it is the local CBO that takes responsibility for the *in situ* multiplication of gene bank accessions, thereby building a more reciprocal relationship.

   The community seed bank (CSB) is a common practice associated with community seed production. CSBs often form a basis for establishing institutional modalities to sustain CBM. Three experiences share variations of one practice – the community biodiversity register in Nepal, the community biodiversity catalogue in Ecuador and the village knowledge centre in India. We realize that this practice is complex; it appears only to function within more consolidated CBM processes. Most of the organizations involved in the experiences that are shared in the first two parts of the book are engaged in six or more practices. Experiences from Nepal, France and Ecuador include 10 or 11 different practices, which is an indicator of the vision of the facilitating organization and its CBOs in the CBM process.

Practices for strengthening the CBOs

We analysed the stakeholders that facilitate the CBM process, and identified what type of organization would in the end sustain the CBM process at community level. The

Table 2.1.1 Overview of practices used by conservation and development organizations as part of the CBM processes described in Parts I and II[a]

CBM practices	CBM experience[b]										
	Number	NP	BR1	BR2	BR3	ET	FR	IN	EC	AS	CA
Diversity/seed fair	8	X	X	X	X	X	X		X	X	
Diversity and culture[c]	10	X	X	X	X	X	X	X	X	X	X
Diversity block[d]	9	X	X	X		X	X	X	X	X	X
Diversity kit	6	X	X		X		X	X	X		
Reintroduction of genebank accessions	7		X	X		X	X		X	X	X
Local multiplication of gene bank accessions	1								X		
Agrobiodiversity and tourism	1								X		
Agrobiodiversity and education	3						X		X	X	
Traditional food fair/recipes	3	X					X		X		
Community seed bank[e]	6	X			X	X	X	X			X
Community biodiversity register[f]	3	X						X	X		
Agrobiodiversity products	4	X			X		X		X		
Community-based seed production	7	X	X		X	X	X	X		X	X
Number of practices		9	6	4	6	6	10	6	11	6	5

a The description follows that given in the chapters, based on the knowledge of the authors concerning the work of the facilitating organization, and therefore it is possible that the overview does not cover all the practices of the organizations concerned; it also excludes those practices that involve value addition (Part IV) and participatory crop improvement (Part V).

b Legend and reference to description of practices in Parts I and II: NP – Nepal; LI-BIRD: Chapters 1.2, 1.8, 2.2, 2.4, 2.8 and 2.9. BR1 – Brazil; the Association of the Micro-Watershed/Santa Catarina State Enterprise for Rural Development and Extension (ADM/Epagri), Guaraciaba, Santa Catarina: Chapters 1.5 and 1.8. BR2 – Brazil; several gene banks of the Brazilian Agricultural Research Corporation (Embrapa): Chapters 1.8 and 2.5. BR3 – Brazil; Semiarid Network in Paraíba (ASA/PB): Chapter 1.6. ET – Ethiopia; Ethio-Organic Seed Action/ Institute of Biodiversity Conservation (EOSA/IBC): Chapters 1.4 and 2.8. FR – France; AgroBio Périgord: Chapter 1.7, 1.8 and 2.8. IN – India; M.S. Swaminathan Research Foundation (MSSRF), Jeypore, Orissa, and Kolli Hills, Tamil Nadu: Chapters 1.8 and 2.8. EC – Ecuador; the National Plant Genetic Resources Department of the National Autonomous Institute for Agricultural Research in Ecuador (INIAP-DENAREF) and the Union of Peasant and Indigenous Organizations of Cotacachi (UNORCAC), Cotacachi: Chapter 2.3. AS – Asia (Bhutan, Thailand and Vietnam); South-East Asia Regional Initiatives for Community Empowerment (SEARICE) and partners: Chapter 1.6. CA – Central America (Guatemala, Honduras); the Collaborative Programme on Participatory Plant Breeding in Central America (PPBMA) and partners: Chapter 2.6.

c Culture and diversity covers diversity theatre, diversity poetry and several cultural, spiritual and religious expressions.

d Diversity blocks also include variations of participatory varietal selection and on-farm trials.

e In Central America, the community seed bank is referred to as the community seed reserve.

f In India, the community biodiversity register functions through the village knowledge centre, and in Ecuador through the agrobiodiversity catalogue.

results of this analysis are summarized in Table 2.1.2. In almost all the experiences, the CBM process results in the strengthening and/or establishment of CBOs. In two cases, the CBM process is jointly facilitated by a CBO with a public agency (Guaraciaba, Brazil; Cotacachi, Ecuador). The experiences from Brazil (Semiarid Network in Paraíba, Brazil; ASA/PB) and France (AgroBio Périgord) show the importance of the networks, with the CBOs themselves facilitating the CBM processes. For other experiences from Nepal (LI-BIRD), India (M.S. Swaminathan Research Foundation; MSSRF), Ethiopia (Ethio-Organic Seed Action; EOSA), Asia (South-East Asia Regional Initiatives for Community Empowerment; SEARICE) and Central America (Collaborative Programme on Participatory Plant Breeding in Central America; PPBMA), conservation and development NGOs, mostly in collaboration with public organizations, are facilitating CBM processes that will eventually be sustained by CBOs.

The experiences from Brazil (Embrapa), Ethiopia (Institute of Biodiversity Conservation; IBC) and Ecuador (National Plant Genetic Resources Department of the National Autonomous Institute for Agricultural Research in Ecuador; INIAP-DENAREF) illustrate where public PGR programmes take on the facilitation of the CBM processes and/or are engaged in practices. The contribution to CBM is defined most clearly for Ecuador in the variation and large number of practices, but also by the fact that the key partner is a CBO (the Union of Peasant and Indigenous Organizations of Cotacachi; UNORCAC), which, upon completion of the project, will sustain CBM in Cotacachi. In Brazil, curators associated with the public PGR programmes are engaged in long-term CBM processes (Krahô community), while in other cases the contribution has been more punctual, aiming to revitalize on-farm

Table 2.1.2 Stakeholders involved in CBM processes and practices and their functions, as presented in Parts I and II

Country	Driving stakeholder	Stakeholders[a]				
		Public agri. dev.	Public PGR prog.	Cons. & dev. NGO	CSO network	CBO
Nepal	LI-BIRD	P	P	D/P		P/C
Brazil 1	ADM and Epagri	P				D/C
Brazil 2a	Embrapa (Krahô)		D			P/C
Brazil 2b	Embrapa (others)		D			
Brazil 3	AS/PBA	P	P	P	D	P/C
Ethiopia	EOSA/IBC	P	D	D		P/C
France	AgroBio Périgord	P	P	P	D/C	P
India	MSSRF			D		C
Ecuador	INIAP-DENAREF & UNORCAC		D			F/C
Asia[c]	SEARICE & partners	P	P	D	P	P/C
Central America[d]	PPBMA & partners	D	P	D	P	P/C

a Public agric. dev.: public agricultural development organization; public PGR prog: public plant genetic resources programme; cons. & dev. NGO: conservation and development non-governmental organization; CSO network: network of civil society organizations; CBO: community-based organization.
b Legend for function of the stakeholders in the CBM process and practices: D: driver or leader the project; C: local organization responsible; P: partner.
c Bhutan, Thailand, Vietnam.
d Guatemala and Honduras.

management through reintroduction. In Ethiopia, an NGO (EOSA) took over the role of the public PGR programme on conclusion of an internationally supported project that facilitated the reintroduction of local varieties, investment in the construction of CSBs, and the establishment of conservation-oriented CBOs. EOSA, with international donor support, continues to support the functioning of the CSBs. Its major strategy for sustaining the CSBs is to motivate regional government programmes to take over their role, illustrating the degree to which the institutional sustainability of CBM practices and processes in Ethiopia is still limited.

CBM work in Brazil (ASA/PB), Ethiopia (EOSA), France (AgroBio Périgord), India (MSSRF), Nepal (LI-BIRD) and South-East Asia (SEARICE) is built upon more than ten years of experimentation, learning, consolidation and scaling-up. It should be realized that these processes are facilitated by either conservation and livelihood development-oriented NGOs (EOSA, LI-BIRD, MSSRF and SEARICE), or by networks of civil society organizations (CSOs; ASA/PB and AgroBio Périgord). This compilation of NGOs and CSOs engaged in CBM confirms the observation made in the *Second Report on the State of the World's Plant Genetic Resources for Food and Agriculture* (FAO, 2010) that on-farm management is implemented by NGOs that are not necessarily associated with national PGR programmes.

CBM processes: seeking consolidation and facilitating social learning

Social learning occurs at a stage when the CBM process has reached a level of consolidation, rather than as a concluding step of the CBM process. As such, the difference between CBM processes that are exogenous (those initiated by external agents) and endogenous (those built upon community-based practices and traditions) is no longer valid. Where CBM processes that were originally exogenous reach consolidation and social learning, we can perceive that the process and practices have been appropriated within community structures and capabilities (i.e. they have become endogenous).

Several of the case studies presented have reached a stage of consolidation, as, for example, can be seen with the three case studies in Thailand, Bhutan and Vietnam described by Doctor (Chapter 1.6). CBM processes are supported by SEARICE and its partners, and they have contributed to up-scaling, taking into account the impact on national and local genetic resource policy frameworks. In Begnas and Bara, the two ecosites of the agrobiodiversity project in Nepal, LI-BIRD and partners have concluded their support as facilitators, and local CBOs are now responsible for sustaining and evolving the CBM process. These experienced CBOs have become consultants for promoting CBM processes in other communities and regions, playing a strategic role as grassroots catalysts and facilitators. In India, the CSBs that are supported by MSSRF in Jeypore, Orissa, and the Kolli Hills, Tamil Nadu, form the foundation blocks for establishing CBOs to engage in a series of CBM practices. In most communities, the CBM process has been consolidated with well-functioning and institutionally sustainable CSBs. MSSRF has a clear policy to only support communities in CBM for a short period of time, given its mandate that goes much beyond that of individual communities. It does, however, monitor progress and evaluation in the villages, and provides the back-up facility with its gene bank. Moreover, MSSRF is committed to continue to provide advocacy and policy support in the context of CBM.

In France, the Maison de la Semence Paysanne is a CBO that targets the farmers in the Aquitaine while at the same time serving as a national network of farmers, CSOs and NGOs. It receives some financial support from the regional government, but is mainly supported and run by member contributions and volunteers respectively. Its strategy is to keep a minimal structure in order to ensure its autonomy from external (public) financial resources. The Seeds of Passion network that consists of more than 200 community seed banks, in the state of Paraíba in Brazil, has reached its size and has been consolidated over a period of more than 20 years. It is sustained by farmers' groups, NGOs and CSOs; its strength can be explained by its institutional embedding in society, and its relative autonomy from public organizations.

By sharing the diversity of experiences, we aim to show the evolution of individual practices and combinations of practices into community structures that are autonomous in their CBM, and which are able to adapt their acquired capabilities, modalities and institutions to address the problems they may face in the future. In this way, facilitating organizations and their CBO partners can achieve the goal in which CBM evolves into a situation where communities manage their agrobiodiversity in a collective and conscious manner. The collective structure is vital for getting to the point where conservation and use are no longer dependent on single nodal farmers, but are instead embedded in collective, and therefore more sustainable, structures. CBOs described in those cases have reached the stage best described as where social learning has become part of their collective livelihood strategy. In reaching such a stage, we consider that the CBM process is contributing not only to the conservation and use of agrobiodiversity, but also to community resilience.

Out- and up-scaling CBM

When we refer to 'out-scaling', we mean mainstreaming in a horizontal manner; thus, in the case of CBM, out-scaling means expanding the use of CBM practices and processes to other farmers and communities, or within similar facilitating or community-based organizations. In contrast, we refer to up-scaling as a vertical process, which embeds CBM within local traditions, practices and skills, and even institutions, at community and farmer levels; and within governmental, and conservation and development organizations, policies and structures, at all levels, from local level to global level. The fact that over the years the Seeds of Passion network in Brazil has grown to its current size and has consolidated its structure, shows its strength and relevance in the context of agriculture in the semi-arid region of north-eastern Brazil, a region that is most often associated with poverty and misery. The efforts made by the network to link with research and engage itself in advocacy have resulted in the establishment of a food-purchasing programme by the government to promote the use of seed of the network. Today, the Seeds of Passion network is influencing current discussions on incorporating local varieties and seed sourced through the network into the same governmental seed distribution programmes for which the Seeds of Passion network was initially established as an alternative.

Other experiences with strong elements of out- and up-scaling can be seen in the activities of LI-BIRD and partners in Nepal. The CBM process, with the CSBs, CBRs, conservation-oriented CBOs and the CBM trust fund modalities, is now being implemented within governmental programmes by various ministries and through other NGOs in many districts. LI-BIRD has been promoting innovation through several PGR policies, creating a policy and legal space for CBM and agrobiodiversity in Nepal. As described by Subedi *et al.* (Chapter 1.2), LI-BIRD is encouraging the use of CBM in southern Asia, and is coordinating a global programme that aims to promote CBM in a context of on-farm management and the enhancement of community resilience. The organization's involvement in the evolution and implementation of the CBM methodology in its sites provides the methodological and institutional capital motivating their assumption of this leading role.

This aspect of out- and up-scaling brings us to defining the boundaries of a community within CBM. The experience of the Maison de la Semence Paysanne in France is relevant in this context. This CSO network operates both with local and national boundaries. First, the CSB and other related practices subscribe to a community with a strong local or territorial foundation. Secondly, the CSB responds to a national community of farmers and their CBOs, which is a group with a shared or common interest. The group involved in AgroBio Périgord in managing the CSB balances their work between those two boundaries. However, if we consider up-scaling and out-scaling as vital to CBM, then shouldn't all communities engaged in CBM, independent from where they are, have to find a balance between the boundaries of their territory and the interests that they share with other practitioners both in the North and South? In a world where farming communities are becoming more and more interdependent, and are at the same time seeking alternatives in local development pathways, the issue of boundaries in CBM reaches beyond agrobiodiversity management. This issue relates to options faced by farming communities concerning which development path they should take, as farmers, to maintain their livelihood, and provide our society with their services in food production, while also taking into account a series of environmental, economic, social and cultural factors.

2.2 Enhancing awareness of the value of local biodiversity in Nepal

Pitambar Shrestha, Abishkar Subedi and Bhuwon Sthapit

Introduction

Enhancing awareness among farming communities is important within a community biodiversity management (CBM) process. Awareness-raising practices increase community participation in managing biodiversity in a systematic way by changing their present mind-set and behaviour. Enhancing awareness is also important for encouraging consumers, development workers, policy-makers and farming communities to make continued use of local crops and varieties, thereby contributing to their conservation (Rijal *et al.*, 2000).

Over the last decade, many CBM practices that involve varying degrees of community participation have been developed in order to sensitize communities on the value of local agrobiodiversity, thus forming a basis for the development of local strategies for the conservation and sustainable utilization of PGR (Sthapit *et al.*, 2008a). In this chapter, we share our experiences of working on CBM with Local Initiatives for Biodiversity, Research and Development (LI-BIRD) in Nepal (Subedi *et al.*, Chapter 1.2). We describe five practices – diversity fairs, rural poetry, rural drama, diversity blocks and diversity kits – that target awareness-raising and contribute to understanding diversity as initial components of a CBM process.

The diversity fair

The diversity fair is a popular practice employed by various organizations to sensitize communities on the value and importance of biodiversity and traditional knowledge (Rijal *et al.*, 2000). The diversity fair has its origin in the Andean countries (Tapia and Rosa, 1993), and since the early 1990s has spread globally as a practice used by organizations engaged in facilitating CBM for conservation and development. In the current volume, Tapia and Carrera (Chapter 2.3) and Dias (Chapter 2.7) share experiences with diversity fairs in Ecuador and Brazil, respectively, in the context of different projects. What both countries shared in common with these projects was their use of the diversity fair as an instrument to raise awareness, and to document and monitor agrobiodiversity, in a process for strengthening communities in their biodiversity management, and as a means of implementing *in situ* conservation (De Boef and Thijssen, Chapter 1.8).

In Nepal, the diversity fair is a collective process and a competitive event during which farmers collect and display plant parts, seed, fruit samples and traditional food items. The farmers share biodiversity-related information and associated traditional knowledge (ATK) either orally, through illustrations, or through displays in their stalls. This practice fulfils multiple objectives, such as creating awareness; locating diversity and custodians; documenting information on ATK; and exchanging seed/planting materials and knowledge. Within LI-BIRD, we find it is useful to organize diversity fairs in the initial stages of a CBM process.

The organization of the fair begins with the formation or identification of participating farmers. Farming communities set the date and venue, and decide on what activities should be carried out for organizing the fair. During a planning meeting, they set up various subcommittees for defining and distributing roles and responsibilities. Stakeholders and community members join in these subcommittees to ensure their ownership over the process. The subcommittees are responsible for the preparation of stalls; management of food and accommodation; and evaluation of stalls. Each subcommittee has three to five members. In this way, we build an initial basis in social and institutional capital at community level for CBM.

The diversity fair can be considered as a stepping stone to CBM for two reasons: first, because it enables us to begin to develop a local biodiversity database, by documenting the biodiversity displayed by fair participants; and second, because the fair encourages the community to assume responsibility in managing its biodiversity. Farmers, farmers' groups, students, teachers, researchers and development workers obtain insights into local biodiversity in the community. Documentation is an important goal of the diversity fair since it contributes to enhancing the understanding of the richness of diversity; identifying ATK and custodians; locating valuable plant genetic resources (PGR) and traditional knowledge; and generating inputs that are vital to the development of a CBM plan. Furthermore, a diversity fair enhances social interactions among community members and promotes the exchange of knowledge among and between farmers, and with researchers and conservation professionals. Farmers obtain new seed and planting materials, thus increasing the biodiversity in their home garden, farms and in the village. Dancing and singing competitions between the farmers' groups can be organized during a diversity fair as a way of promoting the involvement of students, young and elderly people, entertaining participants with songs that bring across the message of the value of local PGR. A food fair may also accompany the diversity fair, facilitating the exchange of recipes, and acknowledging the role of women in their culture and their responsibilities in the preparation of food. The diversity fair provides a good source for assessing the quality of farm-saved and informally exchanged seed, defining in what manner informal seed systems can be enhanced. And finally, the diversity fair offers an easy way for PGR professionals to collect seed of local varieties for gene banks, which may serve for *ex situ* conservation or for further crop improvement.

Rural poetry

Rural poetry, when applied as a CBM practice, is an innovative way of creating awareness of the value of biodiversity, and of documenting traditional knowledge and information in the form of poems. When LI-BIRD applies this practice in Nepal, we invite a group of local and national poets to participate in the creation of poetry on agrobiodiversity. We first share with them the objectives and the different steps of the practice. The poets then visit different households and discuss many aspects of agrobiodiversity with experienced farmers, both young and elderly, addressing the following topics: agrobiodiversity use and value; extent and distribution; ATK and associated information; food culture; and proverbs, myths and sayings. Directly after these discussions, the group converts the local information that has been gathered into poems, emphasizing the value of conservation and the importance of local PGR. A public announcement is made inviting community members to participate and listen to the poems. The poets read their work to a wide audience of community members. They document, enrich and return the information provided by the community in the form of melodious poems and songs. Community members pay great attention while participating in such events as the poems and songs represent their words, feelings, culture and reality.

Rural drama

Drama is a popular way of creating attention regarding any specific issue of public concern. In preparation for the rural drama event, we discuss relevant topics with a group of artists. Such topics include the importance of local biodiversity, the problems faced by the community, or possible harm from the loss of valuable local crops and varieties. After a few days of rehearsal by the artists, we make a public announcement inviting community members to a common meeting place within the community. The artists then present a theme-based play. In this manner, community members feel and internalize the importance of biodiversity in their daily life and they can better imagine the consequences should such a valuable biodiversity be lost. Following the drama, the organizers facilitate a short discussion, during which they ask few short questions to the audience to summarize the key messages and lessons of the drama.

The diversity block

The diversity block is a non-replicated experimental plot of farmers' varieties that is established and managed by the farmers' groups under their own management conditions. The plot size of the diversity block depends on the type of crop, availability of land, management capacity of the farmers' group and the amount of seed they want to produce. Diversity blocks are further discussed in this volume by Kendall and Gras in France (Chapter 1.7), Tapia and Carrera in Ecuador (Chapter 2.3) and Dias *et al.* in Brazil (Chapter 2.5).

In Nepal, we use the diversity block in our CBM activities for several goals, including the following: to test consistency in naming farmers' varieties; to evaluate the varietal performance and collect characterization data; to identify suitable parents for participatory plant breeding; to regenerate seed collected at community seed banks; and to meet the seed demands of local varieties. Besides these, we use the diversity block to create awareness among community members on the value of biodiversity. The diversity block is deliberately established near roads or public places. We erect signs in each plot, to encourage community members to stop and see the differences between the varieties.

While working with local PGR, it did not surprise us that we found the same name being used for different varieties and numerous names being used for the same variety. We find such situations in farmers' varietal nomenclature systems both within and among communities. Therefore, during the maturity stages of the crops, we invite knowledgeable and experienced farmers to differentiate between the varieties; and we document this knowledge for enhancing the community's, as well as the researchers', understanding. The seed produced in diversity blocks (only possible for selfing and vegetatively propagated crops) is kept safely either in the community seed bank or is stored by farmers' groups and distributed in diversity kits over the subsequent year.

The diversity kit

The diversity kit is a set of small quantities of seed of different crops and varieties that is made available to the farmers for informal research and development (Sthapit *et al.*, 2008a). The basic idea of the diversity kit is to deploy the diversity, giving priority to neglected, underutilized, rare and unique crop varieties and species, increasing community access to seed and knowledge, and promoting biodiversity conservation through use. The diversity kit practice is a process which involves the following steps: identification and selection of varieties favoured by farmers through a diversity fair; identification of crops and varieties on the verge of disappearing; selection of varieties from community seed banks and community biodiversity registers, or inventories of biodiversity; establishment of diversity blocks; the production of quality seed of selected varieties; the development of information sheets for each variety containing varietal characteristics as well as the methods of cultivation and use; and finally, the packaging and distribution of seed to farming households in the community. Canci *et al.* (Chapter 1.5) provide a more elaborate description of how the diversity kit in Brazil is embedded within a larger framework of CBM. The diversity kit as a CBM practice enhances the informal exchange of seed and information among farmers. In Nepal, we have promoted the broader use of home garden species through the diversity kit, resulting in an increase in dietary diversity and better income of target families from homestead production. The diversity kit is very useful in a CBM process since it is easy to implement, requires few resources, and takes little time, but can be very effective in building a firm basis in awareness and social organization for CBM.

Conclusion

Community awareness of biodiversity forms an important stepping-stone to a process in which the community will assume collective responsibility for managing its biodiversity in a sustainable manner. CBM practices that aim to enhance awareness are essential for exposing farming communities to biodiversity, and for inspiring them to participate in the conservation and management of PGR in a collective manner. Based on our work and experiences in Nepal, we have compared the practices that are currently being used for contributing to awareness-raising. We demonstrate that practices for enhancing awareness, when embedded in a CBM process, can be effective in contributing to the conservation and use of PGR. Depending upon the availability of time and resources and the interest of the community, various types of awareness-raising practices can be employed. The art of facilitating an effective CBM process lies with the conservation or development organizations, which are responsible for tailoring the appropriate combination and sequence of these CBM awareness-raising practices to meet the needs of the community.

The documentation part of any awareness-raising activity has to be very strong and must be efficiently managed. The information collected is used as the basis for developing a CBM plan, thus moving away from the components on awareness and understanding, towards outlining a process with key steps aiming towards the institutionalization of biodiversity management in the community. Feedback from the awareness-raising practices must be collected from participants during the CBM process to monitor the effectiveness of the practices in achieving their goals. Apart from community members, it is important that local government bodies, researchers, extension workers and policy-makers participate in these practices, and in their continuous monitoring; their contribution is essential for maintaining long-term collaboration, for sustaining the CBM process and for influencing policy-making.

2.3 Practices that contribute to promoting and appreciating Andean crops and identity in Cotacachi, Ecuador

César Guillermo Tapia Bastidas and
Hugo Fabian Carrera Rueda

Ecuador: a diversity of crops and cultures

With its variation in landscapes, crops and human cultures, Ecuador is an important repository of agrobiodiversity. The richness of its biodiversity and its native plant genetic resources (PGR) represent fundamental elements of the Ecuador cultural heritage. Generations of Ecuadorian farmers have contributed to their creation. Unfortunately, however, socio-economic and environmental changes over recent years have caused the erosion of agrobiodiversity, associated knowledge and customs. Nevertheless, indigenous and local farmers still maintain their agrobiodiversity and associated ancestral knowledge concerning the use-values of the varieties, their customs and agricultural management practices, thereby contributing to the conservation strategy of *in situ* on-farm conservation of these PGR.

Cotacachi agrobiodiversity project

Cotacachi is located 80 km north of Ecuador's capital Quito, in the great western plains of Hoya del Imbabura (Figure 2.3.1). The altitude varies from 2600 to 3350 m above sea level and the area extends over 1809 km^2. Cotacachi is recognized as an important site for PGR. This diversity has been shaped by the unique culture and traditions of both indigenous Kichwa people and other ethnic groups living in the area.

In the period 2002–2008, the National Plant Genetic Resources Department of the National Autonomous Institute for Agricultural Research in Ecuador (INIAP-DENAREF) and the Union of Peasant and Indigenous Organizations of Cotacachi (UNORCAC) implemented an agrobiodiversity project in Cotacachi on promoting Andean crops for rural development. Through this project, we aimed to associate strategies for rural development with agrobiodiversity management. INIAP and UNORCAC established partnerships with various national and international institutions, including the Corporation for Export and Investment Promotion and its Sustainable BioTrade Initiative (CORPEI), the Andean Foundation for the Promotion of Sustainable Technologies for Natural Resource Management (FOMRENA), the Union for Economic Cooperation and Development of the People (UCODEP), the United States Department of Agriculture (USDA) and Bioversity International.

The home garden is a critical livelihood component of rural households in Cotacachi. It provides numerous benefits to farming families and acts as a repository of local species diversity that is used for food, medicines, ornaments, fuel and animal feed.

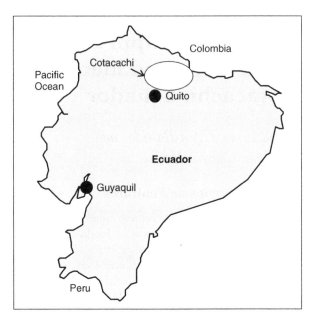

Figure 2.3.1 Map illustrating the location of Cotacachi, in Ecuador.

Farmers maintain their cultural customs (ritual ceremonies) and continue to manufacture handicrafts and tools, through their use of diversity from the home garden.

Practices for recognizing and appreciating the value of diversity

In the project, we decided to respect the Andean cosmovision and legacy of traditional agricultural systems while promoting PGR conservation and development. As such, we associate our work not only with the physical mountainous environment and its cultural aspects, but also with local customs and beliefs concerning Mother Earth, and the important role they play in the maintenance of people's farms and well-being. This starting point guided us towards establishing reciprocal social and institutional relationships, which, over the years, led to the development of a series of practices that contributed to community biodiversity management (CBM). These practices include:

- strengthening the two-way relationship between gene banks (*ex situ*) and farmers (*in situ* on-farm) through several interventions;
- creating an agrobiodiversity catalogue;
- organizing seed exchange and food fairs;
- adding value to local species, crops and varieties and developing market chains for these agrobiodiversity products;
- supporting the establishment of organic producers and rural enterprises;
- linking agrobiodiversity and cultural diversity with tourism;
- developing agrobiodiversity courses for formal education;

- using various publications and communication tools to raise awareness and popularize local diversity and food traditions.

Below, we provide further details concerning the development of these practices in the Cotacachi district, and demonstrate how they benefit both individuals and communities beyond conservation.

Promoting farmers' access to gene bank accessions and use of local crops and varieties

The INIAP gene bank holds 480 accessions collected from the Cotacachi district. These include several cereals, root and tuber crops, fruits, vegetables and legumes. As a means to repatriate and promote their reintroduction, we set up a diversity block on communal land in Cotacachi. Shrestha *et al.* (Chapter 2.2) describe modalities of diversity blocks in more detail. We included 155 maize accessions (*Zea mays*), 111 bread wheat accessions (*Triticum aestivum*), 73 pepper accessions (*Capsicum* spp.), 44 ground cherry accessions (*Physalis peruviana*), 42 figleaf gourd accessions (*Cucurbita ficifolia*), 37 tamarillo accessions (*Cyphomandra betacea*) and 18 achocha accessions (*Cyclanthera pedata*). Community members were involved in assessing and characterizing the accessions. They selected the best varieties based on their own criteria, such as high yield, tolerance to pests and diseases, good eating quality, suitability for agrotourism and ornamental values. For those selected accessions, seed was produced on communal land. We were able to reintroduce and restore the cultivation of dozens of traditional crop varieties. The seed produced was further distributed to other farmers interested in diversifying and enriching their production scheme.

Collaborating with gene banks: community contributions to **ex** situ conservation

During the project activities in Cotacachi, INIAP collected samples of the current local varieties and deposited them in the national gene bank. INIAP characterized and evaluated the accessions, which are now being conserved *ex situ* under long-term storage conditions. This collaboration with the INIAP gene bank also contributed to the *in situ* multiplication of gene bank accessions in their original farming system and/or habitat.

Establishing an agrobiodiversity catalogue

We developed an agrobiodiversity catalogue of the Andean highlands of Cotacachi. Our main aim was to document and better understand PGR and associated traditional knowledge in order to achieve a more effective way of managing genetic resources, including both *ex situ* and *in situ* strategies. The catalogue was published and has been shared among the farming communities and other stakeholders. It reflects the unique diversity of species, crops and varieties found in Cotacachi. For example, it shares the fact that more than 12 maize landraces and 40 common types of beans are maintained by farmers. The catalogue has also been used to identify those species, crops and/or varieties that have a strong association with local liveli-hoods. Consequently, these PGR have been prioritized for follow-up action in the project. The agrobiodiversity catalogue for Cotacachi could be referred to as a kind of community biodiversity register (CBR), as described by Subedi *et al.* (Chapter 2.4); however, its structure is closer to that of an institutional rather than community-based register, when compared with the CBR in Nepal.

Exchanging gene bank materials in seed fairs

Over the course of the project, we organized four diversity fairs, a practice that is described in more detail by Shrestha *et al.* (Chapter 2.2). The fairs assisted us in reintro-ducing gene bank accessions, and facilitated the monitoring of flows of local crops and varieties. The major reason why farmers participate in the fairs is to exchange seed. We organized the fairs in such a way that farmers' groups or households could exhibit their traditional food and culinary diversity. UNORCAC now regularly organizes the diversity fairs. It is a practice that reveals the importance of Andean crops and associated food and cultural heritage to all Ecuadorians. It contributes to awareness-raising, conservation, research, education and agro-tourism; moreover, it enhances local farmers' self-esteem.

Creating artisan food products in support of agrobiodiversity management

The project emphasized the consumption and promotion of local foods, holding workshops to address local nutritional culture, recipe exchange and value addition to local crops through artisanal processing. In training workshops, families from the community, nutritionists and agronomists shared information about the value of local crops, such as their productive, nutritional and culinary properties, and their use in the preparation of artisanal food products for sale. The project supported the estab-lishment of a food-processing plant to produce and package artisanal foods, includ-ing: Andean blackberry marmalade (*Rubus glaucus*), squash seed snacks (*Cucurbita fici-folia*), spicy pepper pastes and dried ground cherries. These attractive, high-quality food products are marketed to tourists visiting the Cotacachi area, the nearby famous Otavalo market, and the capital city Quito.

Establishing agrobiodiversity-oriented enterprises

In order to emphasize the value of the products, our marketing strategy tells the story of the Andean crops and the indigenous farmers who produce them. A rural micro-enterprise, Sumak Mikuy, pays a premium price for the high-quality produce, and any profits made from the business are reinvested into the community. In follow-up to these commercial activities, a network of certified organic producers was established, to guarantee access to the necessary raw materials. These value chain practices led to the publication of an *Agro-culinary Guide to Cotacachi and its Surroundings*, and a *Cookbook of the Traditional and Intercultural Andean Cuisine of Cotacachi*, which further documented and popularized our rich food diversity. The results of the agrobiodiversity project in Cotacachi emphasize the importance of integrating value addition activities for generating economic benefits, as shared by various authors in Part IV.

Promoting agrobiodiversity tourism

The project has contributed to the development of agrobiodiversity tourism in Cotacachi. Rural lodges have been constructed throughout the Cotacachi region, using traditional local materials. These *albergues* are owned and operated by individual families. Overnight stays at the lodges and daytime tours with native guides are coordinated by a community-based tourism agency, Runa Tupari Native Travel (www. runatupari.com), which is associated with UNORCAC. The agency has developed a number of tourist routes and offers tourists a range of local products. Tourists appreciate Andean gastronomy that uses local crops and varieties cultivated in home gardens. Agrobiodiversity tourism provides an important source of additional income for farming families. The gardens, in turn, diversify the families' income and diet, and contribute to the revaluation of the role of native crops in Andean agriculture by the tourists, and also by the farming households themselves.

Integrating an agrobiodiversity course into the formal education system

In our project, we designed an agrobiodiversity course for the formal education system. We developed a guidebook for rural teachers entitled *Planting Seeds of Reflection and Hope*, which is used in 19 intercultural, bilingual primary schools in the indigenous communities of Cotacachi. Our strategy was to link the course with teacher and student clubs. Through the clubs, we were able to create a base of 60 teachers and 400 students. Their efforts helped us to promote and raise awareness of agrobiodiversity to 1200 students and, consequently, to the rural community. An important UNORCAC partner that became an offspring of the project in this process was the Teachers Environmental Association of Cotacachi. This association facilitates and enhances the contribution and involvement of teachers in socio-environmental education and rural development.

Raising awareness of agrobiodiversity beyond farming communities

We used a variety of publications and media for raising public awareness. A cartoon video called *The Most Valuable Treasure in the World* was created with the help of schoolchildren from the Intercultural Bilingual Education Centres of Cotacachi. This cartoon conveys a message to people not directly involved in our project activities, calling for the conservation of Andean crop diversity, and its associated knowledge and values.

Lessons learned

Our project provided benefits to more than a thousand families, either directly or indirectly. It demonstrated the value of establishing functional linkages between *ex situ* gene banks and farming communities. In partnership with several stakeholders, we were able to develop a number of mechanisms for benefit-sharing.

Our efforts resulted in an increase in farmers' access to the germplasm collections of gene banks, enabling farmers to broaden their varietal portfolio. Through our efforts to institutionalize a diversity of CBM practices in the community, we contributed to the empowerment of the rural communities in various ways. We strengthened their ability to manage their agrobiodiversity, establish community institutions, make use of agrobiodiversity and generate benefits from their agrobiodiversity. We can conclude that by the end of the six-year project, in 2008, we had contributed to the enhancement of several assets in the livelihood of the Andean communities in Cotacachi.

The gene bank in turn benefited from the recycling of gene bank accessions in their original ecological and social environment. In addition, they were able to collect new materials, through diversity fairs, monitor genetic erosion and enrich their collections by documenting farmers' knowledge on the accessions. Conservation professionals were able to improve the functioning of their gene banks and to reduce operational costs, as a result of their participation in the agrobiodiversity project in Cotacachi.

The *Second Report on the State of the World's Plant Genetic Resources for Food and Agriculture* identified a gap in linkages between strategies for *in situ* and *ex situ* conservation (FAO, 2010). In this regard, our experiences in Cotacachi can be considered a milestone for filling this gap if other conservation professionals and their institutions follow the same path.

In Cotacachi, we learned that agrobiodiversity conservation can be aligned with the often-considered contrary force of development and community empowerment. We achieve this linkage for example through promoting entrepreneurship in agrotourism and the production and marketing of Andean food products. The collaboration between UNORCAC, INIAP and the private sector can be considered a successful civil–public–private sector partnership.

The diversity of practices introduced by the project increased the self-esteem of the rural communities, motivating them to appreciate and therefore conserve their local agrobiodiversity, culture and traditions, which we collectively refer to as 'native Ecuadorian agrobiodiversity'. This is crucial for acknowledging Andean identity and lifestyle, and becomes a community asset of critical significance during times of socio-environmental changes.

One very important element for the success of this project was the dialogue between the various types of organizations involved. This created the possibility of generating a 'new model' that associates community empowerment with conservation. The project led to a qualitative leap in the development of UNORCAC, an important stakeholder that is now autonomous in the design and implementation of developmental processes. It is no longer an object of research, or the beneficiary of projects run by other stakeholders. This has been a decisive element for allowing the organization to trumpet this cooperative model as an example to other organizations in similar situations in Ecuador and beyond. For further up- and out-scaling this model, combining conservation and development through a local cooperative organization, we aim to analyse UNORCAC's institutional, technical and political capabilities that have made the project into a success. If UNORCAC can sustain its role following the conclusion of the project, we will have been successful in developing an organization and a potential model for 'development with identity'.

2.4 Community biodiversity registers in Nepal

Enhancing the capabilities of communities to document, monitor and take control over their genetic resources

Abishkar Subedi, Rachana Devkota,
Indra Prasad Poudel and Shreeram Subedi

The need to protect associated traditional knowledge

'Traditional knowledge' refers to the body of wisdom, innovations and practices of indigenous peoples and local communities (CBD, 1992b). We see in the world that young generations of farming communities are increasingly reluctant to learn about, or are simply not interested in understanding, plant genetic resources (PGR) and associated traditional knowledge (ATK), which may result in their loss. The risk of losing ATK appears either when an owner of ATK does not pass it on to a recipient (another generation or other interested persons); when the overall situation in which the knowledge is used changes; or when the material concerned is lost. If the information passed on is not useful, or if the related resource is no longer there, the ATK loses its purpose. When any of these conditions or combinations thereof occur, we need to address this situation with a conservation action. After more than a decade of work at Local Initiatives for Biodiversity, Research and Development (LI-BIRD, Nepal), we have learned that a community biodiversity register (CBR) can contribute to the empowerment of community institutions. Through this practice of community biodiversity management (CBM), communities gain a better understanding of their own biological assets and values, and are better able to use those assets for livelihood development while appreciating and sustaining them for future generations.

The CBR is a practice that addresses a range of objectives. A CBR is basically a farmers' information database on biodiversity and traditional knowledge. It documents and monitors ATK and PGR and thereby protects them from bio-piracy (Rijal *et al.*, 2003; Subedi *et al.*, 2005c). In this chapter, we share our experiences in designing and implementing CBRs in a diversity of situations in Nepal.

The design and implementation of a CBR

We developed a CBR in a process that includes three major steps, as illustrated in the time line in Figure 2.4.1.

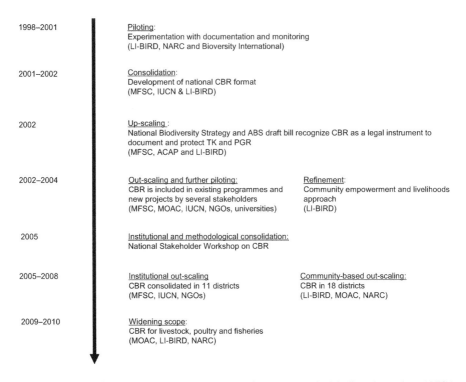

1998–2001   Piloting:
             Experimentation with documentation and monitoring
             (LI-BIRD, NARC and Bioversity International)

2001–2002   Consolidation:
             Development of national CBR format
             (MFSC, IUCN & LI-BIRD)

2002   Up-scaling :
             National Biodiversity Strategy and ABS draft bill recognize CBR as a legal instrument to
             document and protect TK and PGR
             (MFSC, ACAP and LI-BIRD)

2002–2004   Out-scaling and further piloting: Refinement:
             CBR is included in existing programmes and   Community empowerment and livelihoods
             new projects by several stakeholders   approach
             (MFSC, MOAC, IUCN, NGOs, universities)   (LI-BIRD)

2005   Institutional and methodological consolidation:
             National Stakeholder Workshop on CBR

2005–2008   Institutional out-scaling   Community-based out-scaling:
             CBR consolidated in 11 districts   CBR in 18 districts
             (MFSC, IUCN, NGOs)   (LI-BIRD, MOAC, NARC)

2009–2010   Widening scope:
             CBR for livestock, poultry and fisheries
             (MOAC, LI-BIRD, NARC)

Figure 2.4.1 Time line synthesizing the evolution of the community biodiversity register (CBR) as a practice for community biodiversity management in Nepal. ACAP, Annapurna Conservation Area Project; NGO, non-governmental organization; NARC, Nepal Agriculture Research Council; MOAC, Nepalese Ministry of Agriculture and Cooperatives; LI-BIRD, Local Initiatives for Biodiversity, Research and Development; MFSC, Nepalese Ministry of Forests and Soil Conservation; IUCN, International Union for the Conservation of Nature; TK, traditional knowledge; PGR, plant genetic resources.

Source: Based on Subedi *et al.* (2005).

During the piloting stage (1998–2001), we experimented with documenting and monitoring biodiversity and traditional knowledge in the global agrobiodiversity project that is described in more detail by Subedi *et al.* (Chapter 1.2). The need for CBR emerged after carrying out a crop inventory and monitoring the dynamism of PGR at household level in three sites that were each cultivating different crops, the combination of which matched the crops chosen to be addressed by the agrobiodiversity project (Rijal *et al.*, 2003; Subedi *et al.*, 2005c). This investigation provided us with the rationale for documenting and monitoring PGR and ATK; it became the scientific foundation to the development of CBR.

To fulfil Nepal's obligations to the Convention on Biological Diversity (CBD), stakeholders began to explore ways of documenting the national biodiversity and

ATK. The Ministry of Forest and Soil Conservation (MFSC), in collaboration with the Nepal office of the International Union for Conservation of Nature (IUCN-Nepal), explored ways to study and refine the CBR methodology in the context of biodiversity, in one of LI-BIRD's sites. This project constituted the consolidation and development of the CBR (2001–2002). We moved beyond our original PGR focus, and included a wide range of biological resources including forests, wetlands and wildlife.

We used the outcome of this work to out- and up-scale several projects with a wide range of stakeholders (from 2003 onwards). MFSC, LI-BIRD, IUCN-Nepal, the Unitarian Service Committee-Nepal (USC-Nepal) and the Ministry of Agriculture and Cooperatives (MOAC) began to out-scale CBR through various projects or within their regular programmes. Stakeholders used similar or slightly adapted registers that responded to their objectives. Some established CBRs were managed by external stakeholders, with documentation as a major goal, but this modality proved to be unsustainable and was discontinued in 2005. Examples of the other modality, where community-based organizations (CBOs) themselves managed and took responsibility for their CBR, are still operational and benefiting the community and its members.

In 2010, we adapted the CBR for documenting and monitoring animal genetic resources, including fisheries and poultry. As such, the CBR evolved to take a more integral approach to biodiversity. Most importantly, it gradually began to strengthen the capacity of local communities to analyse locally available information and design development and conservation plans to generate social, economic and environmental benefits (Subedi *et al.*, 2005c).

Key steps in the process of developing the CBR

Our experience with CBRs can be synthesized into eight different but interlinked steps that we present below. These steps may vary and can be adapted depending upon the context of any given community and type of biodiversity being addressed.

Step 1: Selecting the area and community

In choosing an area to support the establishment of a CBR, we use the following criteria:

- There must be an availability and richness of biodiversity and/or PGR.
- The community and engaged stakeholders must recognize that diversity is an important asset of their livelihood.
- The community must be interested.
- The area should be representative of an atypical agro-ecosystem of the country.

We use secondary information sources or consultations with relevant stakeholders to gather the information. We form a multidisciplinary team that includes project staff and local government officials, and conclude with a participatory diagnostic survey to verify key characteristics and set a basis for the establishment of the CBR.

Step 2: Informing the community on CBR rationale and objectives

We organize a series of village workshops. Participants represent farmers' groups, natural resource user groups (e.g. forest user groups), schools, youth clubs, local administrative bodies, private sectors and local government extension agents. With the workshop, we start building a common understanding of the CBR, the rationale behind its establishment and way that it operates. Subsequently, we facilitate a multi-stakeholder process during which we design the CBR action plan. Guiding questions in this process are:

- What is the appropriate unit of documentation (household, group, village, or any specific landscape)?
- What local institution has the capacity and long-term interest to locally coordinate the CBR?
- What stakeholders should we involve?
- Where should we register the CBR for legal purposes?
- What mechanisms should we use to link the local CBR database with a national repository?

Step 3: Strengthening institutional capacity

We use several participatory appraisal tools, such as rainbow diagrams, CLIP and SWOT analysis (De Boef and Thijssen, 2007; Chevalier and Buckles, 2011), to define the CBR's institutional set-up. In this way, communities identify local institutions that they consider interested, skilful and legitimate for coordinating or facilitating their CBR. The appraisal defines who will be a member of the committee that will provide strategic support, facilitate collective decision-making, engage in capacity-building and ensure the overall monitoring of PGR and ATK. As an external supporter, LI-BIRD trains, guides and coaches the committee in performing these responsibilities. We also conduct exposure visits that enhance the capacity of committee members and farmers' groups.

Step 4: Defining a specific focus and initial requirements for data collection

We start the actual implementation of the CBR with the development of the register's format. Even before advancing with this, a minimum set of data is required, for which we use the following guiding questions:

- What resource (species, variety or other biological resources) do we have and how do we recognize it?
- How do we utilize it?
- What is its status? Is it abundant or rare? Does this status change over time?
- Why do we need to conserve it?
- Who holds the traditional knowledge associated with the resource, and how is it transmitted from one generation to another?

- Who (men or women; young or elderly) makes decisions concerning the management of this resource?
- To what degree and in what way is the resource shared within and outside the community or beyond?

Once the community has adequately responded to those questions, we can start to collect an initial set of data on the resources identified. Based on our experience working with CBRs in Nepal, we collect detailed information concerning the origin, characteristics, historical background and traditional knowledge of the resource, as well as site and community descriptors, extent of distribution, name and gender custodians, among many other details.

Step 5: Documenting, compiling and validating CBR information

The institution and/or committee responsible for the CBR strengthen the capacity of each specific group, to ensure that they are capable of assuming the responsibility to document their biodiversity and ATK. Committee members coach and monitor the groups in their activities. They further explore how the CBR can be associated to other practices that are part of the community's CBM plan. These include the use of a diversity fair, during which the information recorded in the CBR can be verified, and additional, relevant information can be collected, which may lead to follow-up CBM practices, such as diversity blocks and diversity kits, as described by Shrestha *et al.* (Chapter 2.2). This step is crucial but also resource-demanding for the supporting organization (e.g. LI-BIRD) and the community itself.

Step 6: Analysing and sharing information

We supported communities in their organization of a regular series of village-level stakeholder workshops to discuss and analyse the status and trends of biological resources that can be interpreted from the CBR. Four-cell analysis is an effective participatory discussion tool for identifying abundant and rare crops or varieties in a participatory manner (Subedi *et al.*, 2005c). Through the use of four-cell analysis, the communities can also learn about how many households cultivate certain crops or varieties in large or small areas. The outcome of this discussion is crucial in identifying follow-up CBM practices. Various communication tools, such as posters, pamphlets or radio programmes, as well as other CBM practices, can be used to share the outcomes of this analysis with an audience beyond those households directly engaged in the CBR.

Step 7: Designing and implementing conservation and development plans

During the analysis and sharing of CBR information, committee members facilitate a participatory process to identify priorities for conservation and livelihood improvement. It is crucial that the priorities are based on evidence in the CBR and are endorsed through a participatory and transparent process of decision-making. We have found that for the effective implementation of such CBM plans, the committee needs to take on a role beyond that of managing the CBR. Their role should gradually evolve into one of leadership, where they take a responsibility for facilitating the CBM process and practices; for guiding the community towards collective action; and for ensuring partnerships and collaboration with stakeholders, including local government. Once the institutional responsibilities and social organization have been defined, the successive component is the establishment of the CBM fund, as described by Shrestha *et al.* (Chapter 2.9).

Step 8: Local registration and linkage with a national repository

We learned that in order to make the CBR a legal document, it needs to be formally registered at both village and national levels of government. The committee is responsible for its registration at village level, something that is achieved through the Village-level Biodiversity Conservation and Development Committee (VBCDC). We have also invested in the development of a practical mechanism that links community-based CBR registration to an instrument for documentation at national level. Our argument is that the compilation of information from the various community-based CBRs, managed through VBCDCs, and their linkage to national level, forms a basis for claiming ownership over the use of genetic resources. Such a repository is required for sharing the benefits from any commercial use of the biodiversity and ATK originating from communities. Vernooy and Ruiz (Chapter 6.4) further elaborate on ABS, while also addressing specific aspects of the situation in Nepal.

Success stories of CBR

In Begnas, one of our CBM sites, we observed that CBR is an effective tool for locating unique and rich PGR at both household and community level. Through the CBR, the community were able to identify a single household that maintains more than 20 rice landraces. This information became more relevant when it was known that the community itself was maintaining over 40 rice landraces. When the CBR committee members shared this information in their village, it had an immediate impact. Custodian farmers and farmers' groups that had been conserving unique or abundant diversity were publicly recognized and were awarded with specific grants from the CBM fund.

Farmers realized that only a few households are responsible for the conservation of a large number of local varieties. The material in their hands is thus considered highly vulnerable to genetic erosion. Upon recognizing this situation in the CBM site in Bara, the leadership encouraged 22 farmers' groups to form a CBO, which in turn established a community seed bank (CSB). The CBOs were able to access funds from both local government and other developmental organizations for the construction of their CSBs, using information from the CBR as evidence (Singh *et al.*, 2006).

In Begnas, the community identified unique traits in rice, finger millet, taro and sponge gourd from the CBR, sharing this information with relevant stakeholders. The information was also used in participatory crop improvement, value-addition and market linkage programmes, as a means to create incentives for farmers to continue cultivating these species and varieties.

Within another CBR project, we supported the establishment of the Rupa Lake Conservation and Rehabilitation Cooperative, from 2002 until 2006. An analysis of information in the CBR showed that many wetland resources, such as native fishes, white lotus and freshwater otters, were endangered due to illegal hunting or over use by local communities, particularly by those living upstream. With this insight, the cooperative began a CBM-guided commercial fish-farming venture. Within eight years, the annual income of the cooperative had increased from US$4200 to US$98 000. Each year, the cooperative allocates 25% of the income to conservation actions and ecological services, including the conservation of local fish diversity, the protection of breeding habitats for wetland birds, the conservation block of wild rice species, and other conservation activities targeting species and varieties identified as rare in the CBR and therefore requiring special attention. The cooperative expanded its membership to include households from upstream communities that also benefit from its actions. It promotes the management of local PGR, organic farming and reforestation, has created a scholarship programme for schoolchildren, and provides employment to indigenous fishing communities. In 2010, four years after the conclusion of the project, the cooperative had over 700 shareholders. The CBR experience in Rupa Lake became an important reference for LI-BIRD and other biodiversity stakeholders in Nepal. It showcases how CBR can provide social, economic and environmental benefits, also referred to as payment for ecosystem services.

The CBR has been important in many of our agrobiodiversity sites in supporting and providing crucial inputs to several other CBM practices, including:

- recognizing custodians by providing awards and other means of social recognition during diversity fairs;
- increasing the competitiveness of local varieties through the improvement of specific traits, as illustrated by the case of rice landrace, Jethobudho Pokhareli (discussed by Silwal *et al.* in Chapter 5.5);
- increasing the demand for local diversity by value addition through better processing, packaging and market linkages, and through the establishment of small-scale enterprises (Bhandari *et al.* in Chapter 4.2 provide an example of this, showing how the unique diversity of the citrus species was conserved through a process of value addition and market promotion);
- establishing and supporting the CSB activities (as further described by Shrestha *et al.* in Chapter 2.8 in their global overview of this practice);
- promoting the use of diversity kits and the distribution of seed of rare or unique varieties (as described by Shrestha *et al.* in Chapter 2.2).

Lessons learned and future challenges

The recording of information at household level can be very resource and time consuming, which may deter communities and their partners from sustaining this practice. In order to create a high degree of ownership over, as well as the motivation to sustain the CBR, while ensuring an adequate critical mass of contributing member households, we learned that one CBR per community or village is the most practical and most viable. Biodiversity projects often organize diversity fairs as part of their regular activities. Another way to optimize resources and time is to link the documentation and monitoring activities of the CBR with such events. This will increase the visibility and the use of the CBR practice among members, and will further motivate community members to contribute. A digital database system is an efficient option if the necessary resources and capacities are available. Such an investment ensures the efficient management of the CBR and makes it more attractive to younger community members. Unique PGR and ATK can be documented through video clips, which would maximize their future use. One opportunity is to integrate CBR within digital rural inclusion programmes that are currently being implemented. Another important lesson learned from working with both the institutional and community-based CBR is that the latter modality is not only more appropriate from a development perspective, but is also more sustainable institutionally.

We learned to ensure the legal status of the CBR and to support the development of a framework with national systems for biodiversity conservation. This includes the exchange of PGR for research and development purposes, and the development of mechanisms for access and benefit-sharing (ABS). In Nepal, although provisions have been made for such a mechanism in a draft ABS bill, the practical mechanisms still need to be developed. Our experiences in different contexts of biodiversity management show that the CBR is an experimental ground for developing such ABS mechanisms; stakeholders in charge of this important and difficult task do not need to start from scratch.

In conclusion, we learned that strong socio-political and economic incentives encourage community members to continuously engage in CBM, in which CBR is a rewarding but also demanding practice. We therefore have been able, and will continue, to seek ways to link the CBR to community well-being and welfare. Its association with mechanisms that deal with access to financial or market services is a priority. It is unlikely that biodiversity registration alone is a viable and sustainable strategy for protecting PGR and ATK, whether the government or local institution supports its implementation, or if a CBO is autonomous in managing its CBR. It is only sustainable when the community takes responsibility for the registration and monitoring of its resources, which in turn creates direct and visible benefits for their livelihood.

2.5 Gene banks that promote on-farm management through the reintroduction of local varieties in Brazil

Terezinha Aparecida Borges Dias,
Maria Aldete J. da Fonseca Ferreira,
Rosa Lía Barbieri, Flavia França Teixeira and
Sérgio Guilherme de Azevedo

Background

Brazil is rich in cultural diversity; its population is composed of different ethnic groups, including 219 different indigenous peoples and countless traditional communities. This diversity in population led to the application of diverse agricultural practices. However, rapid industrialization in agriculture has resulted in the loss of plant genetic resources (PGR) in farmers' fields, which in turn affects cultural diversity. This process has had a direct impact on the myths and rituals associated with agricultural practices, transforming their farming routines, and their livelihood in social and economic terms (Dias *et al.*, 2007).

Approximately 35 years ago, the Brazilian Agricultural Research Corporation (Embrapa) set up a system for the conservation of genetic resources *ex situ* in Brazil. Its purpose was, and continues to be, to collect, exchange, characterize and conserve germplasm. Embrapa is aware of the growing demands of *ex situ* conservation and has been gradually organizing itself to respond to the needs of the agricultural sector. As conservation professionals who work with small-scale farming communities on the conservation of PGR, we focus our work at Embrapa on the need to promote social and technological innovation. We aim to motivate indigenous populations and traditional communities to continue their practices according to their local conditions and traditional knowledge, and thereby contribute to the conservation of PGR.

The Brazilian national system of genetic resource conservation is seeking pathways to contribute to the strategy of on-farm management (FAO, 2010) by supporting community biodiversity management (De Boef and Ogliari, 2007). One way of fulfilling this aim is to strengthen the interface between on-farm management and *ex situ* conservation strategies. In this chapter, we document our experiences as conservation professionals and curators of Embrapa gene banks in different locations in Brazil, on the reintroduction of maize, squash and sponge gourd accessions from gene banks to farmers' fields. These experiences illustrate how the relationship between *ex situ* conservation and on-farm management can enforce community biodiversity management.

Case 1: The reintroduction of traditional varieties of maize to the Krahô people

Maize (*Zea mays*) was domesticated in Mesoamerica and introduced into South America by Pre-Columbian populations. Before the arrival of European colonizers, native populations in Brazil were already cultivating and selecting varieties of maize, generating a large genetic variability. The indigenous Krahô tribe comprises 2500 people, who live in 27 villages in the state of Tocantins. Ongoing conflicts with an expanding livestock and agricultural frontier, as well as inter-ethnic contacts, have resulted in the loss of indigenous maize varieties that were once part of Krahô traditional agriculture.

Krahô leaders are used to discussing important issues exhaustively and they remember a time when they cultivated abundant agrobiodiversity in their fields. They also remember the fact that they used to consume indigenous pohypey maize varieties during a ritual abstinence from certain foods, to promote their strength as Krahô people. In 1994, an extension agent informed the tribe about the existence of this indigenous variety, pohypey, in a gene bank. In 1995, the Krahô leaders visited the gene bank of the Embrapa Centre for Biotechnology and Genetic Resources (Embrapa-CENARGEN) in Brasília, accompanied by representatives of the National Foundation for Amerindian People (Fundação Nacional de Índio; FUNAI). During this visit, they identified different traditional pohypey varieties. These maize accessions had been collected by researchers in the state of Mato Grosso during the 1970s, in villages of the Xavante indigenous people. Following their visit, we provided the Krahô with small samples of 200 g of six varieties of indigenous pohypey maize. Through the reintroduction of these varieties we were able to contribute to a cultural revival in the villages, restoring traditional agricultural practices and associated myths and rituals.

In 1996, the Krahô returned small amounts of these varieties to the gene bank for storage. This return was something unheard of in the history of our gene bank at Embrapa-CENARGEN, and it helped to build a relationship based on trust between the Krahôs and Embrapa conservation professionals. Since then, together with the Krahô, we have been developing activities related to the strengthening of community biodiversity management (CBM). We have been engaged in CBM practices such as diversity kits and the identification of guardians of indigenous biodiversity. We join several other stakeholders in the organization of annual diversity fairs that facilitate the exchange of seed of traditional varieties. During this fair, we reward the villages that exhibit and share the largest number and diversity of varieties (Dias *et al.*, 2007).

Case 2: The reintroduction of traditional maize varieties to other indigenous populations

It is interesting to note that the cycle that started with the collection of indigenous maize in the Xavante villages in the 1970s, its *ex situ* conservation and the provision of seed to the Krahô at a later stage, was completed when a Xavante leader visited our gene bank at Embrapa-CENARGEN in 2003. He did not resent the fact that we had been providing the Krahô people with their varieties but said that the Xavantes had also lost their varieties and wanted them back in their fields. Therefore, we multiplied the varieties and later were able to share seed with many Xavante villages.

Our activities with the Xavantes and the Krahô had national repercussions; today, other indigenous populations are seeking Embrapa gene banks, with the aim of recovering their varieties. Recently, at the request of indigenous leaders, the national maize gene bank, Embrapa National Agricultural Research Centre for Maize and Sorghum in Sete Lagoas, multiplied and sent seed of traditional varieties to the peoples of Bororo, Maxacali, Pataxó and Kaingang. This exemplifies the importance of interactions between *ex situ* conservation and on-farm management, because it led not only to the recovery of local varieties but also to the revitalization of indigenous people's rituals and myths, which is important for maintaining their culture and for maintaining plant genetic resources within communities.

Case 3: Squash diversity and its association with ethnic diversity

Thousands of years ago, five species of squash were domesticated in the Americas: *Cucurbita argyrosperma*, *Cucurbita maxima*, *Cucurbita moschata*, *Cucurbita ficifolia* and *Cucurbita pepo*. When the first Portuguese explorers disembarked on Brazilian land, five centuries ago, the indigenous people were cultivating several species and varieties of squash. For the Amerindians, squash was the third most important crop at that time, following cassava and maize. The Portuguese brought some of those squash to Europe, and the Spanish, having colonized other parts of the Americas, brought back local varieties of squash cultivated by the Aztecs, Mayas and Incas. Squash was a success. Its cultivation spread quickly through many countries, reaching Germany and Italy by the sixteenth century. When German and Italian immigrants arrived in Brazil in the nineteenth century, they brought seeds of their own squash varieties, in order to continue cultivating those that had already been incorporated into their culture three centuries earlier.

Nowadays, in many parts of Brazil, small-scale farmers keep their own local varieties of squash, selecting the types of fruit that please them the most. A small number of these farmers play a key role in the dynamics of on-farm management of varieties of several squash species. These farmers act as guardians of these PGR, and their function is to exchange them between relatives and neighbours and to pass them on from generation to generation. The squash guardians are directly associated with the cultural background of their communities.

Below, we share some examples of the association between species diversity and ethnic background in Brazilian people:

- Farmers of Portuguese origin in southern Brazil maintain local varieties of fig-leaf gourd (*C. ficifolia*), the fruits of which have a white and fibrous pulp, which is used in traditional Portuguese sweets recipes.
- Other farmers of Portuguese origin maintain local varieties of *C. pepo*, the fruits of which have hard skin and fibrous pulp, and are called *mogangos*.
- Afro-Brazilians, who live in communities called *quilombola*, have their own local

varieties, especially *mogangos* (*C. pepo*) with its extremely hard skin and a long shelf life. They also maintain local varieties of *C. maxima* and *C. moschata*.

- Descendants of German immigrants maintain *C. maxima* and *C. moschata*. They also maintain ornamental varieties of *C. pepo*. German descendant farmers appreciate fruits that have intense colours and show a large variation in shape. The dried fruits are used in the decoration of their houses.
- Descendants of Italians prefer local varieties of *C. maxima*, the fruits of which are flat and have a firm flesh. They have an orange-coloured pulp. A common feature for all ethnic groups is that genetic erosion has had an increasing effect on their local varieties. This has also been affected by the disappearance of local cultures in culinary and ornamental traditions, caused by the industrialization of traditional agricultural systems, urban expansion and rural exodus, but also by changes in rural people's lifestyles (Heiden *et al.*, 2007; Priori *et al.*, 2010).

Case 4: The reintroduction of squash accessions to small-scale famers in southern Brazil

Demand for seed of local varieties of squash by small-scale farmers in southern Brazil has increased in recent years. Since 2008, we have been making the seed of some accessions available during fairs dedicated to the exchange of seed of local varieties. In 2010, the gene bank of Embrapa Agricultural Research Centre for the Temperate Climate, in Pelotas, distributed three local varieties of squash to farmers in the municipality of São Lourenço do Sul. The farmers had donated the seed to the gene bank in 2005 and 2006 but had lost their varieties due to flooding in 2009. This restoration was extremely important for the community, because these varieties were part of the local food culture (Barbieri and Tempel, 2012).

Case 5: The reintroduction of squash accessions to the Krahô people

In 2006, we organized a training programme on methods of germplasm conservation. Participating Krahô leaders expressed their concern over the shortage of seed and ongoing loss of varieties of squash and pumpkin. On that occasion, we showed them images of the squash and pumpkin varieties maintained in the gene bank of Embrapa National Vegetable Research Centre, in Brasília, and provided them with seed of local squash varieties that they identified from the images. This provided the opportunity for a broader discussion on the existence, rarity, shortage and need for introduction and reintroduction of varieties. Together with the Krahô we identified 46 varieties of interest to the indigenous community. Accessions were identified and distributed.

Case 6: The reintroduction of sponge gourd (*Luffa cylindrica*) accessions in Minas Gerais

Sponge gourds are of great importance in small-scale agriculture because they are used in both human and animal diets. However, they have been facing genetic erosion in farmers' fields. In 2009, we started a project that aimed to promote on-farm management of local varieties of sponge gourds in the Furado da Onça community, in the north of Minas Gerais. We began the project by raising awareness through the establishment of a diversity block, as described by Shrestha *et al.* (Chapter 2.2), consisting of local varieties and germplasm received from our gene bank at Embrapa-CENARGEN. The diversity block was established in the fields of small-scale farmers, using their common management practices, thus without the use of inputs and chemicals. Seven local varieties and 13 gene bank accessions were observed and evaluated using a participatory approach. The criteria used by the farmers to evaluate the local varieties were identified and afterwards used for comparison. The farmers identified one local variety from the community and three accessions as promising. These varieties could be used directly by the farmers for the production of bath sponges and handicrafts (Fonseca Ferreira and De Azevedo, 2012).

Conclusion

The six cases presented in this chapter exemplify the importance of combining strategies of *ex situ* conservation and on-farm management in order to preserve the culture of using local crop varieties, shared by communities of indigenous and traditional people and small-scale farmers. Some of these communities have lost their traditional varieties and have been searching gene banks in order to restore them. This situation is a great challenge for us. We consider it our duty as curators to interact with guardians in a complementary and dynamic manner and feel that it is important to expand the dialogue and, through participatory planning, promote both the understanding of guardians in their original communities, and their recognition and inclusion in national strategies of genetic resource conservation. We consider that such actions, beyond our initial work to reintroduce germplasm in original communities, will contribute to the realization of community biodiversity management, empower these guardians and their communities, and promote the necessary integration of strategies of *ex situ* conservation and on-farm management.

2.6 Community seed reserves

Enhancing sovereignty and resilience in Central America

Mario Roberto Fuentes López and
Sergio Romeo Alonzo Recinos

Introduction

Many important staple crops have been domesticated in Central America, such as maize (*Zea mays*), beans (*Phaseolus vulgaris*), squash (*Cucurbita* spp.), peppers (*Capsicum* spp.) and many fruit species, including pawpaw (*Carica papaya*), avocado (*Persea americana*), custard apple (*Annona reticulata*), guayaba (*Psidium* spp.) and naseberry (*Pouteria sapota*). Maize and beans form the basis of the diet of much of the population. The crops provide 60% of the proteins and 70% of the daily energy needed by the region's rural population.

As a result of a combination of geographical factors (e.g. its location at the confluence of several ocean currents and the effects of periodic climatic oscillations – El Niño/La Niña), the Central American region is prone to extreme and unpredictable climatic events. Their effect on the society and economy of the region is aggravated by a number of economic and social factors (e.g. the low levels of social development and high degree of inequality). The ability of the population to adapt to the impacts of climate change is made more difficult by these socio-economic factors. As a result, the agricultural sector in Central America suffers significant economic loss due to extreme climatic events. For example, 49% of the losses caused by Hurricane Mitch alone were concentrated in the agricultural sector. It is estimated that in times of drought, losses to agriculture can amount to 60%. In Mexico, droughts represent 80% of all the agricultural catastrophes that occurred between 1995 and 2003 (Cifuentes Jara, 2010). In this highly unpredictable environment, the goal of achieving food security is a daunting task that needs multiple and collective efforts.

The Collaborative Programme on Participatory Plant Breeding in Mesoamerica (Programa Colaborativo de Fitomejoramiento Participativo en Mesoamérica; PPBMA), which was set up in 2000, is one such regional effort. The programme is being implemented in the form of national projects in Costa Rica, El Salvador, Guatemala, Honduras, Nicaragua and Cuba, with financial support currently being provided by the Development Fund, Norway. We focus on working with small-scale farmers to maintain, characterize and improve varieties of maize, beans, sorghum (*Sorghum bicolor*) and other crops. We are currently concentrating on the management, conservation and development of agrobiodiversity through participatory plant breeding (PPB). The main goal of the programme is to contribute to achieving sustainable food security. Our principal strategy is to encourage the participation of farmers in breeding (varietal selection, evaluation and seed production) and crop management, thereby contributing to the conservation and utilization of germplasm. We facilitate the sharing of knowledge and the development of skills at regional level, aiming to increase the capabilities of farming communities and stakeholders in a context of adaptation to climate change. Alonzo *et al.* (Chapter 5.8) provide insights into our PPB programme. Within the programme we aim to increase the resilience and sovereignty of farming communities in the management of their varieties, and ensure their seed security in times of change. In this context, we support the establishment and implementation of community seed reserves (CSRs) in the region.

Local mechanisms to cope with change

CSRs are set up as mechanisms for increasing local seed and varietal security during emergency periods that often follow catastrophic events such as floods, droughts and hurricanes. They are established in accessible areas of the communities with the aim of promoting the conservation, multiplication and distribution of seed stocks of portfolio crop varieties under the custodianship of a local committee. They can be immediately accessed and utilized in times of emergency, while in normal conditions they are used strategically as seed stock for supporting production. The CSRs supply communities with the tools necessary for responding to emergency situations, ensuring the continuation of the production processes while guaranteeing food and seed security.

The key steps towards establishing a CSR

Step 1: Assessing vulnerability and risk in a participatory manner

First, we analyse the impacts of extreme climatic events on local crop production and food security. The community identifies and prioritizes their problems. We use this crucial information for the design and development of the operational plan. We then identify the most vulnerable areas and communities, by mapping farmer and stakeholder perceptions on extreme events. We use available secondary information to triangulate and conclude the information gathered for the participatory assessment of vulnerability and risk.

Step 2: Designing the CSR

For this next step, we organize a series of consultative meetings with local authorities, farmers' groups and other relevant stakeholders. These meetings serve to share and discuss the results of the assessment. We discuss the objectives and operational strategies of the CSR, and provide the community and stakeholders with the opportunity to prioritize options for locations, crops, varieties and potential seed suppliers. We use the discussion to estimate the amount of seed to be produced and supplied, and thereby define the size of the CSR.

Step 3: Mobilizing local organizations and farmers' groups

The success of a CSR is directly related to the group responsible within the community for its management. In our project, we experienced several variations in the institutional structure. The CSR can be associated with pre-existing groups that have a good track record. It can also be set up in association with a new group that either represents the community or is associated with local government. We consider it crucial that the group members, or organizations themselves, are engaged in seed production and have the capacity to join in PPB following the establishment of the CSRs.

Step 4: Establishing a CSR committee

We support the establishment of a committee that takes responsibility for maintaining and sustaining the CSR. In addition, the committee establishes the criteria for seed quality, identifies seed suppliers, negotiates conditions for selling and buying seed, and makes sure the system performs well.

Step 5: Identifying seed producers

Within each community, we identify individual farmers who are already known and recognized for quality seed production; these farmers are then invited to become associated with the CSR. We enhance the farmers' capabilities to ensure that the CSR only stores seed of a high quality, in terms of germination, vigour and varietal identity.

Step 6: Estimating the size of the CSR

It is important that the community is aware of the number of available varieties that are strategic for overcoming extreme climatic conditions. Other important information that the community must be aware of is the seed production data of each identified variety; this is to ensure sufficient seed is available for events following extreme climatic conditions. Such data refer to planting and harvesting time, and to the most favourable period for producing an adequate quantity of quality seed. We ensure that each community knows the number of households that requires access to seed of a particular variety, and ensure that the committee is capable of planning the production of an adequate volume of seed of each variety, and of calculating the space and resources required for seed processing and storage.

Step 7: Defining CSR standards

We support the committee in developing a simple document that determines how the CSR operates, defines the responsibilities of the committee, and sets the criteria for choosing seed suppliers and distributing seed in emergency situations. The document is proposed to the community for approval and/or amendments.

Step 8: Ensuring the sustainability of the CSR

The CSR requires significant financial investments. The community itself covers the costs of planning and training, and sets aside financial resources for purchasing seed and maintaining storage silos. It must show that it is capable of continuing the training of community members in seed selection, seed quality management, seed storage, and pest and disease control. These investments and the continued strengthening of capabilities are strategic for sustaining the CSR.

Step 9: Incorporating PPB into CSR modalities

In order to enhance their adaptive capabilities to respond to the changing climatic conditions, we are currently linking the PPB programme with the CSRs. Through this linkage, we enhance the flow of new materials into CSRs, thereby contributing to the development of diverse varietal options for farmers, with the aim of increasing community resilience.

Case 1: Quilinco CSR in Cuchumatanes, Guatemala

The establishment of the Quilinco CSR in 2010 was the first step towards building our knowledge and experience in implementing such activities in the Cuchumatanes region. The region is located in the Department of Huehuetenango, in Guatemala, at an altitude of more than 2500 m above sea level (Figure 2.6.1). In order to create the Quilinco CSR, we invited four groups from four different villages to participate. They identified local varieties and shared the seed of those varieties for multiplication, inclusion in the reserve and dissemination in the community. The CSR currently has 130 accessions of maize, including native varieties, PPB improved varieties, and varieties that are still in an experimental stage. These varieties include white, yellow, pinto, red and black seeded types. The CSR also includes five varieties of common black beans (*Phaseolus vulgaris*) and four yellow varieties of runner beans (*Phaseolus coccineus*). In addition, the community included six species of aromatic herbal plants. Forty male and twenty female farmers contributed to the establishment of the CSR and are involved in running it. They were motivated to join because they are aware of the need to enhance seed security.

The fact that the CSR increases the community's seed and varietal sovereignty, even in times of disasters, makes it a source of collective pride. The seed of those varieties that are considered important in case of extreme weather conditions is stored in the silos of the Quilinco CSR for medium-term periods. Special containers are used to maintain properly identified and documented varieties that represent the agrobiodiversity of Cuchumatanes. The committee responsible for the CSR has set up its own rules for managing and using the CSR.

Case 2: Nueva Esperanza Concepción Sur CSR, Honduras

This CSR was established in 2007, by the two local agricultural research committees (Comité de Investigación Agrícola Local; CIALs) of Nueva Esperanza and El Barro,

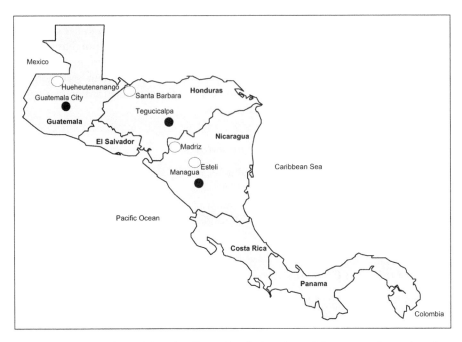

Figure 2.6.1 Map of Central America illustrating the locations of the community seed services discussed in Chapter 2.6 and participatory crop improvement in Chapter 5.8.

in the Department of Santa Barbara, Honduras (Figure 2.6.1). Farmers began to collect and identify local varieties in their communities, storing half a kilo of every local variety with the aim of retaining it for the next sowing season. Thus, they started with a strategy to avoid the potential loss of varieties due to climatic change. This CSR now maintains eight local varieties of maize, three conventionally improved maize varieties, 12 local bean varieties, six conventionally improved bean varieties, five green manure bean varieties, and some other species of local importance. It also includes two PPB improved maize varieties and eight PPB improved bean varieties. CIAL members have started to produce and sell the seed of a wide range of maize and bean varieties. Through this activity, other farmers have started to benefit from the access to and availability of good-quality seed of better-adapted (locally or through PPB) varieties.

Lessons learned

Over the last five years, members of the PPBMA collaborative programme in Costa Rica, El Salvador, Guatemala, Honduras and Nicaragua have been working on the establishment and strengthening of 12 CSRs that directly benefit over 1400 farming families. These 12 CSRs also indirectly benefit up to 2000 farming families by providing them with access to seed in times of emergency. The practice of CSRs is a feasible and cost effective way of ensuring the availability of, and increasing farmers' access to, the seed of local or conventionally improved varieties, or even varieties improved through PPB in times of insecurity. CSRs are similar to community seed banks, as described and analysed in more detail by Shrestha *et al.* (Chapter 2.8). The CSR is a globally important practice that supports community biodiversity management (CBM), thereby contributing to the *in situ* on-farm conservation of plant genetic resources (PGR), while increasing food and seed security at community level. PPBMA has been able to associate conservation aspects of the CSRs with food security; they are both important assets for increasing community resilience. CSRs can be seen as community institutions that foster livelihood development in the context of CBM (Shrestha *et al.*, Chapter 1.3), promote entrepreneurship and contribute to a robust seed system (De Boef *et al.*, 2010), through which they are better able to respond to frequent but unpredictable climatic catastrophes.

Public gene banks regularly face constraints in relation to the availability of financial resources for the renewal of genetic material, putting the PGR maintained in the gene bank potentially at risk. Such situations do not occur when the varieties are used and stored at farm level. We took this aspect into consideration when designing the structure and operational mechanisms of the CSR. They contribute to a form of conservation through which the dynamism that is a vital component of agrobiodiversity is continued. CSRs emerge in such a way as a strategy that embraces the dynamic nature of agrobiodiversity while contributing to its maintenance, and it is this dynamic nature that motivates people to face new problems and continue to find ways to cope with them (De Boef and Thijssen, Chapter 1.8). CSRs constantly revitalize varieties, ensuring genetic variability and pooling genetic materials in ways that allow farmers to respond to the constraints caused by climate change. They ensure the production of food for poor, small-scale farming families and safeguard their control over their seed and varietal stocks.

Note

The authors would like to thank coordinators of the Participatory Plant Breeding Programme for the Mesoamerican Region – Juan Carlos Rosas (Honduras); Javier Pasquier (Nicaragua); Juan Carlos Hernandez and Rodolfo Araya (Costa Rica); and Carlos Reyes (El Salvador) – for their contributions to this chapter.

2.7 Community seed banks in the semi-arid region of Paraíba, Brazil

Emanoel Dias da Silva

Small-scale farmers who live in the semi-arid regions of Brazil have been challenged by their environment throughout history. They use their creativity and skills to observe and learn from nature, and in this way they have developed a livelihood that is compatible with their harsh environment. Associated traditional knowledge develops over time and is actively passed down from generation to generation. Together they constitute the genetic and cultural heritage of farming communities, but are also assets that are ignored or marginalized in normal pathways for agricultural development and modernization. The recovery and use of agrobiodiversity plays an important role in vitalizing this cultural and agro-ecological identity, therefore the conservation of agrobiodiversity cannot be isolated. It needs to be associated with, and appreciate the innovative spirit and autonomy of these small-scale farming communities. The diversity of local varieties that continues to exist in these communities is a product of the capacity of these traditional societies to work with, observe and establish a harmonious relationship with Mother Nature.

In the current chapter, I will share the experiences of the Semi-arid Network-Paraíba (ASA/PB), with which I have been collaborating since 1993. I will focus on our activities in strengthening the ability of communities in the semi-arid region of Paraíba state to sustain their livelihoods, through the establishment of community seed banks (CSBs) in the recovery, maintenance and recognition of local varieties together with their traditional knowledge.

Agrobiodiversity: a vital livelihood asset under threat

The diversity of local varieties is being threatened while the normal pathway of agricultural development spreads around the world. This model is dominant and advanced in Brazil in research and extension services that also target small-scale farming communities. The commercialization and privatization of plant breeding and seed production generates large profit margins at the cost of the loss of plant genetic resources (PGR) from farmers' hands. Exerting intellectual property rights over living organisms, as promoted by genetic engineering, can be seen as way in which monopolistic enterprises can increase their control of the seed market. Hybrid and transgenic varieties are increasingly being sold in large quantities to small-scale farmers. These varieties make farmers more and more dependent on the use of agro-toxins and chemical fertilizers, which undermines their original livelihood strategies that are compatible with their environment.

Government policies concerning seed distribution can also be considered another serious threat. We observe that these policies often contain errors in their structure and execution and, in the semi-arid regions in particular, they tend to reduce the food and seed security of farmers while aiming to achieve the opposite effect. Over the past decades, traditional seed distribution programmes have been distributing only a few varieties but in vast quantities, thus undermining rather than strengthening farmers' livelihoods. Varieties and seed for distribution are selected based on public and private interests rather than on the specific agro-ecological system and cultural demands of beneficiary farmers. Such politically motivated seed distribution leads to dramatic genetic and cultural erosion in Brazil's semi-arid regions; it should be noted that such policies continue and are even further encouraged up to this day.

In spite of the great challenges imposed by the capitalist model and policies that create dependency rather than autonomy, small-scale farmers and their organizations have continued to maintain their knowledge, livelihood strategies and forms of resistance over the years, by using and exchanging farmer-saved seed of local varieties. Such resistance is not only based on agro-ecological and cultural demands, but is also strongly politically motivated by the farming community's struggle to maintain autonomy over their production system and sustain their own livelihood strategies, particularly in coping with a semi-arid environment.

Agriculture in the state of Paraíba

Paraíba is one of the smallest states in Brazil (Figure 2.7.1). Over half of its population lives in its semi-arid region. According to agricultural census data, farm properties of less than 10 ha account for 69% of the total number. However, in the region of Agreste Paraibano, in the central-eastern part of the state, small-scale farming properties form the majority (94%), although they occupy just 56% of the area. In Brazil, small-scale farmers are referred to as *agricultores familiares* (family farmers), since the family or household is the basic unit of agricultural production. Small-scale farmers in semi-arid Paraíba use multiple cropping systems and cultivate different varieties of cowpea (*Vigna unguiculata*), common bean (*Phaseolus vulgaris*), lima bean (*Phaseolus lunatus*), maize (*Zea mays*), cassava (*Manihot esculenta*), sweet potato (*Ipomoea batatas*) and yam (*Dioscorea* spp.). They grow several fruit and vegetable species in their home-gardens, while many of them also engage in the collection of medicinal plants and wild fruits from their surroundings. Small-scale animal production supports their integrated production system (Almeida and Cordeiro, 2002).

ASA/PB: a facilitating network

ASA/PB is a forum that brings together around 350 civil society organizations, including community associations, unions, non-governmental organizations (NGOs), parishes and social movements. These organizations are united in their goals to better coexist with the semi-arid climate, and to strengthen the autonomy and livelihoods of small-scale farmers in the region. ASA/PB was established in 1993 in response to a great social mobilization, which took place throughout the semi-arid region of north-eastern Brazil, in support of effective public policies facilitating coexistence with

Figure 2.7.1 Map illustrating the approximate location of Paraíba in Brazil.

the semi-arid. Since the foundation of the ASA/PB network, we have been in a continuous struggle to find new pathways to increase the autonomy of small-scale farmers and to break free from the normal pathways promoted by developmental policies and interventions that rather result in social dependency. In this struggle, seed is a core theme for ASA/PB. We have been motivating and supporting a network of CSBs for more than two decades, over which time our work and network has become a reference point in Paraíba, the north-eastern region, and Brazil as a whole.

Community seed banks

The need to reassure seed security and sovereignty

Small-scale farmers in semi-arid Paraíba use farm-saved seed of local varieties annually. However, due to the small size of their planting areas, combined with the unpredictable and unfavourable climatic conditions, many households are often unable to replenish their seed stock. Such circumstances lead many households to consume their seed. In response to this threat to food and seed security, government programmes provide these farmers with certified seed of improved varieties. Historically, these programmes have been used to serve political interests, so that seed is simply exchanged for votes during elections.

Another option for farmers is to purchase seed. However, this is most unfavourable, since the price of seed peaks at planting time, reaching up to four times higher than it would be during harvesting. Households that are unable to save or access seed end up having to access what is referred to as *semente de meia* (literally 'seed for a half').

Such farmers 'borrow' seed, agreeing to return 50% of their harvest (the 'half'), resulting in a vicious circle of poverty associated with food and seed insecurity (Londres and Almeida, 2009).

Community seed banks (CSBs) emerged as a way of addressing issues related to food and seed security that result in poverty and lead to a disparity in food sovereignty. At ASA/PB, we developed a mechanism through which households could borrow seed from a CSB and agree to return the same amount plus a relatively low percentage at harvesting time. An informal group or association that is formed within the community is responsible for the structure and procedures for the storage, delivery and return of the seed.

Common mechanisms for implementation

CSBs do not have a rigid operational model. Each bank adopts its own specific procedures. Some common elements include: controlling the flow of seed (loans and returns); monitoring seed quality and storage; and surveying the demand and availability of seed in the rainy season. A seed commission was established within ASA/PB to record information from all the banks in the network in a database, which is updated annually through questionnaires and meetings (Cordeiro, 2007).

The number of members of a CSB varies and it determines the number of species and quantity of seed to be stored. For example, the CSB in the municipality of Vieirópolis has 157 members, while a CSB in Camimbas, located in the interior, includes only eight households. A large bank can store up to 7000 kg of seed, although there are cases where CSBs with large numbers of families maintain just a few varieties of seed and in low quantities (e.g. the Serrotão CSB in the major town of Campina Grande stores just 25 bags of maize and one bag of common beans for its 42 member households) (Cordeiro, 2007).

CSB members define the operational rules of the bank. The CSB of Lagoa do Gravatá, in the municipality of Lagoa Seca, for example, was established in 1998 and its operational rules were decided on during a meeting involving 35 households, which led to the establishment of an association involving members from five other communities. A committee of three persons, appointed by the CSB members, takes responsibility for the day-to-day management of the bank. The assembly fixed a return rate of 20% on the quantity of seed provided to a beneficiary household. In times of drought and loss of production from the fields, the bank can relax the rules so as not to prejudice members and to ensure the continued functioning of the bank (Cordeiro, 2007).

Conservation of local varieties

One of the main functions of the CSB is to provide access to the right quantity and seed of the preferred varieties, as can be seen with the São Thomé CSB in Nova Alagoa. When the CSB was established in 1974, the community maintained the seed of only two varieties, or rather types, of common bean, and depended on external sources for the seed and varieties of all other crops. In 2009, their CSB included three common bean, two cowpea, three lima bean and two local maize varieties, as well as the seed of sunflower (*Helianthus annuus*), pumpkin (*Cucurbita* spp.), sorghum (*Sorghum bicolor*) and pigeon pea (*Cajanus cajan*) varieties.

ASA/PB uses the strategy of integrating conservation aspects into the day-to-day management of the CSBs, bringing agrobiodiversity conservation into the direct domain of food and seed sovereignty. Household members of a CSB are responsible for maintaining samples of varieties included in the seed bank, thus forming an '*in situ* safety reserve'. In some situations, a 'mother bank' is established at municipal or regional level. However, the guiding principle remains that each household maintains its seed stock of local varieties, and the local CSB and 'mother bank' are continuously supplied with farm-saved seed. Shrestha *et al.* (Chapter 2.8) describe a similar mechanism with different levels of seed storage in Bangladesh, where seed huts form the CSBs, and Community Seed Wealth Centres adopt a role similar to that of mother banks operating at the agro-ecosystem level. Pitambar Shresta *et al.* further compare and draw common lessons from a range of experiences with CSBs from around the globe.

In Borborema, located in the Agreste Paraibano region, CSBs are organized at municipal level by the farmers' union. The national NGO, Advice and Services for Alternative Agriculture Projects (AS-PTA), supported the establishment of a regional network of 78 CSBs. Each CSB has its own seed committee. A municipal seed committee has been established in each of the 13 municipalities of Borborema, in association with the municipal charter of the farmers' union. In 2002, the regional 'mother bank' of Borborema was set up in the town of Montadas and is managed by an association of municipal seed committees. The association manages inventories, facilitates the exchange of seed between banks, organizes the receipt and redistribution of purchased seed, and distributes the seed of local varieties from existing banks to new banks (Londres and Almeida, 2009).

Enhancement of capacities and storage facilities

Over the years, member organizations of ASA/PB have been investing in the development of capabilities and infrastructure to improve the storage system of member households and CSBs. We organized workshops to train farmers and staff of our member organizations in the construction of silos and in the construction of drying and processing yards. Farmers constructed zinc silos with storage capacities of 30–500 kg grain, but increasingly use recycled PET 2-litre bottles for seed storage too. In 2009, the storage capacity of the network of CSBs exceeded 140 tons of seed (Londres and Almeida, 2009).

Distribution and network

Paraíba's network of seed banks includes more than 240 CSBs, involving more than 800 families in 63 of its municipalities. Table 2.7.1 provides an indication of the evolution of the CSBs in Paraíba between 1998 and 2008. We have been able to rescue nearly 300 crop varieties. The network of seed banks is an illustration of a practice that contributes to community biodiversity management (CBM), where small-scale famers, with their traditional knowledge and local varieties, assume both individual and collective responsibility for maintaining their genetic heritage. The communities and the network enhance their members' sovereignty over seed, varieties and food; as well as their capacity to cope with their semi-arid environment. In Paraíba, our CSBs

Table 2.7.1 Evolution of the community seed banks in Paraíba since 1998

Season	No. of CSBs	No. of HHs	Total seed stock (tons)[a]	No. of municipalities	No. of silos
1998–99	62	1860	15	n.d.	n.d.
1999–00	129	3838	66	n.d.	n.d.
2000–01	220	6920	100	55	98
2002–03	175	7441	78	51	n.d.
2003–04	205	7170	128	60	437
2004–05	207	7145	161	56	344
2005–06	228	6560	n.d.	61	476
2007–08	205	3730	56	60	558

Source: Londres and Almeida (2009).

CSB, community seed bank; HHs, households; n.d., no data available.

a Total stock of seed available before planting.

strengthen and promote CBM, serving as a model to inspire others in the semi-arid north-eastern region of the country and Brazilian society as a whole.

Participatory research for strengthening the network

In order to strengthen its work, ASA/PB set up a partnership with researchers from the Brazilian Agricultural Research Corporation (Embrapa) and the Federal University of Paraíba. Our aim in establishing this partnership is to demystify the idea that local varieties are inferior to the improved varieties of commercial companies. We conducted participatory varietal selection (PVS) trials in which we evaluated both improved and local maize varieties using both researchers' and farmers' criteria. In all PVS sites, we observed that local varieties performed better than locally available commercial and government distributed varieties. We now use the research results in policy dialogues to promote the acknowledgement and appreciation of local varieties and local people's capacities in seed production.

'Seeds of Passion' festivals: strengthening our identity and facilitating out-scaling

The seed of local varieties in Paraíba is called *Sementes de Paixão*, which means 'Seeds of Passion'. With this name we articulate our relationship with the varieties and with the informal seed system responsible for maintaining and exchanging the varieties we inherited from our ancestors. Our enthusiasm affirms the importance we give to this genetic patrimony for guaranteeing seed security and sovereignty over our food for future generations.

Since 2004, our network has been celebrating Seeds of Passion festivals every two years. We organize the festival to reinforce the identity of the *agricultura familiar* and promote the use of agrobiodiversity. We joined forces with organizations from other states in the semi-arid region of Brazil, and formulated common strategies to reinforce our work and be more effective in reaching our common goals. Sementes de Paixão has become a symbol and cause in defence of sustaining the livelihoods of small-scale farmers. Today, our network includes more than 800 CSBs and *casas de sementes* (family-based banks with local varieties), across semi-arid Brazil. This collective of more than 15 000 small-scale farming families maintains the heritage of hundreds of local varieties and ensures their access to seed of those local varieties in a way that does not prejudice their food sovereignty.

CSBs and policy development

The extensive network of CSBs has gained noteworthy political weight and includes representation at many levels, from families who contribute to and benefit from a single CSB, to clusters of municipal banks, to local government representatives and the state assembly. This facilitates an exchange of experiences, resources and ideas, and provides the opportunity to influence policies (Cordeiro, 2007).

Through this structure and our influence, policies in Paraíba have started to better respond to the demands of small-scale farmers. Between 1998 and 2002, ASA/PB worked with the Paraíba state government to recover the seed stocks and create new seed banks. In 2002, a CSB programme was established under State Law 7.297, enabling Paraíba's state government to acquire seed of local varieties for implementing its public programmes, a situation that had, until then, been the sole privilege of certified seed and improved varieties. This programme has enabled farmers to use local varieties and seed from the CSBs in contracts supplying food to the so-called 'institutional market' – public entities such as schools and hospitals (Schmidt and Guimarães, 2008). Between 2004 and 2010, over 180 tons of food was produced in Paraiba using the seed of 73 local varieties. This provided farmers with a revenue of up to 200 thousand Brazilian Reais (more than US$100 000).

Development of a new pathway in agriculture

With our experience and consolidated network, we realize that informal seed exchange is vital for sustaining the varieties and seed sources that we need. Such seed networks shape a strategy to protect communities of small-scale farmers from a number of threats they face in maintaining their agrobiodiversity. The networks are collective structures that are crucial for gaining political weight to ensure our sovereignty in the conservation of our genetic patrimony. Small-scale farmers have shown that Seeds of Passion are productive in, and suitable for, semi-arid farming conditions. The quality of seed produced by these farmers is no longer challenged. The seed produced is free from chemical inputs and agro-toxins and the fact that it does not include transgenic varieties is gradually becoming an asset for our seed producers. Those who select, store and plant the seed of local varieties are following an alternative and new development pathway, based on the strength and resources of communities of small-scale farmers. As such, we reject the normal pathway that is based on monopolies, industrialization of agriculture and disempowerment of small-scale farmers in the production system. The diversity of local varieties and the great knowledge of this genetic wealth are fundamental elements for the development and strengthening of agro-ecological production in semi-arid Brazil and beyond.

2.8 The community seed bank

A common driver for community biodiversity management

Pitambar Shrestha, Genene Gezu, Saujanendra Swain,
Bertrand Lassaigne, Abishkar Subedi and
Walter Simon de Boef

A backup to informal seed systems

In the early 1990s, many development organizations began to support the establishment of community seed banks (CSBs) with the primary goal of enhancing food and seed security. The normal agricultural development paradigm assumed that all farmers, from subsistence to commercial, would eventually become clients of public and private seed systems, and the formal sectors would ensure their access to quality seed of improved varieties (Douglas, 1980; Frankel and Soulé, 1981). We now realize that this never actually happened. Small-scale farmers in Africa, Asia and Latin America, but also in Europe and North America, use farm-saved seed and rely to a large degree on the informal system for accessing new materials (Byerlee *et al.*, 2007; Louwaars and De Boef, 2012).

Limitations of farm-saved seed and the informal seed systems led to the establishment of CSBs for increasing seed and food security. When farmers lose their seed or varieties, they depend either on quality seed of improved varieties that has been commercialized through agro-dealers, or on the seed of improved varieties that is distributed by governments. However, the varieties are not necessarily adapted to local agro-ecological conditions or specific demands (Abay *et al.*, 2008). The CSBs ensure sufficient seed of the right (local) varieties is available at the right time and for an affordable price, thus enhancing seed security (De Boef *et al.*, 2010). They complement the formal seed sector, in particular for those crops and varieties that are not addressed in seed production and marketing because they are not commercially interesting, or because the market for such crops and varieties is too small (Louwaars and De Boef, 2012). CSBs serve as backup to the informal seed sector, and they motivate farming communities to rely on their own institutions and remain autonomous in their food and seed security. Their success in recovering seed and varietal sovereignty – and thereby food sovereignty – is well illustrated by Dias discussing the Seeds of Passion network in Paraíba, Brazil (Chapter 2.7).

A practice within community biodiversity management

In the context of community biodiversity management (CBM), CSBs are considered a common practice for securing the availability of local varieties (Jarvis *et al.*, 2011), thereby contributing to the on-farm management of plant genetic resources (PGR). In this chapter, we will share examples of CSBs that address seed and varietal security within the framework of CBM. CSBs form the social and institutional capital required to make CBM work as a collective and conscious process in conservation (Pretty and Smith, 2004; Shrestha *et al.*, Chapter 1.3). In addition to the experiences of the Semi-Arid Network-Paraíba (ASA-PB) in Brazil (Dias, Chapter 2.7), and the Collaborative Programme on Participatory Plant Breeding (PPBMA), in Central America (Fuentes and Alonzo, Chapter 2.6), we look at and compare experiences working with CSBs in Bangladesh, Ethiopia, France, India and Nepal, and relate these to other activities in Canada, Mali and Mexico. We try to capture commonalities and variations that exist for the practice, which are summarized in Tables 2.8.1 and 2.8.2. Before sharing the case studies, we outline some common principles. We conclude by drawing some lessons from CSB practice within the larger framework of CBM.

Common principles

The CSBs are collections of seed that are maintained and administered by communities in a central facility or in a structure that is shared among a range of individuals. They are usually established in collaboration with local organizations and, sometimes, national gene banks (Jarvis *et al.*, 2011). A common feature is that non-governmental organizations (NGOs) and civil society organizations (CSOs) play a key role in their establishment and management. Seed is stored in large samples, to ensure that adequate and sufficient planting material is available (to achieve seed security), and/or in small samples, to ensure that local varieties are available for regeneration when varieties are endangered (to achieve varietal security), as shown in Table 2.8.2. Some CSBs have a well-defined relationship with public or NGO gene banks to ensure that a backup system is available under a normal access regime (Bangladesh and Ethiopia), or a black-box arrangement (India). The seed is primarily retained for members or farmer beneficiaries, using informal measures of quality control. The CSBs in France and Canada operate as national networks. The Seeds of Diversity network in Canada facilitates the maintenance and exchange of seed among its members. The national network in France does have a central storage facility, but its members conduct regeneration, as in all other CSBs. Common to all CSBs is that they increase farmers' access to, and timely availability of, diverse types of quality seed and planting materials of locally adapted (not necessary only local) varieties at reasonable prices or conditions. They also ensure farmers' sovereignty over access to seed and varieties (Bangladesh, Brazil, Central America, France, Mexico and India). CSBs can also be considered a practice that contributes to the realization of farmers' rights.

Table 2.8.1 Characterization of community seed bank networks, 2011

Country	Name of facilitating organization	Type of organization	Specific area	Major crops	Year started	No. of CSBs	Average no. of members/CSB
Bangladesh	UBINIG	Private organization	Scattered areas; 19 districts	Rice, leafy vegetables, pulses, oilseeds, spices, uncultivated food plants, fruits, timber plants, medicinal plants	1994	6 + 23[d]	1300
Brazil	ASA-PB	CSO[a]	Paraíba	Maize, cowpea, lima bean, common bean	1993	205	20
Canada	Seeds of Diversity	CSO	National	Vegetables, fruit, grains, flowers and herbs	1984	1	1400
Ethiopia	IBC/EOSA	NPGRP[b]/NGO	Oromia, Amhara and SNNPR[c]	Wheat, chick pea, grass pea	1994	14	680
France	AgroBio Périgord	CSO	Local/national	Maize, several others	2001	1 + 7[e]	280[f]
India	MSSRF	NGO	Orissa, Tamil Nadu	Rice, millets, pulses	1999	38	40
Mali	USC	NGO	n.d.	Sorghum, millet, cow pea, bambara groundnut	1994	7	280
Mexico	Instituto Simone Weil	CSO	Mexico	Maize, vegetables, medicinal plants	2003	10	200
Nepal	LI-BIRD	NGO	Scattered across Nepal	Rice, legumes, gourds, millets, vegetables	2003	14	360

Source: based on an email survey carried out by LI-BIRD, with partners, UBINIG, ASA-PB, Seeds of Diversity, IBC/EOSA, AgroBio Périgord, MSSRF, USC, Instituto Simone Weil and LI-BIRD, in 2011.

a CSO – civil society organization; b NPGRP – National Plant Genetic Resources Programme; c SNNPR – Southern Nationals, Nationalities and Peoples Region; d 6 community seed wealth centres and 23 seed huts; e AgroBio Périgord has a local CSB that serves a national CSB, to which seven other CSBs are associated; f number refers to the CSB of AgroBio Périgord and not its sister CSBs.

Table 2.8.2 Characterization of seed and varietal management aspects, and institutional ground of community seed banks

Country	Name of organization	Mechanism	'Seed bank / loan arrangement'	Varietal security strategy	Link with public gene bank	Specific observations
Bangladesh	UBINIG	Seed and varieties	200% (double seed)	Members continue to cultivate, specialized seed farmers maintain specific varieties, community seed wealth centres (CSWC) and gene bank serves as backup	Functional link with public gene bank as backup	Organized seed conservation – CSWC have linkage with national gene bank, seed huts, specialized seed keepers); focus on promoting autonomy, and sustainable agriculture and uncultivated food
Brazil	ASA-PB	Seed and varieties	120% (flexible community-based)	Members continue to cultivate, community seed bank (CSB) and mother banks serve as backup	None	Emphasis on autonomy in seed and varieties; use of seed of local varieties for institutional market supports CSB network
Canada	Seeds of Diversity (SoD)	Seed and varieties	Exchange among members	Living gene bank, members cultivate accessions, and provide access to members	Informal link, no formal backup	Accessions remain with members; community is composed of persons with shared interest; network is national; sod facilitates access to members
Central America	PPBMA	Seed and varieties	Provision in times of disasters; otherwise sales	CSB contracts out cultivation to maintain the reserve	None	Emphasis on being a backup system in case of climatic disasters for both seed and varieties
Ethiopia	IBC/EOSA	Grain, seed and varieties	Introduction if new, otherwise back-up with dissimilar arrangements	Community-based gene bank, formal gene bank acts as backup	Informal link	CSB associated to legal entity (cooperative); continuous links with and support from NGO; organization is a base for other activities
France	AgroBio Périgord	Seed and varieties	Return of 300% seed for sharing, evaluation and storage	Sharing with as many farmers as possible 'in vivo', and maintenance of rare varieties	Informal link, no formal backup, collaboration in research	All entries are tested for GMOs; community is composed of farmers and amateur gardeners with shared interest; network is local and national; AB Périgord provides access to members only
India	MSSRF	Grain, seed and varieties	100–150% (seed)	One farmer-one variety, in situ community-based gene bank through (black box) backup	Black box; MSSRF-gene bank	Traditional knowledge through village knowledge centre for documentation of farmers' rights
Nepal	LI-BIRD	Seed and varieties	Link with CBM fund	Diversity block, fund beneficiaries grow one variety each	Not yet but in process	'Redbook' of information and associated traditional knowledge maintained; CSB is sustained through links with CBM fund

Source: Based on an email survey carried out by LI-BIRD, with partners, UBINIG, ASA-PB, Seeds of Diversity, IBC/EOSA, AgroBio Périgord, MSSRF, and LI-BIRD, in 2011.

Case studies of community seed banks

Safeguarding rural livelihoods in Bangladesh

Since 1984, Policy Research for Development Alternative (UBINIG; Unnayan Bikalper Nitinirdharoni Gobeshona), a private research organization in Bangladesh, has been supporting a network of 300 000 farmers, called the Nayakrishi Seed Network (NSN). The NSN is an organized effort to promote local seed conservation and sustainable agriculture practices, and build self-reliance in communities. At agro-ecological zone level, UBINIG has established six community seed wealth centres (CSWCs) similar to ASA-PB in Brazil (Dias, Chapter 2.7). The main function of the CSWC is to provide backup seed storage and maintain a compatible passport data system with the national gene bank, thereby facilitating the access to and exchange of seed between *ex situ* gene banks, and linking with on-farm management. These CSWCs maintain up to 2400 accessions of different types of food crops, including uncultivated ones. CSWCs are integrated with 23 Nayakrishi seed huts that are located in various parts of the country, and which use common CSB working principles. These seed huts provide seed to on average 1300 members (Table 2.8.2); returned seed is stored for multiplication and for wider distribution through the CSWCs. Furthermore, the seed huts are linked to over 332 specialized seed keepers (111 of whom are women), located in different clusters of the communities, who make sure that quality seed is produced and that local seed stocks meet the urgent demand of farmers. The involvement of the NSN in seed banking in Bangladesh is contributing to the on-farm management of 2300 rice varieties, a unique success in an era of high genetic erosion of local crop varieties.

Establishing conservation cooperatives associated to CSBs in Ethiopia

In the 1990s, the rich diversity of durum wheat varieties in the central highlands of Ethiopia was almost entirely displaced (95%) by the intensive introduction of improved bread wheat varieties. Farmers became dependent on annually purchased seed, making their agriculture economically and ecologically less beneficial. In response to the common critique that genetic resources in gene banks are not easily accessible to farmers, the Institute for Biodiversity Conservation (IBC) conducted a series of activities in the early 1990s, in which local varieties of several crops were reintroduced (Worede and Mekbib, 1993). The next step was the establishment of CSBs, to enable communities to maintain and use their varieties with autonomy.

While the IBC played a critical role in the initial establishment of CSBs, the NGO Ethio-Organic Seed Action (EOSA) has been providing continuous support to six CSBs in Amhara and Oromia since the early 2000s. EOSA provides technical support and oversight to ensure that seed regeneration through the CSBs maintains the required quality. Currently, it is supporting the establishment of two new CSBs in Oromia. Another five CSBs have been constructed in the Southern Nations, Nationalities and Peoples Region (SNNPR). Regassa Feyissa and colleagues describe the activities of EOSA in more detail in Chapter 1.4.

CSBs in Ethiopia are legal entities with a well-established governing structure and mechanisms for their management. They also act as conservation and seed producer cooperatives. After almost two decades of investment in human resources and infrastructure, and support through institutional and technical assistance, the conservation cooperatives are well organized and sustained (Gezu and Seboka, 2012). The awareness and experiences gained have resulted in the maintenance and use of a higher degree of crop diversity by participating farmers (Bezabih, 2008). These conservation-oriented CSBs in Ethiopia, similarly to those CSBs in other countries embedded within a CBM programme, benefit communities with their organizational capacity and awareness of genetic resources in livelihood development (Engels *et al.*, 2008).

EOSA has been working with the Unitarian Service Committee (USC) for many years, through their Seeds of Survival programme. The programme supported farmers in the establishment of CSBs for safeguarding access to seed of their local varieties. The USC has also been running a similar programme in Mali, and additional programmes are in the process of being set up in other West African countries. These programmes are quite similar to the CSB programme in Ethiopia (Table 2.8.1).

Creating conditions for enhancing resilience and favouring farmers' rights in India

In 1999, the M.S. Swaminathan Research Foundation (MSSRF) established a grain/seed/gene bank as an experiment in Banrangput village, in Jeypore district, Orissa. The aim was to develop a model that would contribute to achieving food security in tribal villages by ensuring the access to, and availability of, the seed of local varieties of several subsistence crops, thereby contributing to their conservation and use. Tribal communities live in areas where they are vulnerable to frequent droughts, flooding and other livelihood threats. Consequently, large numbers of tribal households face regular food shortages; many have no other choice but to consume their seed, resulting in seed insecurity and genetic erosion. Farmers are then compelled to buy seed of improved varieties, which is expensive and often the varieties do not match their farming system. A vicious circle is thus created, linking environmental stress with poverty and social vulnerability.

Over the past decade, MSSRF has been supporting the establishment of 23 CSBs in Jeypore district in Orissa. The majority of the established CSBs have become self-sustaining institutions that store varieties and seed of rice, millets, pulses and oil seed crops. The seed of adapted local varieties is available and affordable. An indirect output of the CSBs is that the aforementioned vicious circle is broken.

MSSRF also supported the establishment of a further 13 CSBs in the Kolli Hills in Tamil Nadu, fostering the conservation and use of several millet species, in particular. The CSBs have benefited more than 1330 farm households in Jeypore and 300 farm households in the Kolli Hills. In order to ensure the sustainability of the community collections, MSSRF supports the CSBs by keeping varieties in a black box arrangement in a gene bank at their headquarters in Chennai.

Besides the conservation and use of local varieties, the CSBs have contributed to the documentation of associated traditional knowledge (ATK). Information on local varieties is documented and further shared through village knowledge centres (VKCs), which digitize the passport data of each member household. VKCs are instrumental in disseminating information on availability of quality seed among tribal villages, which further enhances seed networking and seed security beyond that of individual villages. As such, the VKCs contribute to resilience in the context of availability of seed of local varieties.

In accordance with the Farmers' Rights Act of the Government of India, the documentation maintained in the VKCs supports farmers in registering their local varieties as a farmers' varieties with the Protection of Plant Varieties and Farmers' Rights Authority (as described by Bala Ravi, Chapter 6.3). Consequently, the MSSRF model for CSBs and VKCs increases food, seed and varietal security; contributes to the conservation of local varieties; and enhances community resilience through building social institutions and strengthening the farmers' network. These results could be achieved while creating conditions favourable for the implementation of several aspects of farmers' rights.

Ensuring farmers' autonomy in using local maize varieties in France

The Maison de la Semence Paysanne ('small-scale farmers' seed house') was initiated in 2003 by AgroBio Périgord and Bio d'Aquitaine in France. Kendall and Gras (Chapter 1.7) provide more details on its structure and functioning. Farmers set up this network to create options for them to access, evaluate and exchange local maize varieties that are adapted to small-scale farming and, often, to organic production systems. With this option, they were politically motivated to maintain their independence from large-scale seed companies that advocate hybrid and genetically modified maize varieties (Gaudin *et al.*, 2009). Sovereignty over access to seed and varieties was a key motive for setting up the Maison de la Semence Paysanne.

Currently, more than 250 farmers are associated with the Maison de la Semence Paysanne, contributing to the conservation and continued cultivation of more than 100 composite maize varieties, a dozen sunflower varieties, and varieties from other species (Table 2.8.2). Each year, participating farmers return part of the seed selected from conventional plots, which is distributed to new farmers in subsequent years. To ensure that the material is not contaminated, a sample of each batch is sent to a laboratory for genetically modified organism (GMO) screening. If a seed lot is contaminated, which up until now has not yet occurred, the lot and related plots are destroyed before flowering to avoid further contamination. What is crucial is that through this scheme the network aims to use a dynamic approach to conservation, in comparison with the strict protocols used in gene banks.

Another crucial characteristic of the Maison de la Semence Paysanne is that farmers must become members to get access to varieties. As members, they contribute to conservation and experimentation, and can exchange and use the non-certified seed of most non-registered varieties. The Maison de la Semence Paysanne is thus a network, which avoids the strict seed and variety regulations of France, as described by Kastler (Chapter 6.8), and we therefore consider it to be a CSB in which the community is one made up of farmers with a shared interest. This situation is similar to that of the Seeds of Diversity network in Canada, as illustrated in Tables 2.8.1 and 2.8.2. A key difference between the networks in Canada and France is that in France an actual physical seed bank exists, whereas the bank in Canada is virtual and accessions remain with their 'guardians'. Seeds of Diversity facilitates conservation and exchange, which is also one of the functions of the CSB in France. The definition of community is the major difference between the seed banks in France and Canada, and their sister organizations in Africa, Asia and Latin America, while they share in common many aspects, including the motivation to enhance sovereignty in seed and varieties.

Linking CSBs with the CBM fund to ensure their sustainability in Nepal

The implementation of CSBs in Nepal began in early 1996/1997, through a USC-Canada coordinated project. Other seed banking modalities were developed in 2003, in the Bara district of the central lowlands in Nepal, as part of a global agrobiodiversity project (further described by Subedi *et al.*, Chapter 1.2). The project was inspired by the results of the CSBs in other countries, and motivated by the need to link conservation with access to quality seed of adapted varieties (Shrestha *et al.*, 2006). What was unique in Nepal was the immediate integration of the CSB into the broader CBM processes, as well as its linkage to the CBM fund, which is presented in more detail by Shrestha *et al.* (Chapter 2.9).

In general, poor and low-income farm households in rural areas do not have access to financial services. Consequently, they are not able to make the investments required to increase household income and productivity. CSB members take out loans from the CBM fund, agreeing to cultivate at least one local variety from the CSB for maintenance and multiplication. The farmers then return 50% more seed than they received. This link between the CSB and the CBM fund reduces the financial burden of members, while leveraging their farming skills for conservation services. Although it is minimal compared to other formal financial schemes, the interest generated from the loan is used to cover the costs of CBM practices, for example to set up yearly diversity blocks to compare and popularize local varieties as described by Shrestha *et al.* (Chapter 2.2), and to sustain the CSB. The experience with CSBs not only inspired LI-BIRD to set up another 13 CSBs, but also led other NGOs, and the Nepalese Ministry of Agriculture and Cooperatives, to initiate similar practices across the country. In total, more than 5000 farm households throughout Nepal are now benefiting each year from CSBs supported by LI-BIRD (Table 2.8.2).

The CSB as a driving force of CBM and an asset for community resilience

The CSB practice is older than the CBM methodology itself, with its application dating back to the early 1990s (Cooper *et al.*, 1992; Sperling and Loevinsohn, 1996). CSBs have followed a series of PGR trends that are directly related to the objectives of the organizations, including:

- community-level food and seed security;
- *in situ* conservation and use of PGRFA (on-farm management);
- community custodianship;

- community access to and control over PGR;
- community autonomy and sovereignty over food, seed and local varieties;
- farmers' rights;
- resilience in the context of socio-environmental (climatic) change.

The CSBs are a common practice in CBM programmes and are used in every corner of the world. They are of vital importance for strengthening communities in their social organization, and for promoting and securing the use of local varieties. They contribute to on-farm management and move beyond the dilemma of how to implement *in situ* conservation of PGR (De Boef and Thijssen, Chapter 1.8). The CSB is in this way instrumental to the CBM components of setting-up institutional modalities, consolidating community roles in planning and implementation, and community monitoring and evaluation. We have learned from the experiences shared in this chapter that the establishment of the CSB has led to significant investment by conservation and development organizations in CBM components such as awareness-raising, the creation of social networks and institutions, and the capacity-building of community institutions (Shrestha *et al.*, Chapter 1.3). The experiences of LI-BIRD in Nepal show that CSBs are sustained by their integration with the CBM fund, while in other countries they are sustained by the motivation for seed security and sovereignty. It can be concluded that by linking conservation with access to seed and varieties, CSBs have become a driving force for the CBM process and for enhancing community resilience.

Note

The authors would like to acknowledge with thanks the contribution made by UBINIG in providing them with a case study of their work. Furthermore, they would like to thank all the people who shared their experiences by collaborating with them on the survey used for the writing this chapter.

2.9 Community biodiversity management fund

Promoting conservation through livelihood development in Nepal

Pitambar Shrestha, Sajal Sthapit, Abishkar Subedi and Bhuwon Sthapit

Promoting individual action for a collective purpose

Although small-scale farmers manage their valuable plant genetic resources (PGR), they have yet to benefit from successes in conservation and crop improvement (Bellon, 2006). The empowerment of farmers and communities seems to be a precondition for halting the rapid disappearance of local crops and varieties from farmers' fields (Shrestha *et al.*, 2008). Access to financial resources is one way of contributing to the economic empowerment of farmers, within the larger framework of community bio-diversity management (CBM). The knowledge and skills (human capital) that can be acquired through a CBM process enhance the capacities of farming communities to improve their well-being and livelihoods. As described by Shrestha *et al.* (Chapter 1.3), CBM is composed of a number of components and practices that can be tailored to particular agrobiodiversity contexts. In the current chapter, we focus on the CBM component that is described as the 'establishment of a CBM fund'. This component aims to sustain collective action in conservation and facilitates the payment of conservation services in the interest of the collective, and of society as a whole (Sthapit *et al.*, 2008b).

The initial design and operational modalities of the CBM fund

We developed the concept of the CBM fund in 2003 in the context of the agro-biodiversity project, which is described in more detail by Subedi *et al.* (Chapter 1.2). We experimented with the design of the concept in two of the CBM sites of Local Initiatives for Biodiversity, Research and Development (LI-BIRD), the villages of Kachorwa (in Bara district) and Begnas (in Kaski district). Since then, we have replicated it as a mechanism for matching conservation and livelihood development in more than 21 villages in the country (Table 2.9.1).

Its operational modality is similar to other rural microfinancial schemes. It has its own salient features, such as operating guidelines that address biodiversity management. Community members are responsible for all decision-making, and for developing guidelines to ensure that they reflect local priorities and needs. In designing the CBM fund we have made sure that it promotes local biodiversity-based, small-scale enterprises, and explores ways to establish links with the conservation of rare and

Table 2.9.1 The status of CBM funds established in different villages in Nepal (December 2010)

Development region of Nepal	Number of CBOs with functional CBM funds	Project support (US$)	Community fund (US$)	Community contribution
Eastern	2	6000	300	5%
Central	4	16000	1100	6%
Western	6	38000	450	1%
Mid-western	3	7500	900	11%
Far-western	6	8000	2900	27%
Total	21	75500	5650	7%

Source: Records of community-based organizations (2010).

threatened PGR. We learned that a basic prerequisite for its success is that the farmers' group, or community-based organization (CBO) that is responsible for its implementation has shown itself to be mature and stable, and has a proven record of working in agrobiodiversity management. Other CBM components, from raising awareness and understanding diversity to developing institutional modalities, will ensure that CBM capabilities have been consolidated in the community. Shrestha *et al.* (Chapter 1.3) provide a more detailed overview of the CBM process with its components. The fund can then further drive and sustain an endogenous CBM process. We learned that it brings people together for collective action in contributing to two noble causes: the on-farm management of PGR and improvement of farmers' livelihoods.

Within the formal administrative structure of Nepal, we have established the CBM fund at the level of the Village Development Committee (VDC; the smallest unit of administration in Nepal). The fund is managed by an executive committee that represents all the wards (lower level, local administration), known as the Village level Biodiversity Conservation and Development Committee (VBCDC). Every household associated with the VBCDC is eligible to apply for loans from the CBM fund, though priority is given to households that match specific criteria for poor and marginalized households. The CBM fund is used as a basket fund. Funds received from different sources can be collected and, through its proper mechanisms, can benefit individual households. It further facilitates the implementation of collective conservation actions.

Livelihood improvement

The CBM fund is available for all members of the CBO. The fund does not require collateral to provide loans, but a recommendation from the executive committee is necessary. The loans are accessible because of their low interest rate (<12% per annum), which is lower than the rates of formal and informal financial service providers. We have developed protocols to ensure that priority is given to providing loans to poor and marginalized CBO members. The fund aims to promote their engagement in income-generating activities with small businesses. The protocol includes a participatory exercise in which we rank potential beneficiary members on their socio-economic well-being. We use a number of indicators: the amount of family income, duration of food sufficiency based on their own production, amount of fixed property, and livelihood strategies in general.

We analysed loan distribution data in different project sites from 2005 to 2009. We found that 50% of loan recipients were from a resource-poor category, 63% were female and 66% were members from groups such as Dalits, Janjati, Madhesi and Muslims (Table 2.9.2). These results indicate that the CBM funds provided were successful in promoting social inclusion within the project sites.

Payment of conservation services

Once the fund has been created, the CBOs are able to provide loans according to the agreed guidelines. Those include the commitment to grow at least one rare local variety identified in the community biodiversity register, or from the community seed bank (CSB). Farmers' groups in Kachorwa have been able to reduce the costs for the regeneration and multiplication of local varieties to about US$10 per local variety per year (Shrestha *et al.*, 2011). Each year, farmers' groups in the Western Terai area finance diversity blocks of between five and fifteen local varieties of one crop species. In this way, the farmers' attention and curiosity concerning these local varieties is increased. Such investments have resulted in the cultivation of varieties that were previously at risk of disappearance by an increasing number of households.

Similarly, the CBM fund in Begnas has been able to provide loans to its members for raising livestock on condition that they plant 30 saplings of local fodder tree species. One of the conditions attached to loans for farmers' groups located around Rupa Lake is that they take care of wetland biodiversity, including the maintenance of habitats of wild rice, local fish, birds and white lotus. Such conditions have created a conscious balance between conservation and income generation. Through the management of the CBM fund and their respective conditions for loans, communities are strengthened in promoting individuals to assume responsibilities in collective conservation efforts. Through their organization they are able to mobilize local financial, human, institutional and social capital in the management of natural and biological capital. The CBM fund has evolved into a system for payment of agrobiodiversity conservation services similar to carbon credits.

Sustaining local institutions

Among the reasons for their collapse was their dependency on the external organizations for acquiring resources. Communities often lack the necessary skills and knowledge to mobilize financial assets needed for community development. A more controversial but often valid perspective is that it is not in the interests of many development agencies and their donors to create such capacities as it would move them out of the development business.

The CBM fund is a mechanism from which local institutions can generate regular income in the form of interest, with which they are able to continue their basic conservation and livelihood development work when externally financed projects have been phased out. For instance, Jaivikshrot Sanrachhan Abhiyan in Begnas, and the Agriculture Development and Conservation Society in Kachorwa, have been able to

Table 2.9.2 Summary of direct beneficiaries of the CBM fund in LI-BIRD's project sites, September 2005–2009

Year	No. of sites	Total no. of beneficiaries	No. of beneficiaries by gender		No. of beneficiaries by socio-economic category			No. of fund beneficiaries by ethnicity				
			♀	♂	Rich	Medium	Poor	D	J	OM	BC	M/O
2005	2	62	50	12	6	30	26	9	9	21	22	1
2006	3	271	186	85	45	120	106	24	82	47	109	9
2007	6	336	207	129	35	155	146	17	167	57	92	3
2008	7	531	332	199	59	197	275	73	186	74	190	8
2009	9	577	347	230	44	196	337	50	236	78	198	15
Total number		1777	1122	655	189	698	890	173	680	277	611	36
Total frequency		100%	63%	37%	11%	39%	50%	10%	38%	16%	34%	2%

Source: Based on CBM fund establishment and mobilization records of local institutions in LI-BIRD project sites (2010).

D, Dalit; J, Janjati; OM, other Madheshi; BC, Brahmin Chhetri; M/O, Muslim and others.

mobilize more than US$12 000 and US$8000 respectively. Both organizations have their own staff and run programmes from income-generating activities. They do not need external financial support to run their regular CBM activities.

Benefit-sharing of genetic resources

The international regime on access and benefit-sharing (ABS) states that local communities and indigenous people are the true custodians of genetic resources and associated traditional knowledge (ATK). They have the rights to make decisions on various aspects of genetic resources management, including the rights to define access to genetic resources and claim an equitable share of the benefits arising from their use. Many national governments, including Nepal, have not yet developed a legal, administrative or organizational framework for implementing ABS (Paudel *et al.*, 2010). Nevertheless, the CBM fund emerges as a community-based mechanism for the implementation of ABS. Moreover, as a practice, it is compatible with the conservation and sustainable use of biodiversity, which is an objective that is associated with ABS. The CBM fund mechanisms contribute to benefit-sharing, as well as to conservation and livelihood development. The results in Nepal are an inspiration for other countries and development organizations.

Lessons learned

The establishment of a CBM fund is an important component for contributing to the sustainability of CBM. It enhances the empowerment of farming communities in their organization; and strengthens their collective capacity to define and invest in their conservation and livelihood development agenda. Within the context of conservation, the fund creates a local incentive structure that supports the on-farm management of rare and valuable PGR, facilitates livelihood development and contributes to the social inclusion of poor households. The allocation of resources by the CBM fund, as developed in 21 villages in Nepal, motivates communities to take collective responsibility in the management of agrobiodiversity. The fund creates a platform for discussing various issues related to farming, marketing, social equality and even family matters of the associated members, thereby contributing to social cohesion. Governments and donors may wish to support local communities in establishing CBM funds as a strategy to reach poor people and also as an instrument for mobilizing local financial capital for the on-farm management of PGR. We learned that it is crucial that CBM fund guidelines are formulated in a participatory manner within each specific context, to ensure that poor and marginalized people are the major beneficiaries. To avoid the possibility of misappropriation of funds that could create conflict among the community members, we also learned that it is essential to invest in human resources for the effective implementation of transparent account-keeping systems, with proper check and balance measures. Another challenge with the CBM fund is to make appropriate arrangements for scaling-up within institutional and legislative frameworks at national and lower levels, which will further allow out-scaling to cover wider geographical areas.

Part III

People, biodiversity and landscapes

3.1 People, biodiversity and landscapes

Introduction

Walter Simon de Boef, Nivaldo Peroni and Natalia Hanazaki

Synopsis

When we describe agricultural biodiversity as the 'complex patchwork of dynamic relations between people, plants, animals, other organisms and their environment', our attention is drawn to the 'people' who develop and use, and thereby maintain agricultural biodiversity for their livelihoods, in all its levels (De Boef and Thijssen, Chapter 1.8). Plant genetic resource conservation strategies and community biodiversity management (CBM) primarily focus on local crops and varieties, as can be seen in many of the case studies discussed in this book. As a result of this, we tend to ignore dynamic relations between farming communities and agro-ecosystems, or, more specifically, the landscapes. In the current section, the authors share a number of relevant concepts and insights that approach agrobiodiversity at both landscapes and species levels, and which are supported by a series of case studies from Brazil, complemented by those from Ethiopia and India, that all focus on the relationship between people, biodiversity and landscapes.

Ethnobotany is a field of science that focuses on the relation between people and plant diversity at species level, although it increasingly addresses this relationship beyond this level (e.g. landscape and genetic diversity). Hanazaki *et al.* (Chapter 3.3) highlight opportunities concerning how, through participatory learning and action research, ethnobotany can support development processes linking conservation and development. Such research has the potential to facilitate efforts made by traditional people to protect their rights over landscapes and biodiversity and thereby sustain their livelihood. Assis *et al.* (Chapter 3.2) analyse the unique attention given to traditional people in biodiversity conservation in Brazil, providing socio-cultural and political insights into this specific relationship. Peroni *et al.* (Chapter 3.4) introduce 'domesticated landscapes', as well as 'cultural' and 'ecological keystone species', which are relevant concepts for addressing the relationship between people, biodiversity and landscapes. These concepts constitute scientific inputs in the development of the CBM methodology for community-based landscape management. The authors describe the concepts, and present two distinct case studies from Brazil: the *faxinal* collective management system in the Araucaria Atlantic Forest in the state of Paraná and the conservation and use of the umbuzeiro tree (*Spondias tuberosa*) by farmers in the semi-arid plains of north-eastern Brazil.

Building on these introductory chapters, Part III continues with four case studies. Reis *et al.* (Chapter 3.5) detail the traditional landscape management system of *caívas* with araucaria *(Araucaria angustifolia)* and yerba mate *(Ilex paraguariensis)* in the Atlantic Forest of the state of Santa Catarina in southern Brazil. Feyissa *et al.* (Chapter 3.6) illustrate how the unique relationship between natural and semi-domestic populations of coffee *(Coffea arabica)* has been vital for the conservation of the Afromontane forests in Ethiopia. Market-based incentives guide the conservation of natural coffee populations in those landscapes, which are in fact the centre of origin of this global commodity. A case study from India brings us to the Western Ghats, where a non-timber forest product, garcinia *(Garcinia gummi-gutta)*, has turned into a booming business because of its recently discovered valuable traits (Ramesh *et al.*, Chapter 3.7). The authors describe how several CBM practices contribute to a more sustainable management and use of this biological resource and its associated forest landscape. Zank *et al.* (Chapter 3.8) conclude Part III, introducing the sustainable development reserve (SDR), which is a protected area in Brazil that creates a legal structure for CBM in the context of landscape management. The authors describe how this protected area is being proposed to respond to the needs of a traditional farming and fisher folk community that is struggling to maintain its livelihood, which is strongly associated with the *restinga* landscape of the Atlantic Forest in southern Brazil. In this way, the concluding chapter brings us back to the heart of this section, exploring ways to recognize, value and sustain and, as expressed in this case study, protect the relationship between people's livelihoods, their biodiversity and landscapes from major 'development forces'.

In this introductory chapter, we synthesize key aspects of the relationship between people, biodiversity and landscapes, such as the role of keystone species for landscape management, which become drivers for CBM. Furthermore, we look at the conservation of agrobiodiversity at landscape level, where the focus on people and their traditions is vital. Participatory action and learning research, associated with scientific fields such as ethnobotany, are instrumental in the design of conservation strategies building upon traditional people's management systems and strengthening their livelihoods. We conclude the chapter by taking a closer look at the protected area of the SDR, with its prospect of serving as a unit to formalize CBM at landscape level, drawing lessons from this Brazilian case of relevance for discussing units for CBM at global level.

People and agrobiodiversity

With regards to agrobiodiversity at the level of crops and varieties, boundaries cannot easily be defined among communities because of continuous in- and out-flows of varieties through the seed systems. At species and landscape levels, boundaries of agrobiodiversity can be well defined by territories shared by local people. We follow practitioners of the collaborative management of natural resources and adapt their definition of a community as a human group that shares a territory and is involved in different but related aspects of livelihoods, such as the management of natural and/ or biological resources or landscapes; the production of knowledge and culture; and the development of management and productive practices (Borrini-Feyerabend *et al.*, 2007). Interpreting this definition does not mean that agrobiodiversity at those levels is static; several cases in this section illustrate dynamism at landscape and species levels. If we consider CBM as contributing to the *in situ* conservation of landscapes, plant species and communities of those species (at genetic level), we need to ensure the continuation of the dynamic interactions between those levels of diversity, the people and the environment. In these cases, it is then logical that the entry point for CBM is not a specific species but an ecosystem or landscape. We do, however, realize that such landscapes are no longer found in areas that have been transformed into monoculture agricultural systems consisting of just a few crops and varieties by mainstream forces of a globalized agro-industrial food market system. Accordingly, examples of those communities that continue to maintain landscapes with associated agrobiodiversity and traditional lifestyles are found in those areas that are less attractive for such transformation.

We find these areas close to fragments or remains of forests, often in hillside or mountainous areas such as the Western Ghats in India and the Afromontane forests in Ethiopia. In Brazil, those areas are found among the remnants of the Araucaria Atlantic Forest and in the fast disappearing restinga, the sand dune vegetation that is located along the Atlantic coastal areas of Brazil. As elaborated by Peroni *et al.* (Chapter 3.4), these landscapes can be understood as a historical, cultural and evolutionary expression of the relationship between people and their environment. Farming communities associated with those landscapes manage them consciously, targeting the use of particular biological resources. For example, *faxinais* and *caívas* in southern Brazil are forest landscapes in which non-timber forest products are gathered through traditional management systems. A similar relationship between forest landscapes, species and their products can be found between the Afromontane forests in Ethiopia and coffee. In north-eastern Brazil, the umbuzeiro tree (*Spondias tuberosa*) is relevant for its cultural, economic and ecological functions. It is considered a sacred tree as it serves as a symbol for the landscape, livelihood and culture of the people.

The restinga landscape in Brazil is being threatened by urbanization and other development forces along the coast. In some parts of restinga landscape, its conservation also depends on its connection with the livelihoods of traditional communities, who cultivate local manioc varieties, gather fruits from butiá palms (*Butia catarinensis*) and collect medicinal plants. These examples illustrate how landscapes including agrobiodiversity can be expressions of the livelihood of traditional people, who, by maintaining these landscapes, provide a service to society.

Traditional people, biodiversity and conservation

It is in the context of this relationship between landscapes and traditional people that our attention is drawn to the livelihoods and lifestyles of these people. Assis *et al.* (Chapter 3.2) describe how the ever-expanding agro-industrial frontier in Brazil threatens landscapes and local cultural values. It is therefore logical that the conservation of biodiversity is directly related to guaranteeing a space for traditional people, who emerge from invisibility, avoiding further marginalization that goes hand in hand with the perceived single model of globalized agro-industrial development. It is important to understand that this focus of attention on traditional people is motivated not just for the sake of conservation, but also as a way of legitimizing the culture and territory of traditional communities, as well as the strong association they have with their surrounding natural environment. This special attention has led in Brazil to a better understanding and gradual recognition of the role of traditional people and their livelihoods in contributing to the conservation of landscapes, and as such is part of a larger debate on conservation and development. The relationship between biodiversity conservation and traditional people may seem to be a construction unique to Brazil, with its history of settlers and agro-industrial frontier, but it has been endorsed by critiques of this development model and its implications on traditional people and their management of natural and biological resources (Borrini-Feyerabend *et al.*, 2007).

During the implementation of the global study on CBM and empowerment, researchers from India and Nepal visited several CBM sites in Brazil, and concluded that the association between traditional people and conservation is well articulated in Brazil. The researchers observed that ancient agricultural societies of India and Nepal provide stark contrast to the settler and frontier agriculture of Brazil; however, despite this contrast, the marginalization of traditional people and the requirement to address them in association with biodiversity conservation, are characteristics shared by the three countries (Assis *et al.*, 2010a).

Landscape as an expression of agrobiodiversity

In order to better understand the relationship between people and landscapes, Peroni *et al.* (Chapter 3.4) describe concepts of domesticated landscape, cultural and ecological keystone species. Examples of these keystone species include coffee in the Afromontane forest of Ethiopia, the umbuzeiro in the north-eastern semi-arid plains of Brazil, and both araucaria and yerba mate that are maintained through the faxinal and caíva traditional management of the Araucaria Atlantic Forest in southern Brazil. Traditional people contribute to the formation and sustenance of landscapes with a particular ecological and cultural identity. In discussing the CBM methodology and its practices, we therefore need to address the managed landscapes with some degree of domestication, as expressions of agrobiodiversity as a whole. CBM reveals itself as a process for supporting local governance in the context of the conservation and use of biodiversity at landscape level, and where this relates to traditional people, this means contributing to their empowerment in maintaining their livelihood assets.

CBM in the context of species and landscape management

The CBM processes that support the people in the Atlantic Forests, the Western Ghats and the Afromontane mountains vary, as can be seen in the examples shared in this book, which mainly target food crops and local varieties. Since the collective management of biodiversity is already a way of life for the Ethiopian farmers and their coffee forests, for the faxinalense people and their Araucaria Atlantic Forest, and for the farming communities and their restinga, CBM in these contexts can be referred to as an endogenous process, as described by Shrestha *et al.* (Chapter 1.3). Therefore, actions carried out by external conservation and development organizations have a different focus, sequence and logic. Such endogenous CBM processes are supported in different ways, as illustrated by various experiences shared in the book, for example, where an exogenous CBM process is initiated by enhancing awareness of the value of local crops and varieties, and by establishing community structures and modalities for their conservation and use.

A characteristic that distinguishes the cases in Part III from those that address annual crops is that these communities manage their biological resources at landscape level over a long period of time, so the association covers multiple generations. Moreover, the resources are often managed within traditionally defined territorial boundaries. Another distinct feature relevant to CBM is that the case studies in Part III primarily relate to perennial species (e.g. araucaria, yerba mate, butiá, umbuzeiro, coffee and garcinia). All these species and their associated products, with the exception of garcinia, have strong cultural values. The species vary in their degree of domestication. Coffee is fully domesticated, though the case study focuses on those semi-domesticated coffee populations in their centre of origin. Yerba mate is partially domesticated, though in the faxinais and caívas more genetically diverse populations are maintained. The faxinalense communities gather *pinhões* from ethno-varieties of araucaria trees that have to some degree been domesticated, in the same way as the landscape in which they are maintained. We do not consider umbuzeiro and butiá as domesticated species, even though their cultural and market values are important. The landscapes and their specific species supply biological resources for food and income through value addition and commercialization.

Landscapes, species and markets

All perennial species addressed in Part III are of important economic and, in most cases, cultural value; they sustain the community's interest in the landscape as an expression of their identity and as an asset for income generation. As such, the conditions at landscape level are favourable for working with CBM. More so than for annual crops and species, sharing access to the forests and their resources is in the interest of communities as well as for the conservation of genetic resources. If the social and cultural capital for livelihood development can be enhanced through CBM practices and processes, this would contribute to the conservation of the landscape and its biodiversity.

Products such as coffee beans, pinhões, yerba mate leaves, and umbu and butiá fruits are traditional and have the potential to reach premium markets, and as such the cases discussed in Part III can be linked to those issues addressed in Part IV on agrobiodiversity, markets and chains. Reis *et al.* (Chapter 3.5) discuss the case of pinhões and yerba mate from the Araucaria Atlantic Forest, the natural origin of which is increasingly appreciated by rural and urban consumers, who are willing to pay premium prices for such products. Economic growth in Brazil and its emerging classes of conscientious consumers creates opportunities for this type of product diversification, a pattern that can also be observed in India and Ethiopia. For a global commodity such as coffee, forest coffee from the centre of origin has the potential to obtain 'multi-label' premium prices. However, this requires sophisticated systems of product certification and traceability, which in the remote areas of Ethiopia creates many challenges. With regards to forest coffee in particular, the fact that it is at its centre of origin means that when such market incentives emerge, management practices should be such that the genetic diversity of the coffee is maintained and that, in the end, these coffee forests become and remain *in situ* conservation sites. In the case of umbu fruits, their products are sold in local towns, at markets in São Paulo and even through fair trade channels in Europe. Value addition and commercialization encourages communities to appreciate the umbuzeiro tree and thereby its associated landscape.

For garcinia in India, the market dynamics are very different as it is a highly valued product that is driven by new export markets. The garcinia project described by Ramesh *et al.* (Chapter 3.7) uses a series of adapted CBM practices, through which they enhance community awareness of the forest and its resources, and thereby contribute to the strengthening of community-based forest management, in conditions that are rapidly changing. As in similar exogenous CBM processes, the project invests in social and institutional capital that includes the promotion of more sustainable practices in garcinia processing, advancing the domestication of the species through the identification of elite materials for planting in home or forest gardens.

In the case of most of the species, labels for organic production, fair trade or certificate of origin create incentives for communities to market their forest or 'landscape' products. If CBM strengthens communities in their social organization, they should be prepared for, and capable of, facing the specific requirements set by those premium markets. The perspective that emerges is that the livelihoods of traditional peoples can be sustained because of their association with those products that contribute to the sustainable management of those landscapes. Conscientious consumer behaviour creates opportunities for 'pro-traditional' and 'pro-biodiversity' products that consequently create incentives for CBM.

Scientists and traditional people

Several of the case studies are the result of interdisciplinary research involving scientists with backgrounds in ethnobotany, ethnoecology and forest genetics, as well as anthropology, geography and development studies. What motivates their interest and brings them together is the interface between people and biodiversity at landscape level. A number of the studies presented in this section document the relationship between traditional people, landscapes and species such as umbuzeiro, araucaria, yerba mate, coffee and butiá. More detailed studies address the degree to which landscapes and their particular species have been domesticated, such as Afromontane, araucaria and restinga landscapes. Insights into both species and landscapes provide inputs for the design of conservation strategies. These strategies investigate the impact of traditional or modern management practices on particular species, thereby gaining insights into how to promote more sustainable practices that result in the maintenance of viable populations.

Scientists who aim to contribute to more sustainable management practices become engaged in participatory learning and action processes. Supported by this collaboration with scientists, traditional people, who are often perceived as marginal in society, are now beginning to be recognized; moreover, they are gaining self-esteem. Participatory research, which often includes CBM practices, facilitates action and learning, and contributes to the emancipation of those communities, which then becomes a precondition for recognizing their role in contributing to biodiversity conservation as custodians.

SDR: a 'laboratory' for new professionalism and new governance

In Brazil, the sustainable development reserve (SDR) has been created as a new conservation regime that includes rather than excludes traditional people, and fosters the relationship between biological and cultural diversity, and between nature and people as described by Zank *et al.* (Chapter 3.8). The SDR matches the intentions of CBM in the context of traditional people managing their landscape in a collective manner, building upon common practices of the collaborative and joint management of natural and biological resources (Borrini-Feyerabend *et al.*, 2007). However, the SDR is still in its infant stage, with only limited and initial experiences; additional research is needed to build evidence for the further development of its structure and modalities in Brazil, and to consider its application in a more global context.

Scientists can act as facilitators and catalysts in learning and action processes, designing and implementing SDRs, and adapting their modalities to a diversity of conditions. Consequently, the role of the scientists in addressing the link between traditional people, biodiversity and landscapes in such situations goes beyond what is common in normal professionalism (i.e. the documentation and enhanced understanding of the relationship between people, plants and landscapes). Since the relationship between traditional people, biodiversity and landscapes has many facets, such a process requires an interdisciplinary approach and long-term commitment. As described in the cases on the faxinais, caívas and restinga, the collaboration between a university and traditional communities matches a new professionalism and contributes to a new governance (De Boef *et al.*, Chapter 7.1). Such collaboration engages students and scientists in learning and action, in response to a social demand; explores, in a collaborative manner, the development of the SDR for meeting the demands of the communities associated with those landscapes; and provides responsible governmental agencies with grassroots and evidence-based insights. Through the involvement of scientists, these new protected areas can be turned into laboratories for participatory action and learning, thereby embracing the adaptive management approach that facilitates social learning in order to drive policy and management (Gunderson *et al.*, 1995; Berkes and Folke, 1998). In such processes of social learning, adaptive management positions scientists as catalysts and facilitators in the further development of these legally protected areas, which are important in Brazil and beyond – considering the SDR a global laboratory.

Outcomes of this type of collaboration may directly benefit the communities, which gain insights and capabilities about how to prepare technical arguments to justify the creation of an SDR (i.e. agency). Such a management plan, a formal prerequisite for establishing an SDR, was in the past often conducted by outsider professionals. However, if the SDRs are not designed in a participatory, involved manner, the traditional communities will continue to depend on those professionals during implementation. The engagement of scientists in such action has the potential to generate insights that can contribute to the further refinement of the requisites and modalities of the protected areas and, if required, can provide evidence-based inputs into its improvement, or even design alternative protected areas and/or modalities. Such a contribution thus goes beyond collective agency; it puts both community-based and scientific evidence in the design of the SDRs and in the development of modalities for the implementation of SDRs (i.e. structure). Consequently, such participatory learning and action research contributes to the empowerment of traditional people in landscape management both in agency and structure (De Boef *et al.*, Chapter 7.2). The case studies presented in this section not only provide tangible insights into the relationship between people, biodiversity and landscapes, they also explore the ways in which the capabilities of traditional communities can be strengthened to motivate them to assume their responsibilities in the sustainable management of landscapes; they share experiences of how to encourage and recognize communities in those efforts; while at the same time contributing to the (re-)design of protected areas and their modalities to recognize and sustain that relationship.

3.2 Traditional people and the conservation of biodiversity in Brazil

Ana Luiza de Assis, Sofia Zank, Nivaldo Peroni and Natalia Hanazaki

Traditional people in Brazil

Traditional people are characterized as having strong ties with their territory, and they depend in many ways on local natural and biological resources for their livelihoods. In Brazil, we have used the expression 'traditional people' since the 1980s to refer to the inhabitants of protected areas. This may seem artificial, since traditional people should not just be living in those conservation areas only. However, it becomes more logical when trying to understand the struggle of traditional people to maintain their livelihood and customs in association with efforts to conserve biodiversity. This apparently artificial expression meets the demands of many organizations that strive towards the two distinct aims of achieving biodiversity and nature conservation and recognizing the specific and often traditional livelihoods associated with natural and biological resources (Vianna, 2008).

It is easy to link traditional people and their livelihood to biodiversity conservation when we address community biodiversity management (CBM) in this book. The chapters include several experiences from Brazil, and they are compared with experiences from other countries in Latin America, as well as in Africa, Asia and Europe. In these comparisons, it is clear that some aspects of CBM in Brazil, as part of the New World, are very distinct. The structure of society in Brazil, for example, differs greatly from ancient cultures such as those of Ethiopia and India, or even France. In Africa and Asia, the term 'indigenous people' is largely synonymous with traditional or local people, whereas in Brazil, 'traditional', 'indigenous' and 'local' people each can refer to a different or differently defined social or ethnic group. To further increase understanding on the variations in groups in Brazil, as addressed in various chapters in this book, we show their location in the country in Figure 3.2.1, and provide a summary in Table 3.2.1.

We use the common terminology of the book when referring to small-scale farmers, which in Brazil would usually be referred to as *agricultores familiares* (literally translated as 'family farmers', or also referred to as smallholders). Many of the groups addressed in the chapters that concern Brazil refer to agricultores familiares. In cases where they are recognized as traditional or indigenous people, their specific denomination is used, such as Krahô, Xavantes, Faxinalenses or Caiçaras. In this chapter, we further explain the historic and social background of indigenous and traditional people, as well as of small-scale farmers in Brazil. We aim to show that the association between traditional people and conservation is consistent with the Brazilian development path.

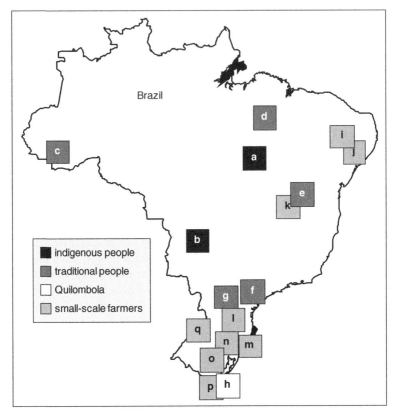

Figure 3.2.1 Location in Brazil of the communities referred to in chapters in the current volume.

The expanding agricultural frontier and consequential evolution of distinct social groups

The occupation of Brazilian territory took place during successive waves of colonization that departed from the coast to the interior. Not wishing to ignore our pre-Columbian history, we focus here on the context of the social and historical construction of the term 'traditional people', beginning with the year 1500, and with the discovery of what we now call Brazil. Portugal ruled the territory as a colony for almost three centuries, in a period that is characterized as the first exploratory phase. Jesuit institutions organized missions to Brazil, with the aim of indoctrinating indigenous peoples; and pioneer expeditions swept the territory in search of minerals and other resources, also appropriating some indigenous people for slave labour. Throughout this period, runaway slaves and pioneers settled inland and survived by using the local resources. The settlers also used African slaves for labour, some of whom eventually escaped or were freed.

Colonization, and the diseases and massacres it brought with it, led to a huge decrease in the indigenous population, and many of those who did survive lost their original way of life. It is estimated that at the time of colonization, four to five million

Table 3.2.1 Characterization of the communities from Brazil referred to in chapters in the current volume

Ref[a]	Chapter no.	Author(s)	State	Location	Specific denomination	Description
a	2.5	Dias et al.	Tocantins		Krahô	indigenous
b	2.5	Dias et al.	Mato Grosso do Sul		Xavantes	indigenous
c	3.2	Assis et al.	Acre	various	rubber tappers	small-scale farmers and extractivists
d	3.2	Assis et al.	Maranhão	various	Babaçueiras	extractivists
e	3.2	Assis et al.	Minas Gerais	various	Vazanteiros	riverside small-scale farmers and fisher folk
f	3.2	Assis et al.	São Paulo	Ilha de Cardoso	Caiçaras	small-scale fisher folk and farmers
g	3.3	Peroni et al.	Paraná	Central-south Paraná	Faxinalenses	small-scale farmers
h	2.5	Dias et al.	Rio Grande do Sul	Tavares	Quilombola	Afro-Brazilian
i	2.7	Dias	Paraíba	various		small-scale farmers
j	3.4	Peroni et al.	Pernambuco	Altinho		small-scale farmers
k	2.5	Dias et al.	Minas Gerais	Porteirinha		small-scale farmers
l	3.5	Reis et al.	Santa Catarina	Northern plateau		small-scale farmers
m	3.8	Zank et al.	Santa Catarina	Imbituba		small-scale fisher folk and farmers
n	5.3	Santos et al.	Santa Catarina	Serra Geral		small-scale farmers
o	5.3	Santos et al.	Rio Grande do Sul	Ipê and Antonio Prado		small-scale farmers
p	2.5	Dias et al.	Rio Grande do Sul	Tavares	Azorean	small-scale farmers
q	5.7	Ogliari et al.	Santa Catarina	Guaraciaba		small-scale farmers, German and Italian descendants
q	1.5	Canci et al.	Santa Catarina	Guaraciaba		small-scale farmers, German and Italian descendants

a Reference to the location in Brazil in Figure 3.2.1.

indigenous people inhabited the Amazonian region alone (Clement, 1999). Brazil today has between 450 000 and 700 000 indigenous people among its almost 200 million inhabitants. These indigenous people belong to 230 ethnic groups and speak 195 different languages (Cunha, 2009), many of which are threatened with extinction. Indigenous territories are, according to the Brazilian constitution, inalienable public lands, owned by the federal state, but indigenous peoples have the right of permanent possession of their traditional territories, as well as exclusive rights over their natural resources. Today, indigenous people in Brazil are considered to be the descendants of its native people, according to the Brazilian constitution, yet the process of recognition of indigenous territories is still controversial.

When Brazil gained its independence in 1822, the occupation of its territory and its economic orientation reinforced the production of commodities, such as sugar cane and coffee, and led to further exploration of the country's resources (e.g. timber and gold). In the early frontier of development, groups of mainly Portuguese and Spanish but also French, English and Dutch settlers gradually occupied areas of north-eastern Brazil. During this period, these settlers continued to use African slaves for labour, a system that was inherited from the Portuguese and continued until 1888. Today, several inland indigenous or settler communities, descendants of those refugees or settlers, are considered traditional people, having developed a livelihood and culture that is strongly associated with their environment; Caiçaras are traditional people with this background (Diegues, 1999). Also, Quilombola communities, the descendants of runaway or freed slaves, are commonly found in many regions of the country.

Quilombolas are considered to be one of Brazil's traditional peoples because they have maintained elements of their African origin in their culture, society and spiritual life, but also in their customs and in the way they manage their natural and biological environment. They enjoy territorial rights similar to those granted to indigenous people. A total of 1624 communities have already been officially recognized as Quilombola, but civil society organizations estimate that there are approximately 3000 Quilombola communities in Brazil. To this day, only 180 communities have received territorial rights from the federal state (Comissão Pró-Indio de São Paulo, 2011).

In the late nineteenth century, many European immigrants came to Brazil to work on coffee plantations in the south east. These were followed in the first half of the twentieth century by groups of immigrants who came from countries affected by the two world wars, such as Italians, Germans, Japanese and Spanish. The federal government encouraged these settlers or *colonos* to settle as farmers, granting them land titles for native forestland formerly inhabited by indigenous people. The colonists transformed this land into agricultural land, which is occupied to this day by small-scale farming households engaged in crop and livestock production. The agricultural frontier in those areas is relatively recent, and current small-scale farmers, such as those in Guaraciaba, Santa Catarina, are only the second generation of German and Italian immigrant farmers occupying and transforming these lands from original Atlantic Forest into agricultural landscapes.

Brazil's history has resulted in a population that is described as multicultural and mixed (Fernandes-Pinto, 2010). We consider many different groups within the Brazilian population as traditional people. They share in common the fact that they do not fit into the mainstream development process, and usually they have a strong link with a traditional territory. Consequently, many of these groups have been occupying spaces or landscapes that, within the hegemonic development point of view, are considered demographically empty (Coelho-de-Souza, 2010). Today, it is estimated that Brazil has 4.5 million people who are considered to be part of traditional communities. This fractionally less than 2.5% of the population is estimated to occupy about 25% of the country (Fernandes-Pinto, 2010).

Definition of the term 'traditional people'

With the expansion of the agricultural frontier in the twentieth century, we have witnessed many conflicts over access to land and territories. In many cases, traditional people have lost their livelihoods as a result of these conflicts, having no other option than to move to major urban centres. These conflicts are embedded in our history and to this day they continue to impose the status of poverty and marginalization on those people who have maintained, and often wish to continue, their traditional lifestyle and livelihoods.

The socio-political debate over traditional people and its association with conservation is rooted in critiques of the dominant development model. These topics emerged alongside the formation of socio-environmental movements during the gradual return to democracy in the 1980s and 1990s. This debate in Brazil, as in other Latin American countries, cannot be separated from a process of democratization and the building-up of civil society. It contributed to an international debate that resulted in 1992 in the Convention on Biological Diversity, which addresses the link between biodiversity and associated traditional knowledge. To some extent, the recognition of traditional people developed simultaneously with environmental policies. Since the debate was, and continues to be, closely related to the Brazilian development model, which is associated with its agricultural frontier, the process of consolidating the recognition of traditional people, and its translation into conservation regimes, is complex and therefore slow. However, we can share some experiences that demonstrate that significant and important advances have been made, and these may serve as important and innovative references beyond Brazil.

Rubber tappers and their struggle for recognition of their customary land rights

An important event in the mid 1980s, involving a group of Amazonian rubber tappers, influenced the debate on traditional people and biodiversity conservation. Rubber tappers in the Amazonian state of Acre, neighbouring Bolivia, have developed a livelihood based on rubber extraction from plantations that have been incorporated into the natural forest, through an agroforestry system practised since the early twentieth century. In the 1980s, the rubber tappers' access to land was threatened by the agricultural frontier expanding to meet the demands of the livestock industry. As a result, the trade union of rubber tappers was engaged in a long and violent struggle to maintain access to their territory. They not only fought for their land rights, but also became a national symbol of conservation for the Amazonian rainforest.

The rubber tappers proposed the establishment of reserves that would belong to the government, which would grant the tappers collective and unique users' rights (Cunha, 2009). The National Institute for Colonization and Agrarian Reform (INCRA), which is the Brazilian agency responsible for agrarian reform, addressed the first legal disputes. In 1989, the Brazilian Institute of the Environment and Renewable Natural Resources (IBAMA) was established for managing environmental conservation areas. Part of IBAMA's responsibilities, including the management of such reserves, was later transferred to the Chico Mendes Institute for Biodiversity Conservation (ICMBio). Through its name the institute honours Chico Mendes, a leader of the rubber tappers who was killed during the struggle in 1988. The first Brazilian extractive reserve (Resex), Alto Juruá, was formally established in 1990. That same year, three more Resex were created, all of them in areas where traditional and collective rubber extraction is carried out (Cunha, 2009). The Resex became a catalyst for linking the recognition of traditional people with the conservation of Amazonian rainforest and beyond.

In 2007, among the emerging conservation regimes a National Policy for the Development of Traditional People and Communities (PNPCT) was enacted by Federal Decree No. 6040 (Brazil, 2007). This policy defines traditional populations as 'culturally diverse groups that are recognized as such, who have their own forms of social organization, occupy and use lands, and use natural resources as a condition to their cultural, social, religious, ancestral and economic reproduction, and who use knowledge, innovations and practices that are generated and transmitted by tradition'.

Traditional people, biodiversity and conservation regimes

Traditional people are increasingly identified with playing an important role in conservation. Diegues (1999) and Hanazaki *et al.* (2010), for example, emphasize that biodiversity is also part of a cultural and social construction, in which the species and the landscapes are objects of knowledge, domestication and use. Besides this, they play an important role as sources of inspiration for myths and rituals of traditional societies (Pilgrim and Pretty, 2010). The relationship between biodiversity and traditional people can be seen, for example, in the maintenance of, and even the increase in, biodiversity in tropical forests that are closely related to traditional practices of shifting cultivation (Peroni and Martins, 2000; Peroni and Hanazaki, 2002).

The National System of Conservation Units (NSCU – Federal Law No. 9985/2000) acknowledges the presence of traditional populations in some categories of protected areas. One of its objectives is to protect the livelihoods of traditional people living in areas targeted for biological conservation. The law guarantees respect for their livelihood through their effective participation in management councils; it further guarantees traditional peoples' participation in the formulation of management plans for protected areas. Among several other categories of protection, it proposes two that have a high level of integration between the conservation and sustainable use of biodiversity: the extractive reserve (Resex) and the sustainable development reserve (SDR). The SDR is a protected area for sustainable use by its traditional people, which reconciles biodiversity conservation with natural resource management, mediated by local culture and local ecological knowledge, as further elaborated by Zank *et al.* (Chapter 3.8). These two conservation categories are considered to be globally innovative because they link social and ecological interests in an intrinsic manner, and institutionalize community conservation areas (Fernandes-Pinto, 2010).

The diversity of traditional people

Traditional people have diverse ethnic and cultural backgrounds, and are currently characterized by the biological or natural resources with which their livelihood is associated (Cunha, 2009). The first to be recognized as traditional were the rubber tappers in Acre. Another group that is associated with a biological resource are the Babaçueiras or babaçu palm extractivists, in the transitional biome between the Amazonian rainforest and the cerrados (savannah) in Maranhão (north-eastern Brazil). Other examples can be seen with the Vazanteiros, riverside people whose livelihoods are associated with the São Francisco River, in the semi-arid zone of the north-eastern region; and artisan fisher folk along the entire coast of Brazil. In addition to these groups are the small-scale farmers who are associated with specific ethnic groups, such as Quilombolas; or who have strong associations with particular landscapes (Vianna, 2008; Hanazaki *et al.*, Chapter 3.3), such as Caiçaras (fisher folk/farmers of the Atlantic Forest coast in the south-eastern region). The Faxinalense people in Paraná are also recognized as traditional; these are small-scale farmers who use collective areas for livestock and extractivism in native Atlantic Forest in Paraná (Souza, 2007). Peroni *et al.* (Chapter 3.4) address the role of the Faxinalense people in the domestication of landscapes, and the conservation of araucaria and yerba mate.

The term 'traditional people' is deliberately used in a broad manner in Brazil, but still has a clear and precise conceptual definition (Cunha, 2009). One common feature is that they all interact with ecosystems and struggle with maintaining a particular form of natural and biological resource appropriation and landscape interaction, which are vital for sustaining their livelihoods (Fernandes-Pinto, 2010). They also share a history of making a low environmental impact, and of being interested in maintaining or recovering territories, often providing important environmental services (Cunha, 2009). The concept of traditional people in Brazil carries a strong political connotation, associated with a collective identity, and the historical relationship of a particular group with a given territory.

The current phase: towards self-recognition

In several cases, the category of traditional people can be considered an exogenous construction (Vianna, 2008). However, these populations are gradually appropriating and transforming the label of traditional people. A similar process has occurred with other labels, for instance: 'indigenous', 'aboriginal', 'tribal' and 'black'. Labels that were once full of prejudices are becoming powerful voices in social mobilization (Cunha, 2009). The struggle for recognition is only the start of a process of emancipation. Traditional people are slowly appearing from invisibility and marginalization (Cunha, 2009). To be traditional is the result of a process of self-constitution that is closely related to empowerment. During such a process, traditional people reaffirm their cultural identity, their vocation and desire for environmental conservation, besides legitimizing their own way of social organization.

Traditional people are also entering the market by partnering non-governmental organizations and international companies, selling goods with high added value because of their traditional knowledge and environmental services (Cunha, 2009). Today, the rubber tappers in Acre, for example, are engaged in value-addition programmes, trying to secure better prices for native and more natural rubber. The traditional small-scale farmers in southern Brazil obtain a premium price for their yerba mate, as described by Reis *et al.* (Chapter 3.5). It is clear that the concept of traditional people is in a process of evolution and constant change, for which we should consider the identity of different groups that are heterogeneous and very diverse but that all together constitute our Brazilian national identity.

3.3 Opportunities for ethnobotany to contribute to community biodiversity management

Natalia Hanazaki, Elaine Mitie Nakamura,
Bianca Lindner and Walter Simon de Boef

The increasing importance of ethnobotanical studies in plant genetic resource conservation

Ethnobotany is the scientific study of the dynamic relationships that exist between people and plants (Alcorn, 1995). Ethnobotanists aim to document, describe and explain complex relationships between cultures and plants, focusing primarily on how plants are used, perceived and managed by human societies. This includes the use of plants for food, clothing, currency, rituals, medicines, dyes, construction and cosmetics. While ethnobotany originally addressed interactions between people and diversity at the level of plant species, the current scope of this research field now addresses aspects of plant diversity at system, landscape, and also genetic levels. Moreover, it emphasizes the links between conservation and management, thus the human role of biodiversity, and, together with associated traditional knowledge, it creates a space for the better use of our expertise in the development of a new professionalism in plant genetic resources, as further elaborated by De Boef *et al.* in Chapter 7.1.

This new approach to biodiversity conservation involves carrying out action research in a manner that acknowledges and builds upon the livelihoods, knowledge, rights and customs of traditional people to contribute to the design and development of innovative conservation regimes. Within such new professionalism, research and context unite aspirations to contribute to biodiversity conservation and the empowerment of traditional people. For example, Assis *et al.* (Chapter 3.2) elaborate on the historical context in Brazil, where a logical association exists between supporting biodiversity conservation and traditional people in their livelihood development. In the current chapter, we provide some historical background on the field of ethnobotany, and explore ways in which participatory tools and research methods can improve our work in community biodiversity management (CBM).

Ethnobotany: the historical context

When the term 'ethnobotany' was introduced in the late nineteenth century, it focused on the description of artefacts that had been gathered to form collections in museums or universities. Since then, ethnobotany has evolved to include discussions on the importance of fieldwork, and an understanding of the local communities that interact with plants. During the 1990s, numerous manuals and texts on methodologies and techniques in ethnobotanical studies were published. It is also possible to identify an almost utopian vision in the 1990s, in which ethnobotany was linked to finding solutions for local development. In this sense, Alcorn (1995) pointed to the scope and objectives of ethnobotany as being committed to a developing world, emphasizing its integrative focus and its potential for public policy planning in the sphere of conservation of biodiversity and beyond. Alexiades (2003) points out that today, boundaries between local and global, conservation and development, basic and applied, and public and private sectors are more diffuse. Consequently, he argues that ethnobotany holds a strategic position that facilitates the dialogue and exchange between academics and other social actors to create mechanisms that generate new knowledge and new forms of dialogue in various dimensions.

A major change in ethnobotany was the shift from a descriptive scope to one that contributes to conservation in a more integrated manner, thereby appreciating and encouraging the use of local knowledge in the realm of designing policy and development agendas (Alcorn, 1995). This reflects a growing recognition of the role of local ecological knowledge in conservation strategies (Cunningham, 2001; Hanazaki, 2003; Albuquerque, 2010; Hanazaki *et al.*, 2010). Ethnobotany can play an important role in providing access to this knowledge in order to combine local knowledge and scientific strategies for the use and conservation of biodiversity, and in facilitating the quest to understand the processes between people and plant resources (Albuquerque, 2010).

By its very nature, ethnobotany recognizes and values the traditional knowledge of plant resources. However, most ethnobotanical studies still highlight the traditional knowledge of plant resources without taking into account the importance of issues such as management and domestication. In Brazil, this field of study emphasizes knowledge relating to, and uses of, medicinal plants in approximately 64% of publications, occasionally including research on patterns of use and management of plant resources. Other topics studied are investigations into the knowledge, use and management of plant resources in general; their cultural significance, origins and flows; the loss of knowledge; and ethnotaxonomy (Oliveira *et al.*, 2009). Some studies directly emphasize the importance of traditional knowledge on genetic resources, particularly with regard to management practices and maintenance, and for broadening the use of agrobiodiversity.

Ethnobotanical studies reveal, document and assess the value and the unique importance of traditional knowledge and the high number of varieties maintained by small-scale farmers, as is shared in a case study from the state of São Paulo. Peroni and Hanazaki (2002) conducted an ethnobotanical assessment of the inter- and intra-specific diversity of crops under shifting cultivation systems managed by traditional people known as Caiçaras in the southern region of the Atlantic Forest, in the state of São Paulo (see Figure 3.2.1; Chapter 3.2). They revealed 261 varieties from 53 crop species, and estimated that more than 30% of varieties/species had been lost within a period of 30 years. The studies show how itinerancy cycles of fallow within traditional shifting cultivation systems increase diversity in human-managed landscapes. Local farmers were able to point out several factors that contribute to a reduction in managed diversity, including restrictive environmental laws, rural exodus, an increase in tourism, and the overall change of livelihood activities of traditional people. The loss of crop diversity indicates an urgency to design conservation strategies and development policies that aim to maintain the biodiversity and knowledge associated with the human-managed landscapes of these communities, and prevent further loss.

Ethnobotanical methods and tools

Interdisciplinary techniques, including methods that combine the areas of geography, biology and anthropology, are useful for learning about the local management of plant resources. Qualitative research methods have a wide use in ethnobotany, including semi-structured interviews, participant observation and a variety of participatory tools (Tuxill and Nabhan, 2001). Participatory tools of visualization and triangulation are used to increase the reliability of the study being conducted, and to enhance the active role of traditional peoples and their partner communities in the research process. If applied in an appropriate manner, and if the ethnobotanists are professional in their facilitation, the tools can enhance the knowledge of both the scientists and traditional peoples (De Boef and Thijssen, 2007). Knowledge and experiences of local people are not only documented, but are also discussed and included in joint plans concerning the management and conservation of local resources. The research now serves to increase awareness of traditional knowledge and biodiversity; to foster an understanding of the local situation and the need for joint actions; and to define strategies for the development of community-based institutions for managing biological resources and associated traditional knowledge.

In this context, participatory learning and action research (PLAR) is a prominent methodology that can help to meet this challenge. In PLAR, researchers and communities drive the research in a participatory way, while being guided by the initial demands of the communities. It is valued because of its reiterative structure, which involves several learning cycles, and works with stratified groups of stakeholders in an effort to link scientific and local knowledge systems. In addition, it is valued for its long record in supporting natural resource management. A key factor concerning participation in the research is the way in which the tools are applied. However, the most important factor is the relationship between researcher, community and stakeholders in the design, implementation and evaluation of the research; and this is relevant not only to ethnobotany, but to any other study in which scientists work with local and traditional communities.

The establishment of sustainable development reserves by communities

Participatory research methodologies increasingly guide ethnobotanical research towards enhancing the links between the conservation of biodiversity and social development (Oliveira *et al.*, 2009). However, participation is still often viewed and used as a tool for data collection, thus maintaining the researcher in the driver's seat of the research process. Few cases are known where participation is seen as the inclusion of the researcher in the processes of local communities. Two other chapters in this book demonstrate how researchers position their research within the struggle of traditional people to sustain their access to a territory that is associated with their lifestyle, and which could be institutionalized by the establishment of a conservation unit. Zank *et al.* (Chapter 3.8) share the case of a traditional community in Imbituba, Santa Catarina, where ethnobotanical studies on the management of manioc, butia palm (*Butia catarinensis*) and medicinal plants, within a defined area, have provided insight into the role of the community in maintaining not only these species and genetic resources but also the specific landscape. The participatory nature of such ethnobotanical studies enhances the self-recognition of the community, and can also contribute to legal processes concerning access to the land and recognition of the conservation unit. Peroni *et al.* (Chapter 3.4) and Reis *et al.* (Chapter 3.5) illustrate the case studies of ethnobotanical research in the context of, respectively, the faxinalense traditional people in central-southern Paraná and traditional small-scale farmers in northern Santa Catarina. Ethnobotany contributes to the documentation of traditional knowledge of land use systems and keystone species in the landscapes, and promotes the development of strategies for the collective management of the landscape. Both these areas contribute to the struggle of traditional people to sustain access to their territory and maintain a traditional lifestyle, thereby contributing to the conservation of the landscapes of which they are a vital part. In both areas, ethnobotany has become an important tool for strengthening communities in their biodiversity management, while supporting them in their social and even political aspirations.

Participatory research methodologies are being used more and more every day, although their application in ethnobotany is still emerging. In several cases, participation is still viewed and used as a tool to aid data collection, with few examples in which participation is seen as the inclusion of the researcher in the processes of local communities. This is the result of a scientific environment that is still guided by a normal professionalism directed by positivist paradigms. Thus, there is no effective return to the community, because purely scientific questions have little relevance to local communities (Wilmsen, 2008).

However, those following more constructive or participatory paradigms in ethnobotany seek and develop expertise that contribute to the emancipation and empowerment of traditional people in a larger context of biodiversity conservation. In the context of Brazil, as described by Assis *et al.* (Chapter 3.2), an interest in biodiversity conservation is linked with the need to recognize and support traditional people and their livelihoods. Ethnobotanical research studies carried out in response to specific requests from traditional people seeking the support of scientists creates opportunities for ethnobotanists to embed their research within the larger context of conservation and development. In this sense, researchers can contribute vital elements in the design of community-based management plans required for the establishment of sustainable development reserves.

3.4 The domestication of landscapes and cultural keystone species in a context of community biodiversity management in Brazil

*Nivaldo Peroni, Ulysses Paulino de Albuquerque,
Ana Luiza de Assis and
Ernani Machado de Freitas Lins Neto*

The domestication of landscapes

Domestication is understood as involving different human actions at distinct levels in the organization of biodiversity, from species population level to the level of ecological communities and landscapes (Clement, 1999). The human management of a particular landscape can, for example, result in a more domesticated environment. Desired outputs of that management include an increase in the occurrence of useful species, an altered demographic structure of species, and increased ecosystem productivity (Clement *et al.*, 2010). Anthropogenic forests fit within domesticated landscapes (Balée, 1989) and we have much evidence of large-scale changes of landscapes in the Amazonian rainforest caused by human management (Fraser *et al.*, 2011). A domesticated landscape can be understood as a historical, cultural and evolutionary expression of the relationship between people and their environment, occurring at a specific location that is not necessarily natural or pristine (Balée, 2006; Johnson and Hunn, 2010).

Management practices undertaken by traditional people can result in the formation of landscapes units with a particular ecological and cultural identity that is recognized as distinct within a heterogeneous landscape. We can also refer to such units as 'cultural ecotypes' (Johnson and Hunn, 2010). They may be a source or reservoir of biodiversity, with many components of cultural significance at landscape, species or genetic levels and may be a source of income leading to the economic recovery of a species. In these cultural landscapes, the species present may exhibit a distinct degree of domestication; perennial species are especially interesting because they may show the effects of individual and/or collective management carried out over a long period of time. Such cultural ecotypes are relevant when designing integrated strategies for the conservation and use of biodiversity, in which we can see a direct relevance to the development of community biodiversity management (CBM) processes that address several components of agrobiodiversity.

Cultural keystone species

Cultural keystone species are those species on which people depend for their livelihood. They play a unique role in shaping and characterizing the identity of the human populations that rely on them. Within the context of a domesticated land-scape, some species are salient, in the sense that they contribute to the ecological structure of the landscape. They may influence the composition and structure of an ecological community as a whole, and are referred to as 'ecological keystone species'. In the management of a domesticated landscape, where natural and cultural components are interconnected, an intricate relationship exists. Cultural keystone species are of greater importance for the functioning of a community than expected by their abundance (Garibaldi and Turner, 2004; Platten and Henfrey, 2009).

Since domesticated landscapes can be spatially structured and biologically diverse, they may be considered complex historical inscriptions, resulting from complex forms of management. Several recent approaches aimed at the conservation of plant genetic resources (PGR) and social development through the sustainable use of agro-biodiversity use a community focus for analysis and action. Some examples of these approaches include: community biodiversity management; community management of natural resources, social and community forestry; community management of wildlife; co-management or cooperatives; buffer-zone management; multi-purpose participatory community projects; and indigenous resource management in communal areas, among others (Kellert *et al.*, 2000).

Domesticated landscapes, cultural keystone species and CBM

In this chapter, we look at those approaches that move beyond the level of diversity at the genetic and species levels, focusing on long-term managed landscapes, with some degree of domestication, as one of the expressions of agrobiodiversity. A link can be made to CBM, which reveals itself as a collective strategy for supporting local govern-ance and promoting community empowerment in the context of conservation and use of biodiversity in domesticated landscapes. Consequently, CBM practices may address or target the conservation and use of landscapes and cultural keystone spe-cies. The overall goal of such efforts may be the continuation or even strengthening of the historical process of managing the landscape, its species and genetic resources.

In Brazil, human ecologists and ethnobotanists have investigated and contributed to CBM processes in which the concept of cultural keystone species is rooted within domesticated landscapes. We are engaged in this type of research at the Laboratory of Human Ecology and Ethnobotany of the Federal University of Santa Catarina, and we share our experience working with these concepts in Paraná, in southern Brazil. The Applied Ethnobotany Laboratory of the Federal Rural University of Pernambuco con-ducts similar interdisciplinary research, but it focuses more on important cultural and economic species. Its experience in carrying out such research in the semi-arid region of north-eastern Brazil is shared in this chapter. In both case studies, we concentrate on aspects where such studies can contribute to the development of CBM approaches.

Faxinal landscapes managed as collective areas in Paraná

Faxinais (plural of faxinal) is the name given to common, shared land in the state of Paraná in southern Brazil, as indicated in Figure 3.2.1 (Chapter 3.2). In general, users of faxinais are engaged in silvo-pastoral activities that include the extraction of non-timber forest products and the rearing of animals in open fields. The communities separate their crop fields from the common forest with fences and ditches. Through this collective land use system, forest areas with remnants of the original Araucaria Atlantic Forest are conserved and used.

The people associated with this system are called the faxinalense people. Even though the communities are made up of descendants of European immigrants who arrived over a century ago, as explained by Assis *et al.* (Chapter 3.2), they are considered traditional people because of the strong connection they have with their natural and biological resources. Their cultural identity is associated with their livelihood and forest use that is governed by consolidated community rules. They extract pinhão seeds from the araucaria trees (*Araucaria angustifolia*) and leaves from yerba mate plants (*Ilex paraguariensis*); both products are primarily commercialized. In addition, community members raise animals (cattle and local pig breeds) in the common areas, where a customary system regulates animal ownership and rearing conditions. This silvo-pastoral system contributes in particular to the maintenance of the landscape, but also to the conservation of populations of araucaria, and even protects the local pig breeds as their collective rearing reduces chances of inbreeding.

The forest areas constitute remnants of the original araucaria Atlantic Forest and are regarded as domesticated landscape units. Araucaria and yerba mate are considered to be cultural keystone species as they characterize the livelihood of the faxinalense people. Formal and informal community agreements govern the management of the faxinal, the common forest area. These agreements address the use of the forest area for animal grazing, as well as the extraction of both the pinhão seeds and yerba mate leaves. Reis *et al.* (Chapter 3.5) describe the system and its practices in more detail, and look at caívas forests in the state of Santa Catarina.

The history of the establishment of the faxinais is closely linked to the management and production of yerba mate. Yerba mate was of great symbolic importance to the indigenous people who originally inhabited the region. It is considered a culturally unique species, used in the preparation of *chimarrão* tea, a drink of cultural relevance to rural as well as urban people in southern Brazil. The maintenance of the Araucaria Atlantic Forest is ecologically relevant for native yerba mate because of its requirement for shade, which can only be obtained by maintaining the forest canopy. Furthermore, the native populations of yerba mate have been valued as a non-timber forest product (NTFP) and have been traditionally managed by local people in the states of Paraná and Santa Catarina (Reis *et al.*, Chapter 3.5).

The pinheiro (araucaria tree) is a symbol of the region, just as is yerba mate. With its unique structure, the species characterizes this type of Atlantic Forest and its people. Its seed, the pinhão, is an important food resource and was of vital importance to the indigenous people who inhabited the region before the arrival of the European immigrants. Pinhão is important both for human consumption and livestock feed in the faxinais. Today, it is an important source of income for the faxinalense people; for example, in the faxinal of Bom Retiro, located in the municipality of Pinhão, Paraná, some households depend on the sales of pinhão for up to 70% of their income.

Although no consensus exists on the taxonomic nomenclature of araucaria varieties, traditional people identify and recognize varieties in their ethno-taxonomic terms. In the faxinal of Bom Retiro, the community identifies at least three ethno-varieties: San José, caiová branco (white caiová) and caiová vermelho (red caiová). The presence of varieties in a tree species with a long life cycle is highly relevant for demarcating particular ares of the araucaria forest that could be viewed as domesti-cated landscape under faxinal management, in which the pinheiro can be considered both a cultural and ecological keystone species.

Currently, the state of Paraná recognizes faxinais in its conservation regime as areas where Araucaria Atlantic Forest is conserved (Löwen-Sahr and Cunha, 2005). The fact that these areas maintain a biodiversity associated with the livelihoods of the faxinalense people has inspired and created opportunities for the development of CBM strategies that associate landscape conservation and livelihood development while recognizing and sustaining more traditional lifestyles.

Umbuzeiro: a sacred plant from north-eastern Brazil

The umbuzeiro (*Spondias tuberosa*) is a wild tree species endemic to the semi-arid plains of north-eastern Brazil. It is a small tree about 6 m high, and has a short trunk with a cup-shaped umbrella canopy about 10–15 m in diameter, which provides dense shade. Its shallow roots reach just a metre in depth and have a body (structure) known as xylopodium, locally referred to as *batatas de umbu* (umbu potatoes), which consists of spongy tissue that stores water, mucilage, glucose, tannin, starch and acids. In times of severe drought, the tubers are eaten. The word *umbu* is based on the indigenous (Tupi Guarani) word *y-mb-u*, which loosely translates as 'tree that gives a drink'.

The fruit of the umbuzeiro is called umbu; its pulp is almost watery when ripe, which makes the fruit very perishable. The fruits are much appreciated by both the people and animals. Goats are particularly fond of umbu and they often accompany local people during harvesting. Umbu is rich in vitamin C and has a characteristic sour taste. As well as being consumed raw, it is also used in culinary preparations, such as ice cream, jellies, jams and *umbuzada*, which is a delicacy prepared with milk and sugar that is very much appreciated by the people. As an opportunity for income generation, several cooperatives in the region have been set up for the production of such products, and for marketing the products in major urban centres in the country, and possibly even for export to Europe.

Although its distribution covers major parts of the semi-arid savannah or caat-inga biome in north-eastern Brazil, the umbuzeiro grows more abundantly in the dry interior of the states of Pernambuco and Bahia. As in the case of the pinheiro in the south, the umbuzeiro is considered to be more than just a tree: it is a deep expres-sion of the people's culture and livelihood. People refer to the umbuzeiro as a 'sacred plant'. This perception is not related to magical-religious or ritualistic aspects of the plant; it relates to the fact that the umbuzeiro flowers and bears fruits during the dry season, making it a source of food and income in times of scarcity. In the heat, it pro-vides people and animals with a place of comfort and shade (Cavalcanti *et al.*, 2000; Albuquerque and Andrade, 2002). The age of the trees, many of them are known to be older than 100 years, contributes to their sacred status.

In the semi-arid region of Pernambuco (see Figure 3.2.1, Chapter 3.2), we observed local practices that preserve this species and shape the local landscape around it.

Fruits are principally collected from trees that grow in managed areas as opposed to areas of native vegetation, which is an indication that the local farmers collect fruit from trees that they have known for years. Umbuzeiros in managed lands produce an abundant supply of fruits, which are generally more appreciated by local people. The umbuzeiro is a source of food for both humans and animals, with all its other possible uses having only secondary importance to the community.

Albuquerque and Andrade (2002) observed that while native food sources can serve as important dietary complements as well as sources of subsistence in difficult times, rural communities of the semi-arid north-eastern region of Brazil are gradually changing their lifestyle and associated diet. Despite the abundance of the umbu fruit, and the potential for income generation through the commercialization of the fruits or other umbuzeiro products, our assessment in the region showed that local people have stopped harvesting this resource commercially, because farmers direct their attention almost exclusively to the cultivation of maize and beans, which are widely promoted by rural extension programmes. One of our observations indicates that the knowledge and use of the umbuzeiro is well established but the interest of the some rural communities in this biological resource is diminishing, thereby reducing the strength of their relationship with the landscape and this culturally and economically important species. The umbuzeiro is still used for domestic purposes, but the interest of rural households is diverting away from the species, thereby reducing their cultural and economic appreciation of the domesticated landscape (Lins Neto *et al.*, 2010). With this gradual loss of interest in umbu and its potential use for income generation, the communities lose opportunities to improve the quality of their lives by maintaining the species and its associated landscape. However, it should be recognized that in other areas, as indicated above, several initiatives aim to implement umbu value-addition and marketing activities through cooperative groups. Such initiatives, embedded within a larger framework of CBM, may enhance the relationship between communities and the umbuzeiro, in managed and native stands, thus contributing to the conservation of this tree, which is important to the landscape and culture of the people, while supporting local people's livelihoods.

CBM: opportunities for the conservation of domesticated landscapes

The CBM processes that support the people in the faxinais and semi-arid regions differ from most of the other experiences shared in this book, which target food crops rather than landscapes. Since the collective management of biodiversity is already a way of life for the faxinalense people, the CBM process in the faxinais can be referred to as an endogenous CBM process and structure. Other distinct features include the focus on perennial species and their association with animals (pigs and goats) within landscapes with some degree of domestication, and the fact that the landscapes provide food for subsistence, as well as resources for value addition and commercialization. The faxinal stands out with its sophisticated system of common property management. One critical point that we have learned in both cases is that the domesticated landscape, with its cultural keystone species, creates the boundaries for the CBM processes. Another characteristic that distinguishes these CBM cases from others that are associated with annual crops is that communities manage their biodiversity at landscape level over a long period of time, and so the association covers multiple generations.

We have learned from these experiences that individual perennial species, such as the araucaria, yerba mate and the umbuzeiro, which are of important cultural and economic value, are also vital for sustaining the community's interest in the landscape as an expression of their identity as traditional people or communities. As such, we find working with agrobiodiversity at the landscape management level ideal conditions for working with CBM, combining the conservation of genetic resources with the enhancement of social and cultural capital for livelihood development.

3.5 Caívas and their contribution to the conservation of Atlantic forest landscapes in Brazil

*Maurício Sedrez dos Reis, Camila Vieira da Silva,
Andrea Gabriela Mattos, Alex Anderson Zechini,
Adelar Mantovani and Nivaldo Peroni*

Domestication of landscapes

Multiple uses of distinct environments have historically favoured many past and present human populations in the discovery of plant resources. Various studies have been carried out on the impact of human activities on changes in species composition or their demography in landscapes. Historical ecology seeks to understand these relationships by looking at such transformations of landscapes as mediated by human actions (Balée, 2006).

The historical ecology perspective complements that of focusing on the domestication of landscapes, as described by Peroni *et al.* (Chapter 3.4), as it enhances our understanding of how human activities have transformed landscapes and to what extent these landscapes have increased the productivity of specific biological resources. The domestication of landscapes results from human management of biotic and even abiotic components of the environment. Such a process can take many generations, but may contribute to an increase in the availability of useful species, change the demographic structure of individual species and maximize the productivity of the environment (Clement, 1999).

Landscape conservation: a unique strategy for sustaining domestication

In Brazil, many biological scientists focus their research on the threatened Atlantic Forest. However, studies rarely focus on its historical ecology and just a few species and associated landscapes have been studied from the domestication perspective (Reis *et al.*, 2010). Little focus is given to species that continue to be domesticated by humans today. For such species, traditional people employ management practices that may also be of relevance to the conservation of those landscapes. We focus here on the interaction between people and their landscape and, in particular, on their management of cultural keystone species, thereby looking for transformation at species and landscape levels. Such an approach appears to be the most logical as it targets the dynamic relationship between people, biodiversity and landscapes.

Caívas: managed patches of the Araucaria Atlantic Forest

The mixed ombrophilous forest (MOF), also called the Araucaria Atlantic Forest, is one of the most endangered ecosystems of the Atlantic Forest. Yerba mate (*Ilex paraguariensis*) and araucaria (*Araucaria angustifolia*) are typical of this forest type. It is estimated that MOF remnants occupy just 1–4% of their original area, and are currently spread over small, dispersed fragments (Guerra *et al.*, 2002).

In the northern plateau of Santa Catarina (see Figure 3.5.1), small-scale farmers manage existing MOF fragments, which form part of an area that is used for cattle rearing, extraction of firewood, commercial extraction of yerba mate and *pinhão* (araucaria seed). The extraction activities complement other commercial farm activities that include maize and soybean cultivation, and cattle husbandry for dairy production. Local farmers refer to these managed forest fragments as caívas, landscape units or patches within the matrix of a fragmented forest. In the composition and structure of their plant communities, these caívas reflect past and present human influences. Araucaria and yerba mate predominate in the caívas, together with several species of the Myrtaceae family, such as the araçás (*Psidium* spp.) and the guavirovas (*Compomanesia* spp.).

From an ecological perspective, caívas are heterogeneous landscape units, and they range in composition from areas with dense herbaceous vegetation, dominated by yerba mate, to forested areas in the advanced stages of secondary succession. The management of these areas is quite variable but may include the following three basic elements, in varying degrees of intensity: the extraction of yerba mate, cattle rearing and seasonal management.

Yerba mate

Yerba mate is an arboreal species of the Aquifoliaceae family, which has stimulating properties and is consumed as *chimarrão,* a typical tea drink common in southern Brazil, the northern provinces of Argentina (Corrientes and Missiones), Paraguay and Uruguay. The yerba plant is an evergreen shrub or small tree, which grows to

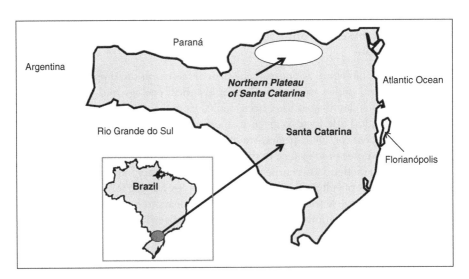

Figure 3.5.1 Map showing the location of the Northern Plateau of Santa Catarina.

about 15 m tall. Its leaves are often called *yerba* in Spanish or *erva* in Portuguese, both of which mean 'herb'. The leaves contain caffeine and related compounds, are harvested commercially and are consumed as a hot or cold tea. Variations in consumption and processing techniques have evolved over time from those initially practised by the various groups of indigenous peoples (in particular the Guarani) that previously inhabited this part of the continent. Today, the use of yerba mate partly characterizes the way of life of many rural and urban human populations who live throughout the southern inland plains of Brazil and its neighbouring countries, where the daily consumption of chimarrão or mate is common. Drinking chimarrão or mate provides an impetus for gathering with relatives and friends in southern Brazil, northern Argentina, Paraguay and Uruguay. Sharing is ritualistic and has its own set of rules, which vary from community to community within the region.

The economic exploitation of yerba mate was already significant in the eighteenth century, even before the start of commercial timber exploitation. Currently, yerba mate can be cultivated as a homogeneous crop, using selected varieties, but the leaves may also be extracted from the caívas, where extractors gather the leaves of 'native' yerba mate directly from forest populations.

In areas where yerba mate is harvested from managed forests, the harvesting of leaves, called *poda* ('pruning'), is carried out at intervals of 2–4 years, depending on the management strategy adopted by the producer. In addition, many producers tend to practise the management technique of *roçar* ('cleaning') in areas of yerba mate to promote leaf growth, or to facilitate the management of the productive area. In general, farmers do not cultivate crops within the yerba mate areas. Within the forest, the management of each yerba mate area is usually carried out by the same individuals for many years. Some producers claim that their yerba mate plants, the leaves of which are continuously harvested, are more than 100 years old.

Despite the higher productivity of homogeneous plantations, producers focus on native yerba mate plants. This preference arises from the higher prices that producers get in the market for the mate extracted from native plants. According to the palate of chimarrão consumers in southern Brazil, native yerba mate has a less bitter taste than that cultivated in plantations (Vieira *et al.*, 2003). Thus, producers who still extract yerba mate from the understorey of araucaria forest fragments have a more valuable product than that from homogeneously cultivated plantations. Currently, the price difference can amount to a premium of 70% more for the native yerba mate.

In the northern plateau of Santa Catarina, the average property size is 45 hectares, approximately 23 hectares of which are covered by forest. Almost half of the properties possess forest areas from which producers extract yerba mate. We can divide the landscape into two main types, based on the management practices undertaken in forest fragments. The first type is called traditional simplified management, in which yerba mate is exploited using simple management practices and minimal human intervention. The second type is traditional caíva management, in which yerba mate is exploited with local management practices, cattle rearing in the understorey and intense human interventions.

The degree of intervention in the landscape, depending on which of the two management types is applied, results in a difference in forest structure and species density, with implications on the arboreal forest layer. In areas where traditional simplified management techniques are applied, the average density of individuals of all species is 5976 plants per hectare (plants/ha), with yerba mate presenting a density of 1067 plants/ha. For traditional caíva management, where management practices are more intense, the density of individuals of all species is 805 plants/ha, with yerba mate presenting a density of 324 plants/ha.

The high densities of yerba mate in landscapes managed by human populations show that it is a species that has remained within the forest fragments. Hence, once sustained, the management of native yerba mate favours the conservation of forest fragments, thereby contributing to *in situ* conservation through the management of yerba mate in caívas as a plant genetic resource.

Araucaria

The most common uses of the araucaria are primarily associated with timber procurement (Guerra *et al.*, 2002). Timber exploitation increased from the early twentieth century until the 1970s. It was followed by the arrival of European immigrants in the south, resulting in extension of the agricultural frontier, as described by Assis *et al.* (Chapter 3.2). The timber exploitation of araucaria made a significant impact on the Brazilian economy and contributed to the devastation of major areas of the Atlantic Forest, especially in the southern states. As a result of its predatory exploitation, associated with the partial destruction of its habitat, araucaria is considered to be an endangered species according to the official list of endangered Brazilian flora species.

The seed of the araucaria, the pinhão, can be eaten and used to form a significant part of the diet for indigenous peoples, such as Xokleng and Caingang (Bitencourt and Krauspenhar, 2006). Nowadays, it is considered to be a delicacy and is greatly appreciated by both rural and urban populations.

Producers have reported significant variations in productivity over the years. An understanding of this variation may be useful for the design of interventions targeting the conservation and continued use of the species. However, much still remains unclear on the production of pinhão (Reis *et al.*, 2010). Estimations of productivity vary from 45 kg/ha of seed (Vieira da Silva and Reis, 2009) to more than 4000 kg/ha (Solórzano-Filho, 2001). This variation may be related to differences in estimation methods, age of the individuals, soil fertility and annual variations in productivity.

Few studies have investigated the structure of the pinhão supply chain, or tried to improve it. Studies carried out in the states of Santa Catarina and Paraná indicated that the pinhão value chain involves pinhão producers and collectors, intermediaries, wholesalers and retailers (Santos *et al.*, 2002; Vieira da Silva and Reis, 2009). An investigation into this value chain showed that the price of pinhão fluctuates both during the harvest and between harvests. The variation in price is influenced by the availability of pinhão, the productivity of which is irregular and unpredictable, and by demand for the product, which appears to increase in the event of a severe winter. Understanding how ecological aspects of the species and the variations that exist among araucaria populations or varieties influence the supply chain is important in order to align the conservation of this plant genetic resource with the effective rural development of the farming communities that manage the species in their caíva landscape.

Forest landscape conservation and its integration with araucaria and yerba mate

Araucaria and yerba mate are typical in the everyday lives of households in the northern plateau of Santa Catarina area and other regions where the MOF is the primary forest type. For yerba mate, this means particular care is given to the species populations, favouring their growth and productivity. Some households in the northern plateau give names to araucaria individuals that refer to the history, unique structure and seed capacity production of an individual tree. Assis *et al.* (2010b) conducted an analysis of the MOF as a domesticated landscape and identified araucaria and yerba mate as its cultural keystone species. The results indicated that both species are among the most valued, demonstrating high cultural indicator levels in all aspects of the analysis. Much ethnobotanical information has nevertheless been lost, as many farmers no longer claim to use names for individual araucaria trees as they used to (Vieira da Silva and Reis, 2009).

Symbolic aspects of the two species are important, such as the identification of regional landscapes, as delineated by the presence of araucaria, the consumption of the pinhão, and the preparation and consumption of chimarrão, using yerba mate. Both species have a unique place in local culture, i.e. there is no other plant species that replaces either of them in the preparation of chimarrão or pinhões, or their commercial exploitation for this purpose. This exclusivity, and the continued appreciation by both rural and urban people in southern Brazil, creates a favourable environment to support the management of the caíva landscape, facilitating the conservation of the Araucaria Atlantic Forest and its two keystone species, but it also contributes to farmers' livelihoods, which are directly associated with, and vital for, the continuation of this landscape. With the current economic growth in Brazil, it is likely that markets for products with a higher value, such as quality yerba mate and pinhões, will grow. However, conservation efforts should promote regional products within markets that tend to favour industrialized and global products. In order to encourage farmers to maintain their remaining forest areas, better market conditions for their products may encourage them to alter their management practices in favour of araucaria and yerba mate. These alterations will in the long run provide additional income, and ensure the maintenance of viable populations of both species and a landscape that retains fragments of the once abundant Atlantic Forest.

3.6 Community management of forest coffee landscapes in Ethiopia

Regassa Feyissa, Genene Gezu, Bayush Tsegaye and Kassahun Tesfaye

Coffee genetic resources in Ethiopia

Ethiopia is the primary centre of origin and diversity for arabica coffee, which is the country's most important crop in terms of its economic contribution. The Afromontane rainforests of south-western and south-eastern Ethiopia are home to wild populations of *Coffea arabica*. The original coffee habitats and major coffee-growing areas of Ethiopia are illustrated in Figure 3.6.1. In Ethiopia, very few forests are maintained, mainly due to high population pressures. The forests that remain are basically those in which we can find wild populations of coffee. The presence of coffee and its important requirement for shade has contributed to forest conservation (Senbeta *et al.*, 2007). This ecological service of the coffee production system for biodiversity conservation can be considered even more valuable if we look at the fact that the Afromontane

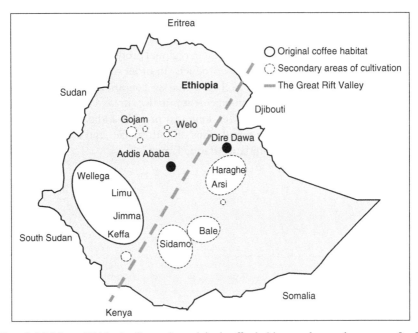

Figure 3.6.1 Map of Ethiopia, illustrating original coffee habitats and secondary areas of coffee cultivation.

rainforests, in which we find the wild coffee populations, include over 700 vascular plant species, comprising about 11% of Ethiopia's total flora (Senbeta, 2006).

Wild coffee plants are usually found in patches in the understorey of the forest. Traditionally, local communities either harvest the berries of those wild populations without much interference in the forest, or manage the forest to increase coffee production. Analysis of the genetic diversity of coffee populations collected from different coffee ecologies confirms the presence of a genetic diversity that is specific to each of them (Tesfaye, 2006). More than 130 landraces have been identified in these coffee ecologies (Teketay and Tigneh, 1994). In order to conserve coffee genetic resources we need to focus on the variation among and within these coffee ecologies.

One particular problem, from a biodiversity point of view, is that the application of traditional management practices has a negative impact on the structure and composition of communities and species diversity in the forest (Senbeta, 2006). At present, in cases where the intensification of coffee production is promoted by the government, the balance between the conservation of forest ecology and coffee production has been destabilized. Furthermore, the pressures of a growing population are causing the transformation of coffee forests or production areas into crop-land for food production.

Most arabica coffee can be found in Ethiopia at altitudes ranging between 1500 m and 1900 m. However, in the very wet south-west of the country, we can also find forest coffee at an elevation as low as 1000 m, while in other regions we find it growing in forest gardens as high as 2500 m (Edwards, 1991). The phenotypic diversity of coffee in Ethiopia is manifested in leaf shape, colour and size, in bean shape and size, in branching habit, as well as in tree shape. The inherent genetic traits and conditions that occur within different coffee ecologies influence the type of coffee flavour and natural qualities (Labouisse *et al.*, 2008).

Coffee production systems

Coffee production in Ethiopia is based on forest coffee (FC), semi-forest coffee (SFC), small-scale farmers' forest garden coffee (FGC) and large-scale commercial coffee plantations (CCP), which covers a wide range of complex landscapes. The first three production systems are of a traditional nature, in which predominantly small-scale farmers are responsible for the management of the coffee production. The fourth, CCP, is a modern production system. The high population density of the highlands of the country does not allow much space for the establishment of new, intensively managed commercial coffee plantations.

Forest coffee system

Since wild coffee plants grow as understorey plants in the forest, local farmers are able to simply pick the coffee berries from those plants. Forest coffee requires very little in the way of management interference in the forest for improving production. In the FC system, the floristic composition, diversity and structure of the forest is little modified or affected by human interference. The occurrence of coffee plants in coffee ecologies is defined by a series of natural and, to some degree, human influences

(Senbeta, 2006). Estimates of the contribution made by this system to the national production vary between 5% and 10%, as indicated in Table 3.6.1. The FC system was originally assumed to host the largest genetic diversity of (wild) coffee plants (Senbeta *et al.*, 2007).

Semi-forest coffee system

In the SFC production system, coffee is grown under natural conditions. The plants are wild, but human selection and the management of their environment affect their distribution and even their genetic make-up. In order to improve productivity in the SFC system, farmers regularly remove understorey trees, shrubs and lianas. Even though the production systems form a continuum, the human impact on plant composition and the reduction of plant density distinguish the SFC system from the FC system. Coffee plants become dominant in the SFC system because management suppresses tree regeneration and reduces tree density, leading to the disappearance of some forest species. In the forest structure, this modification leads to the formation of a tall tree canopy for shade and a coffee canopy layer, without the original intermediate canopy layers. These management practices result in a reduction of 25% in individuals per hectare as compared with the FC system. It could be said that within the SFC system the communities manage or rather domesticate the forest landscape to increase production and the productivity of the coffee trees. We can thus consider that the SFC constitutes a landscape that is more domesticated, since its productivity has increased (Clement, 1999; Peroni *et al.*, Chapter 3.4). The SFC system comprises about 35% of the national production.

The SFC system has a high level of genetic diversity, despite or rather because of human intervention. This has been shown in a comparative analysis of FC and SFC systems (Tesfaye, 2006). A reason for this higher diversity is that the removal of small trees and understorey shrubs leads to a reduction in genetic erosion at coffee seedlings stage, since there is less competition with plants of other species. Seed flow among wild coffee populations is enhanced during such thinning and gap-filling practices. Seed and pollen flows result in an increase in the genetic diversity of the population in

Table 3.6.1 Characterization of coffee production systems in Ethiopia

Production system	Coffee genetic resources	Type of producer	Contribution to national production
Forest coffee (FC)	Native plants and barely managed habitat	Extraction under common property regime	5–10%
Semi-forest coffee (SFC)	Native plants in a domesticated landscape	Extraction under common property regime	35%
Forest garden coffee (FGC)	Local varieties (landraces) in mixed farming	Small-scale farmer cultivation	45%
Commercial coffee plantations (CCP)	Improved varieties in monoculture	Large-scale plantations	15%

Sources: Senbeta *et al.* (2007); Labouisse *et al.* (2008).

this type of managed forest. However, as soon as coffee management intensifies and begins selecting for some specific coffee genotypes by favouring more uniform stands, the original increase in coffee genetic diversity is halted (Senbeta *et al.*, 2007).

Forest garden coffee and commercial coffee plantation systems

There are two further coffee production systems in Ethiopia. In the FGC system, farmers select local coffee varieties for cultivation in small, shaded stands in the area surrounding the household residence. The FGC system is a completely managed agro-forestry system where coffee is cultivated in association with other crops such as ensete (*Ensete ventricosum*), fruit trees and chat (*Catha edulis*). Trees create the shade, albeit in a much lower density than in the SFC system. The diversity in crops is high, but compared with the SFC the diversity of the species accompanying the coffee, and the degree of diversity of the coffee plants themselves, is reduced. As indicated in Table 3.6.1, this small-scale farmer coffee system accounts for a major share of the national production. The remaining system of large-scale commercial coffee plantations is relatively less significant. Under the CCP system, improved commercial cultivars with high productivity are cultivated under intensive management on private and state-owned farms.

Challenges to the traditional coffee production systems

Between 1973 and 2005, the conversion of forest-land into crop-lands in the Afromontane rainforests led to a reduction in FC and SFC systems from 71% to 48% of the original coffee stands (Wakjira, 2007). This land-use change led in turn to an increase in CCP systems and the replacement of many of the original diverse coffee populations with genetically narrow-based (improved) varieties. As such, deforestation and the promotion of modern commercial coffee plantations threaten the natural gene pool, which is rich in variation and in harmony with its landscape.

The traditional production systems (FC, SFC and FGC) face many other less obvious challenges. The lack of attention given to them is important in terms of technical assistance and extension. At the same time, disagreements exist on technical issues, such as on the number of trees that need to be thinned out in the SFC system. Interest in maintaining the forest cover while increasing coffee productivity is also an issue. The traditional systems are not recognized as having a good potential by government and concerned bodies. The push for modernization of coffee production systems ignores and does not invest in the potential of the traditional systems.

A major threat to the traditional coffee production system and the genetic diversity it harbours is the pressure exerted on the Afromontane forests by population growth. A conversion of the FC into SFC and FGC production systems eventually results in the loss of coffee genetic resources, loss of biodiversity and the ecological services of coffee forest. The fact that both production and biodiversity conservation goals can be achieved in the traditional systems has attracted the attention of some bodies within the Ethiopian government. Nevertheless, in order to support the traditional systems, a paradigm shift is required from those research and extension agencies that continue to focus their attention on the production and export of large quantities of coffee, through intensive, highly productive and less quality-oriented systems.

Conservation of coffee genetic resources and its challenges

Different approaches and techniques may be applied to maintain coffee genetic resources. In designing coffee genetic resource conservation strategies it should be realized that the mid- or long-term storage of coffee seed under sub-zero temperatures is problematic since the seed does not survive desiccation to a low level of moisture content. Field gene banks and different *in vitro* techniques are used for *ex situ* conservation of coffee germplasm. Experiences of the Ethiopian coffee field gene bank show that some coffee types are sensitive to temperature, which is one of the factors that reduce the viability of the collection in the field gene bank. When coffee collections from higher elevations were maintained at lower elevations, for example, some accessions suffered stunted growth and chlorophyll disintegration, resulting in the production of physiologically deformed coffee beans. It may be concluded that it is not possible to maintain coffee collections from different agro-ecological conditions in one location, making the field gene bank an institutionally and logistically difficult and demanding conservation strategy.

Since forest coffee is sensitive to symbiotic relationships with other plant species and microorganisms, and the disruption of this relationship usually causes changes in the viability of the genotype, *in situ* conservation in original forest habitats or ecologies emerges as the most viable strategy. The conservation of the forest coffee landscape allows for the interaction of multiple species, the continuum of evolutionary processes and the continuation of the relationship between coffee genetic resources and the managed forest. In this manner, *in situ* conservation is achieved by maintaining a dynamic pattern of interactions between coffee genetic resources, the forest environment and its associated people (De Boef and Thijssen, Chapter 1.8). It can be assumed that the more dynamic these interactions are, the more diverse in biochemical, physiological and behavioural traits the coffee genetic resources will be. The simplification of human–ecological interactions and habitat structure, as is usually the case if accessions are maintained in field gene banks, may cause a speedy decline in the diversity of phenotypic and genotypic traits. As such, maintaining coffee under reduced ecological complexity may not enable the conservation of the full range of diversity it exhibits in forest habitats and landscapes.

Conservation strategies can allow for the continued domestication of the forest coffee landscapes, thereby contributing to the domestication process of coffee as a cultural keystone species, as discussed by Peroni *et al.* (Chapter 3.4). Interdisciplinary approaches that involve fields such as population ecology, community ecology, landscape ecology and ethnobotany need to be adopted in efforts to conserve the forest coffee landscapes that are essential for the conservation of forest coffee genetic resources.

For the conservation of the cultivated coffee landraces, conservation practices associated with *in situ* conservation on-farm are more appropriate. We need to strengthen the capacities of farmers in maintaining and using coffee landraces in the forest gardens where the coffee genetic resources have developed their own distinctive properties. This shows that complementary strategies are required for the *in situ* conservation of the different coffee genetic resources associated with their original production system.

The regulation of forest access: customary systems under threat

De facto local communities are often the primary custodians of coffee genetic resources. Their management is highly site-specific, culturally influenced, diverse and is based on the maintenance of the forest structure and its species diversity. Their traditional knowledge of the local ecology, forest landscape, its species and coffee diversity, is based on a culture of the shared notion of kinship, taboos and rituals. This culture-based management results in shared community values concerning the landscape and the resources.

Traditionally, forest resources are perceived as common property that is governed by open-access regimes. Many coffee forests in Ethiopia fall under such regimes, where exclusion in some cases is difficult and common access can sometimes lead to conflict. These traditional access regimes are subject to local social conditions. Ownership of the resources is not absolute; it involves the relationship between the resources and communities, where individuals or groups have rights over the resources that are socially determined in traditional common property regimes at community level. In this respect, coffee genetic resources in the FC and SFC production systems are intimately tied to local communities, and the fate of the diverse coffee gene pool partly depends on the continued relationship of the communities with the landscape. However, as time and situations change, this relationship is weakening, as are social and communal entitlements concerning the use of the resources.

The strengthening of forest coffee management: experiences from Harrena

Much of the forest coffee in Ethiopia is an open-access resource, to which individuals and groups have free access for different kinds of use. Coffee can be harvested for household use and for sale. Customary regimes are closely observed when accessing the forest resources. However, access to the resources sometimes goes beyond those systems that regulate the use of the resources. This happens mainly when individuals or groups migrate to natural forests for seasonal coffee harvesting. The Harrena coffee forest is one of the few remaining natural forests that are exposed to such access by seasonal gatherers. Usually, coffee that is harvested through open access does not meet the desired level of quality. Open access can also be the cause of harm to standing trees. These gatherers harvest both ripe and unripe cherries together, giving a mixture during drying which spoils the quality. Since local markets primarily are oriented towards quantity rather than quality, prices do not encourage proper coffee harvesting and processing.

This situation necessitated a change in attitude and management practices, and also in the structure of the coffee market. In 2006, Ethio-Organic Seed Action (EOSA), in collaboration with partners, took the initiative to enhance community ownership over resources and promote the sustainable development of Harrena forest coffee through market incentives. The market incentives generated economic benefits that helped to improve the quality of coffee through the promotion of proper harvesting and processing techniques. Local communities were assisted in social and market organization, and were provided with training in, and the facilities for, harvesting and processing. The project also facilitated the establishment of a secure premium price for Harrena forest coffee.

After the local communities had been trained in best practices, the harvested coffee was awarded premium grades. For the first time in history, coffee from Harrena was finally recognized and thereby rewarded for its quality. We conducted laboratory analysis on the cup quality of the coffee from three different micro-ecologies of the same forest, which exhibited specific flavours as spicy, ginger and mocha. Recognition of coffee from their forest excited the communities even more than the premium price; it motivated them to take more responsibility to safeguard the forest. This approach has served as a useful lesson that has encouraged many, including policy-makers, who have concerns over the threat to the forest coffee resource base in the country.

Challenges and opportunities for premium market coffee

Coffee from natural Ethiopian forest is organic by default. There is potential for linking such coffee to niche and international markets, which can reward communities with a price that could be an incentive for sustainable management and for the use of traditional management systems (FC, SFC and FGC). It is necessary to upgrade the techniques for coffee harvesting, processing and packaging associated with the management practices of each production system. These investments are a prerequisite to ensuring access to premium markets. Product labelling such as environmental certification, geographic origin and fair trade have great advantages but also require a high level and quality of organization by the producers, as well as the enhancement of the structure in the value chain to assure consumers of the reliability of quality as well as provenance. This could be a strategy for communities to be rewarded for their custodianship through the continued management of forest coffee landscapes and associated traditional coffee production systems, thereby contributing to the continued process of coffee domestication and evolution in its original and unique habitat in the Afromontane rainforests of Ethiopia.

3.7 Promoting the community management of garcinia genetic resources in the central Western Ghats, India

Vasudeva Ramesh, Narasimha Hegde and Bhuwon Sthapit

Community-based forest management

Community biodiversity management (CBM) is a community-driven participatory approach to the management of biological resources. Elements of CBM can be recognized within existing common property and natural resource management regimes, and as such a number of the customs and traditions related to natural resource management that are followed in villages in India qualify as CBM practices. Consequently, when conservation and development organizations aim to implement a CBM process under such circumstances, it is referred to as 'endogenous CBM' (Shrestha *et al.*, Chapter 1.3).

Throughout history, different communities across India have been managing usufruct forests by applying CBM practices. Contemporary systems of formal forest management in India use CBM practices without even making reference to it. The state government of Karnataka has been implementing Joint Forest Planning and Management (JFPM) since 1990. JFPM involves participatory approaches in the form of 'co-management', in which non-governmental organizations (NGOs) mediate between rural communities and the forest department (Sundar, 2000). According to this system, stakeholders have a joint responsibility to plan and manage forest areas, but they also share any benefits arising from forest protection (Fisher, 1995).

Tropical fruit diversity project: focus on garcinia in Karnataka

As scientists representing a university (Vasudeva Ramesh, College of Forestry, Sirsi), an international organization (Bhuwon Sthapit, Bioversity International) and a NGO (Narasimha Hegde, Life Trust), we are collaborating on the project 'Conservation and Sustainable Use of Cultivated and Wild Tropical Fruit Diversity: Promoting Sustainable Livelihood, Food Security and Ecosystem Services', which is being funded by the Global Environment Facility (GEF) through the United Nations Environment Programme (UNEP). The project is being implemented in 18 communities in India, and six communities each in Indonesia, Malaysia and Thailand. The College of Forestry, Sirsi, is collaborating on the project at our site in Karnataka, with two NGOs – LIFE Trust and EcoWatch. Bioversity International and the Indian Council of

Agricultural Research (ICAR) provide technical support. The project aims to conserve tropical fruit tree genetic resources *in situ* and on-farm by strengthening the capacity of farmers, user groups, local communities and institutions to sustainably apply good practices and to secure benefits.

The project site in Karnataka focuses on *Garcinia* spp. Although the species has not been completely domesticated, its trade has in recent years reached commercial levels, including international markets (Vasudeva and Hombe Gowda, 2009). This commercial status poses several challenges which need to be addressed within community forest management when working with CBM practices in the context of garcinia, an increasingly valued non-timber forest product (NTFP). This combination of fruit tree genetic resources, NTFP and forest management creates a challenging environment for participatory action research using the CBM approach.

The Western Ghats and garcinia production systems

The Western Ghats, a chain of hills that stretches 1600 km along the western coast of southern India, is a globally recognized hotspot of biodiversity (Vasudeva *et al.*, 2006). The project site is located in the Uttara Kannada district, which is situated in the central part of the hill chain (Figure 3.7.1). The district has varied geographical features, with thick forests, perennial rivers, abundant flora and fauna, and a coastline

Figure 3.7.1 Map showing the approximate location of the Uttara Kannada district, in Central Western Ghats, Karnataka, India.

of about 140 km. More than 80% of the area is covered by forests, while agricultural lands comprise just 12%. The tropical climate of this region is strongly influenced by the monsoons. During the monsoon season, the region receives one of the heaviest rainfalls in the world, with an annual average of 2835 mm. However, the western coastal and crest-line areas of the district receive rainfalls exceeding 4000 mm.

People who live on the edge of the forest in the Western Ghats depend primarily on forest resources for their livelihoods. Forest resources are recognized as a critical source of income. For instance, over 40% of the households in communities in our project area gather NTFPs in a traditional manner, mainly collecting fruits of the species *Garcinia gummi-gutta*. In a recent survey, we learned that households earn on average US$770 a year based on garcinia collection (Hegde and Vasudeva, 2010). Garcinia resources have become extremely important as a source of income, and as such are an important livelihood asset in the central Western Ghats.

Garcinia diversity and its traits

Garcinia is a large genus of the family Clusiaceae, consisting of over 35 genera and 800 species, 40 species of which produce edible fruits. *G. mangostana* and *G. indica* produce well-known edible fruits. Thirty-five species of garcinia are known in India, of which seven species are endemic to the Western Ghats region and five are commercially exploited. The pulp of the fruits of *G. gummi-gutta* and *G. indica* is used as a flavouring agent in culinary preparations. Their seeds are a rich source of edible fats, and the fruit rinds of *G. indica* are used in the preparation of a popular beverage. Since both species possess curative properties, especially for stomach and heart ailments, local healers use them in folk medicine.

G. gummi-gutta has become an important and valuable NTFP species because of a newly identified pharmaceutical property, hydroxycitric acid, which is believed to reduce fat accumulation in humans (Singh *et al.*, 1995). The marketable product is extracted from the dried fruit rind. The increased commercialization of processed garcinia fruit has motivated many subsistence-level households to get involved in entrepreneurship directed towards the international export of this new commodity.

CBM activities for promoting garcinia conservation and sustainable use

Enhancement of awareness of genetic diversity and establishment of community platforms

We are currently working with three rural communities in the central Western Ghats, reaching over 1000 households. Over the course of the project, we have organized a number of village workshops to create community awareness of issues around the diversity of garcinia. We used participatory tools, such as four-cell analysis, to identify the common, threatened and rare garcinia types, and then decided on strategies of community action aimed at their conservation and sustainable use. We involved representatives of local stakeholders, such as collectors, processors and members of cooperative societies and women's self-help groups, thus creating community platforms for sharing traditional and scientific knowledge, and facilitating the identification of rare and elite types of garcinia.

Diversity fair

In 2010, we held a national seminar on garcinia genetic resources at the College of Forestry, Sirsi and, in tandem with the seminar, organized a diversity fair to inform and engage farmers on various issues related to the conservation and utilization of plant genetic resources (PGR). The seminar and fair served to showcase variations found in garcinia and share innovative products developed by farmers (Vasudeva *et al.*, 2010). Over 60 farmers from different parts of the district participated, as well as forest managers, policy-makers, scientists and horticulturists from across the Western Ghats. During the fair, participants displayed about 25 types of *G. indica* and *G. gummi-gutta*. The farmers were able to exhibit their talents, skills, products and genetic diversity; indeed, the fair became an instrument for heightening the farmers' pride in their garcinia resources. Interactions with diverse stakeholders, including scientists, researchers, private entrepreneurs and policy-makers, helped the farmers to learn from other people's experiences, and enhanced the visibility of our project.

Discovery of the valuable white type of garcinia

Interestingly, during one of the village workshops, farmers identified two morphotypes of *Garcinia indica*. The red fruit type is common and generally fetches a price of about US$1.60/kg for dried fruit rinds in the market. Farmers recognized an off-type with pale yellow fruit coloration, which is popularly referred to as 'white'. Local healers prefer it because of its unique pharmaceutical properties. Dried fruits of the white type can fetch more than double the price, at US$3.55/kg. During a planning session, we agreed to conduct a large-scale grafting scheme in our project so that farmers could plant this white type in their forest gardens. Unfortunately, very few white fruit type trees are available in nature; so far we have only discovered 20 white type *Garcinia indica* trees in the entire district of Uttara Kannada.

To increase the numbers available, we trained members of the self-help groups to produce grafted white fruit type seedlings on a large scale and we further encouraged them to cultivate them in their forest gardens.

Local innovation for overcoming problems with the propagation of rare, elite types

One of the important issues discussed during the fair was the failure of grafted seedlings of garcinia to show normal tree growth. Since the species is dioecious in nature, farmers have taken to planting grafted saplings obtained from an elite type to ensure the adult plants possess the desired elite traits and bear fruits. Unfortunately, many of these grafted seedlings have a bushy habit and do not form straight-growing trunks, which is not good for productivity. This bushy characteristic only expresses itself five or six years after grafting, which reduces the motivation of community members. During the fair, one innovative farmer suggested that cutting the top portion of the tree results in the development of many straight-growing stems. He claimed to have obtained many (thousands of) scions this way that had ultimately grown into straight stems. We documented this experience and promoted its further sharing, which had a great impact. Currently, we are recording many such experiences and are exploring further nuances to develop a more sophisticated propagation system based on traditional and scientific knowledge. Such participatory action research contributes to improving the use of garcinia genetic resources.

Establishment of an informal group of grafting experts

Farmers have a flair for recognizing, growing, innovating and exchanging genotypes of several indigenous fruit tree species. Mr Dattatreya Hegde is one such farmer from a remote village in Uttara Kannada district. He thoughtfully allocated his diversity-rich garden to fruit crops 15 years ago and now has some 35 varieties of mango and seven types of *Garcinia indica* spread out across an area of 6 hectares in serene Western Ghats forest-land. Today, almost 20% of his household income is based on selling fruits and grafted seedlings. Every year, he provides training to several farmers on grafting techniques. He is a true custodian of fruit tree genetic resources.

In the garcinia project, we identified and acknowledged several custodians like Mr Hegde. Currently, we are setting up an informal forum of people skilled in grafting. Forum members voluntarily contribute to the identification of rare, special and traditional types of tropical fruits. The forum now includes ten farmer experts, and has already produced over 1000 grafted seedlings. The grafted plants are purchased by the project and distributed to the local communities. Through the organization of skilful and entrepreneurial farmers, our project will be able to make a significant contribution to the conservation of fruit tree genetic resources.

Mitigation of the ecological cost of garcinia fruit processing

The dried fruit rind of *G. gummi-gutta* has become an important commercial product. The ripe fruits are collected during the months from June to August. The fruits are very acidic, which makes them highly susceptible to fungal infestation, so they must be processed immediately after gathering. The fruits are cut in half and the seeds are removed; the cleaned fruits are then spread out on a raised mesh above a fire (Hegde and Vasudeva, 2010). Sun drying is impossible because fruit maturity coincides with the peak of the rainy season, when it rains for more than 25 days a month.

Farmers use three traditional techniques for drying. In the first, the rinds are spread out on a bamboo mat above an open fire, while the second involves drying the rinds on an iron mesh above an open fire. The third technique involves the use of a traditional oven, which is built with mud walls on three sides and an iron mesh top, on which the rinds are spread out to dry. The time and quantity of fuel wood required differs for each of the methods. For the two traditional open-fire systems on average about 22 kg of fuel wood is required to obtain 1 kg of dried rind. However, when a traditional oven is used, the fuel wood requirement is reduced by 50%. We tried out dryers that had been designed by the Centre for Sustainable Technologies (CST) in Bangalore and found them five times more fuel efficient when compared with the open-fire system (Hegde and Vasudeva, 2010).

The cost of producing dried fruit rind using traditional open-fire systems is estimated to be around US$0.75/kg of dried fruits. The use of CST dryers can reduce this cost to just US$0.15. Despite this, nearly 99% of the garcinia collectors in the central Western Ghats continue to use the open-fire system, probably because fuel wood is available free of charge from forests. Clearly there is a huge hidden ecological cost involved in using the traditional method. Fruit processing places a substantial burden on the ecosystem when taking into account the carbon emissions of the open-fire systems (Hegde and Vasudeva, 2010). In order to reduce this ecological cost, the UNEP–GEF tropical fruits project is encouraging the use of community dryers. The project has created a trust fund in each of our three project communities, following a structure similar to the one discussed by Shrestha *et al.* in Chapter 2.9. The fund provides accessible loans to communities for the purchase and installation of community dryers under conditions favourable to environmental protection and garcinia genetic resource conservation. With this intervention, the communities are now moving towards an ecologically sound and economically viable way of processing garcinia resources.

CBM practices that support the sustainable use of NTFPs

Rural communities retain customary rights to collect NTFPs from the wild, thereby contributing to the maintenance of usufruct forest patches in the central Western Ghats. We used an approach inspired by CBM to implement the garcinia project, which focuses on the conservation and sustainable use of garcinia genetic resources. In a relatively short period of time, using a series of adapted CBM practices, we have enhanced the awareness of the community on the forest and its resources, and have strengthened the capacity of rural communities to value and use traditional knowledge. Through the organization of various levels of groups, we were able to enhance the social and institutional capital related to forest management, and promote more sustainable practices in garcinia use. In this way, we were able to adapt the CBM approach to a context of NTFPs, contributing to the further domestication, and promoting the sustainable use, of garcinia. Our work in the sites has become an inspiration to many others in the Western Ghats, and to our partners in tropical fruit tree genetic resources projects in Asia and beyond.

3.8 The sustainable development reserve

An option for securing livelihoods in Imbituba, Brazil

Sofia Zank, Ana Luiza de Assis, Marlene Borges, Nivaldo Peroni and Natalia Hanazaki

Sustainable development reserves and community biodiversity management

Conservation policies and strategies can contribute to securing farmers' livelihoods in areas with important features for biodiversity conservation. In Brazil, the sustainable development reserve (SDR) is a protected area that reconciles biodiversity conservation with sustainable natural resource management in a manner that is mediated by its traditional people, with their local culture and local ecological knowledge. In the current chapter, we look at the development of SDRs in Brazil and discuss their potential for securing the rights of traditional people and local communities over access to territories on which their livelihoods depend. We refer in detail to one case in particular, concerning the establishment of an SDR in Areais da Ribanceira, in the municipality of Imbituba, in the state of Santa Catarina.

Biodiversity conservation and traditional people

For many years, biodiversity conservation was based on a model that ignored the role of its human populations. Until the mid twentieth century, the aim of establishing a protected area was primarily that of protecting the flora and fauna. However, this typical model of biodiversity conservation generated a series of socio-cultural conflicts (Diegues, 2000; Tuxill and Nabhan, 2001; Teixeira, 2005). Assis *et al.* (Chapter 3.2) elaborate on how the critique of this model in Brazil has led to a better understanding, and gradual recognition, of the role of traditional people and their livelihoods, and as such is part of a larger debate on conservation and development.

In Brazil, these intense discussions are associated with those of a more global level that link conservation with poverty alleviation (Adams *et al.*, 2004), emphasize the role of traditional people in the conservation of landscapes (Fisher *et al.*, 2008) and promote the development of approaches to enhance the relationship between biological and cultural diversity (Maffi and Woodley, 2010; Pilgrim and Pretty, 2010). In Brazil, the permanence of traditional people in protected areas has been considered an issue since the 1980s (Vianna, 2008), as demonstrated by the well-known struggle of the rubber tappers and environmentalists in the state of Acre. The rubber tappers' struggle resulted in communities regaining their former access to land and natural resources, and was instrumental in the development of a conservation unit referred to as an extractive reserve (Resex) (aspects of this case are further explained by Assis *et al.* in Chapter 3.2). The SDR was created within a new conservation model that includes rather than excludes traditional people, and fosters the relationship between biological and cultural diversity, and nature and people. Such a model places the communities of traditional people or local communities as central to the management of the landscape and natural resources (Diegues, 2000), thereby creating a legal unit for community biodiversity management (CBM).

SDR: the background to its development

In 2000, the National System of Conservation Units (NSCU) acknowledged the roles and responsibilities of traditional people in protected areas, including categories for sustainable use (Teixeira, 2005). The NSCU defines the goal of the SDR as being – loosely translated – 'to preserve nature while at the same time ensuring that the necessary conditions and resources are in place for the reproduction and improvement of livelihood assets, and the exploitation of resources by traditional people; and to enhance and maintain the knowledge and techniques developed by these people in environmental management' (MMA, 2002a, 2002b). Traditional people are not only recognized as partners of conservation agencies, like the Chico Mendes Institute for Biodiversity Conservation (ICMBio), but also as holders of rights over their territories and the benefits arising from the resources they manage. In this way, the NSCU provides the communities with better guarantees and protection through a legal regime for conservation. Furthermore, it enables traditional people to exercise their rights within a context of access and benefit-sharing (ABS) over genetic resources. Albuquerque *et al.* (Chapter 6.5) address the Brazilian ABS system in more detail.

Is the SDR able to secure livelihoods of farmers and fisherfolk in Areais da Ribanceira?

Description of the context

Restinga, or sand dune vegetation, dominates the coastal area of the municipality of Imbituba, in the state of Santa Catarina in southern Brazil (Figure 3.8.1). Colonizers and pioneers from the Portuguese Atlantic islands of the Azores began to settle in this area in 1715. Over the years, the settlers and their descendants, who also share Afro-Brazilian and indigenous backgrounds, developed their own particular livelihood of small-scale farming and fishing, based on their relationship with the natural environment, for which they are considered traditional people in the Brazilian context (as further elaborated on by Assis *et al.* in Chapter 3.2).

A characteristic feature of the area is that the communities have been engaged in the hunting of the southern right whale (*Eubalaena australis*) since the eigtheenth century. In southern Brazil, the hunting of this whale continued up until 1973, when it had become almost extinct in the area. In 2000, the central-southern coast of Santa Catarina was declared an Environmental Protection Area with the aim of creating a safe habitat for the southern right whale, and of contributing to a form of tourism protective of its habitat.

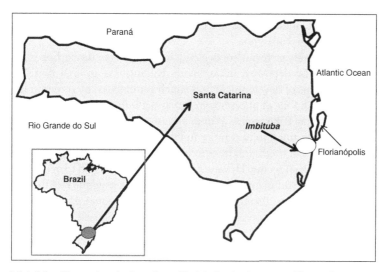

Figure 3.8.1 Map illustrating the location of Imbituba, in the state of Santa Catarina.

Until the 1960s, traditional communities in the coastal region ensured their livelihoods through a combination of agriculture, fishing and hunting (Seixas and Berkes, 2005). Farmers used the *coivara* system ('slash and burn') in the production of manioc and some other crops, allowing for several years of fallow, following which they would burn the branches of regenerated native shrubs before cultivation, to clear and fertilize the farm-land. They cultivated a range of manioc varieties, including both sweet (aipim) and bitter (mandioca) varieties. In addition, the farmers maintained the dense restinga patches from which they extracted fruits from the butiá palm (*Butia catarinensis*) in particular, as well as medicinal plants (Zank and Hanazaki, 2012). The land-use system of coivara, with its associated diversity in manioc varieties, and the traditional management of the restinga patches and their butiá palms, has, over the centuries, contributed to the landscape mosaic of Areais da Ribanceira. Until the late 1970s, the traditional livelihoods of communities in the area continued to reflect the strong associations between the restinga, small-scale farming and extractivism, combined with artisanal fishing.

Urbanization: a threat to traditional livelihoods and the associated landscape

In Imbituba, urbanization began in the 1950s with the construction of a commercial port and the establishment of several ceramic industries. It was further accelerated by the completion of the express highway (BR 101) that links the coastal area near Imbituba with the metropolitan areas of Porto Alegre and Florianópolis. Imbituba, known for its beaches, has developed into a surfing resort that attracts many tourists. As in many such areas along the Brazilian coast, the development of tourism has resulted in intensive land speculation, associated with a vast influx of financial capital and changing lifestyles. Consequently, the livelihoods of traditional people have changed, in particular for those communities closest to the ocean, like the small-scale farmers and artisan fisherfolk (Seixas and Berkes, 2005). The establishment of a carbon steel industrial complex in the port zone of Imbituba further accelerated the process of urbanization.

The local municipality government decided that waste from the carbon steel industrial complex could be deposited in Areais da Ribanceira, an area that is considered marginal in terms of development and which is inhabited by traditional people. Accordingly, 3 million tons of phosphogypsum and a million tons of iron oxide were deposited in Areais da Ribanceira. Within the master plan for industrial development, the area was destined to host more industries. In recent decades, many farming households have lost their land titles following a formal process of expropriation initiated by the industrial sector. However, households continued to occupy the area and use it for crop cultivation and extraction activities. Despite the fact that they had lost their formal land titles, they continued with the traditional practices of manioc cultivation and butiá extraction associated with their livelihoods. These traditional communities provide an ecological service, sustaining the characteristic domesticated landscape through their special relationship with the natural environment.

ACORDI: a community organization leading the struggle

Faced with the threat of being denied access to, or excluded from Areais da Ribanceira, the community has been living with insecurity for years. Decades after the formal expropriation of their lands, they continue to cultivate and maintain their unique landscape. In order to ensure continued access to the territory, the Areais da Ribanceira community established the Rural Community Association of Imbituba (ACORDI) in 2002. The acronym ACORDI is also linked to the Portuguese word *arcorde*, which means 'wake-up'.

Practices supporting community management

A manioc fair has been held annually since 2005. The idea for the fair came from the recognition that it was necessary to publicize the community's struggle to maintain their customs and land rights. It is also an important way of generating income for ACORDI and the community members involved. The idea of creating an SDR to resolve territorial issues arose during the first fair and, over the course of the following fairs, this theme continued to be discussed within ACORDI and with its partners and supporters. In 2010, ACORDI opened a community manioc flour mill and processing centre next to its community centre. The centre allows farmers to process their manioc collectively at a lower cost when outsourcing it to the few families in Imbituba that continue this work.

Project to transform Areais da Ribanceira into an SDR

In August 2005, based on the assessment of several options for securing their livelihoods, through ensuring community access to the territory, ACORDI decided to develop a proposal for creating an SDR in Areais da Ribanceira. In April 2006, ACORDI began the process of submitting such a proposal, in collaboration with the National Centre for Traditional Populations (CNPT). The CNPT was set up within the Brazilian Institute for the Environment and Renewable Natural Resources (IBAMA), with the specific aim of providing such services to communities of traditional people. The area proposed for the SDR includes restinga, agricultural lands and dense forest patches. What is crucial for the establishment of an SDR is that the communities of farmers and fisherfolk associated with ACORDI are using the area in a sustainable manner both in farming and in the collection of butiá and medicinal plants, for sustaining their traditional livelihoods. The total area proposed for the SDR covers approximately 4000 hectares, a part of which overlaps with the Environmental Protection Area of the southern right whale.

Current events in relation to community access to land

In July 2010, a court order, responding to a legal request by the company that had initiated the process of expropriation and that had acquired ownership over a large part of Areais da Ribanceira, issued the farmers with an order to expel them from the area. The court order left the community with access to just 12 hectares, in stark contrast to the 4000 hectares foreseen in the SDR. The order has caused the farmers further insecurity, as they fear losing their investments in agriculture. The residences of four farm households were demolished, crops were destroyed, one of the access roads to Areais da Ribanceira was closed, and areas were fenced off to hinder farmers' access to their crop- and forest-lands. ACORDI members fear the loss of all territory and the demolition of collective structures such as the community centre and manioc processing plant. ACORDI is guiding the community and is putting up fierce resistance to what they consider to be the violation of their rights to sustain their traditional livelihood. The situation has been further aggravated by the installation of a cement factory in Areais da Ribanceira in an area that is considered to be inappropriate for the establishment of industries of that size.

To strengthen its position, ACORDI has developed a close collaboration with a range of institutions, including the Landless Workers' Movement (MST) and the Federal University of Santa Catarina (UFSC). The loss of much of their territory in July 2010 strengthened the mobilization and the struggle of farmers, but it has increased insecurity and farmers are therefore reluctant to invest in crop-land for fear of loss. This situation requires the urgent resolution of the conflict, the provision of guarantees concerning the community's access to land, and recognition of the role of custodians in the maintenance of their territory.

Areais, the SDR and CBM

The establishment of ACORDI, the organization of the manioc fairs and the construction of a community manioc mill and processing centre have been fundamental for strengthening the collective identity of the vulnerable community from Areais da Ribanceira. These practices will only result in the conservation of the landscape and its plant resources if they achieve formal custodianship over their territory. Over the past decades, but most recently in particular, the traditional and informal management of the landscape within the dominant model of urbanization and industrial development has been disregarded. The establishment of an SDR in Areais da Ribanceira will allow farmers to maintain traditional practices for managing the landscape with its traditional manioc varieties, butiá palms and characteristic restinga forests. The protected regime will ensure the conservation of the territory, community members' livelihoods and their access to biological resources. With the creation of an SDR, they will be eligible for public programmes that support their livelihoods, which will further enhance the autonomy of the community in the management and conservation of its landscape with its biological resources, and in the maintenance of the traditional lifestyle associated with its natural environment, which are the main objectives of the Brazilian government for developing SDRs. The establishment of the SDR would create a legal instrument for recognizing the farming community's custodianship over Areais da Ribanceira. It would also be a subsequent step in a long-term learning process of participatory management of the landscape and its biological resources, for the local community, but especially for activists, conservationists, environmental lawyers, academics and policy-makers.

The collaboration with UFSC has helped to enhance the self-esteem of community members through documentation and in publications by recognizing their role in the development of the landscape and its species. The creation of the SDR would bring justice to the community of small-scale farmers and artisan fisherfolk in Areais da Ribanceira, and would formally recognize the role of such traditional people in the sustainable management of our environment.

Part IV

Agrobiodiversity, livelihoods and markets

4.1 Agrobiodiversity, livelihoods and markets

Introduction

Walter Simon de Boef, Marja Thijssen and Monika Sopov

Synopsis

A common feature among conservation and development organizations supporting community biodiversity management (CBM) is that they strengthen the capabilities of farming communities in marketing what we refer to as agrobiodiversity products (i.e. those based on the local crops and varieties that farmers cultivate, and on non-timber forest products that they gather from their agricultural landscape). In many cases, value addition is incorporated into the CBM processes to increase the income generated from this specific livelihood asset of farming households or communities. Market-based actions have become part of the CBM methodology to create incentives for farming communities to continue to use local crops and varieties (Jarvis *et al.*, 2011). The chapters in Part IV share some experiences from Asia and Europe; a number of chapters throughout this book approach the value addition and marketing of agrobiodiversity products in Ethiopia, Ecuador and various regions of Brazil as an individual component of a larger process of CBM or participatory crop improvement.

Experiences are initially divided into two areas, according to whether agrobiodiversity products are approached either at the level of species or variety. Bhandari *et al.* (Chapter 4.2) introduce the first case study, sharing the experience of supporting groups of women farmers in value addition and marketing products of two different citrus species in Nepal. King *et al.* (Chapter 4.3) present a case study from India, where tribal farmers in the Kolli Hills, in Tamil Nadu, are able to generate income through value addition on several millet species. The authors show that when marketing traditional and healthy products from what are referred to as neglected and underutilized crops, groups of often poor women farmers face unfair competition from subsidized food products of major crops, such as rice and wheat, which undermines the ability of such small-scale farmers and their agrobiodiversity products to take a fair share of the market. Other chapters in the section share efforts made to explore the potential of locally known and appreciated varieties. Chaudhury and Swain (Chapter 4.4) share the case of Kalajeera rice, in the Jeypore Tract in India, highlighting the fact that the value addition of products of local varieties may be supported by their participatory genetic enhancement. Through marketing the grain of this variety at regional premium markets, the poor tribal farmers that are known as the custodians of rice genetic diversity are recognized and receive direct 'monetary' benefits.

Another case study shares the experience of farmers in the Lubéron region, France, who are engaged in the dynamic management and marketing of a local wheat variety,

Meunier d'Apt. Farmers have joined together with artisan millers and bakers to protect their product through the application of a certificate of origin that associates them and their product with the Lubéron regional natural park. This case study shows that small-scale farmers, in a marginal production environment, are able to sustain their livelihood by matching a traditional and local product to modern lifestyles that appreciate original, organic and locally produced food (Kastler and Moÿ, Chapter 4.6). The aspect of protection as a means to guarantee specific markets is addressed in more detail by Thomas and Anh (Chapter 4.5), who describe the market potential of local rice varieties in Vietnam that were improved through participatory genetic enhancement, echoing the case of Kalajeera rice in India. The producers need to be well organized to ensure that the quality of the product can be discerned by the customers, and also that the product can be distinguished by customers from those of its competitors. It is at this point that many of the CBM projects face the following dilemma: should they strive for an optimal or maximum market potential?

Agrobiodiversity products are characterized by their uniqueness and, owing to their specific adaptation, they often have limited production potential and will never be produced in large quantities. For these reasons such species and varieties are addressed in CBM processes; we need to target specific, well-defined, and not necessarily large markets.

Agrobiodiversity and livelihoods: a new focus in market development

The case studies in this section approach value addition and marketing with the aim of contributing to the livelihoods of small-scale farming communities as part of larger CBM processes. In a synthesis of the case studies, we identified the following five common topics: agrobiodiversity as a livelihood and market asset; local and specific markets; social organization as a market advantage; challenges in entering the value chain; and the custodian's position in the value chain and market. In Figure 4.1.1, we illustrate the relationship between CBM and these five topics, which are subsequently discussed in more detail, and linked to relevant experiences shared in the book.

Agrobiodiversity as a livelihood and market asset

Market development is incorporated in the CBM methodology because of its potential to contribute to the conservation of agrobiodiversity. Markets may be accessed for specific species, enhancing their value through emphasizing their association with a unique landscape or production system. An example can be seen with the marketing of the 'tea' leaves of native yerba mate plants that are harvested in collectively managed forests of the faxinais in Brazil, and which obtain higher prices in nearby urban markets when compared with the cultivated product (Reis *et al.*, Chapter 3.5); or where coffee beans gathered from natural coffee populations in the Afromontane forests in Ethiopia target global premium markets (Feyissa *et al.*, Chapter 3.6). Products may be specific varieties, such as local varieties of rice, for example Kalajeera in the Jeypore Tract in India, Tam Xoan in the Red River Delta in Vietnam (Thomas and Anh, Chapter 4.5), and Pokhareli Jethobudho in Kaski

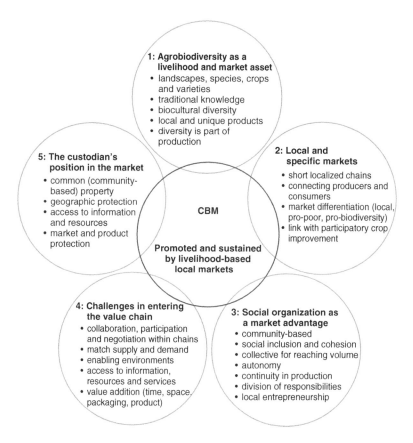

Figure 4.1.1 Value addition and marketing of agrobiodiversity products for promoting and sustaining community biodiversity management, as characterized by five topics.

District in Nepal (Silwal *et al.*, Chapter 5.5). Other examples address the processing of unique species and varieties by farming groups or associated artisan food producers into special products, such as:

- flour or bread made from the Meunier d'Apt traditional wheat variety from Lubéron, France (Kastler and Moÿ, Chapter 4.6);
- chuk and squash, traditional products made from citrus species, processed by women farmers' groups in Ghanteshwor, Nepal (Bhandari *et al.*, Chapter 4.2);
- Andean blackberry marmalade and spicy pepper pastes, produced by members of the conservation cooperative in Cotacachi, and marketed at tourist markets in nearby Otavalo and Quito in Ecuador (Tapia and Carrera, Chapter 2.3);
- umbuzada, a delicacy based on the fruits of the umbuzeiro tree and prepared with milk and sugar, marketed in nearby rural towns in the semi-arid region of north-eastern Brazil, but also reaching people with north-eastern backgrounds as far afield as São Paulo (Peroni *et al.*, Chapter 3.4).

All these products are produced by small-scale farmers, often groups of poor women and/or tribal farmers, and are based on unique and often rare agrobiodiversity. The product's characteristics and the uniqueness of the genetic resource base have major implications on how we approach the market and characterize the relationship between the market and those farming communities. From a distant point of view, the cases shared in this book all emerge as product development driven by conservation and development objectives (i.e. supply driven rather than responding to market demand), which, using common marketing logic, may reduce the viability of the whole enterprise. However, these cases are distinct from many other experiences where chains are built on market demand. In the case of agrobiodiversity products, they enter the market with the push of a project or subsidy, rather than as a response to specific customers' demands (i.e. market pull). It is indeed true that what is driving the marketing of products is an interest in contributing to agrobiodiversity conservation, livelihood development and the alleviation of poverty, rather than meeting the existing demands of potential customers. We know only too well that within the organization of the globalized agro-industrial food market system, with its agricultural system of monocultures and uniformity that is based on just a few global crops, farmers have been transformed into the sole producers of primary materials. Our goal is to link agrobiodiversity and livelihood development through CBM, creating alternatives to the global agro-industrial food market system by stimulating local products for local markets, with perhaps traditional and even new products, and by targeting consumers who recognize and appreciate those products and their provenance. In this way, we show that CBM does not stand alone in a context of conservation and development; it is linked to a much larger social and political debate concerned with defining the role of an estimated 500–800 million small-scale farmers and their communities in agricultural production and food market systems.

Table 4.1.1 compares the models. We realize that the comparison may be exaggerated, but it positions the diversity of experiences where farming communities, supported by conservation and development organizations, are engaged in processes of value addition and marketing their local agrobiodiversity products, as part of a larger discussion on the organization and structure of the global agro-industrial food market system versus the local livelihood-based food market system.

Local and specific markets

Most of the case studies seek opportunities in niches of local, regional or, in some cases, national or global markets. Structures or incentives for investment are created through CBM programmes, and those external organizations that facilitate this process must have a clear vision of sustainability, thus avoiding artificial, project or subsidy-based marketing mechanisms. These organizations must therefore seek partnerships, particularly with the private sector or marketing bodies within government structures, and invest in community organizations and their capacity-building, linking them to viable business partners, and avoiding in this way unsuitable marketing mechanisms. Of key importance to farmers' organizations engaged in the production, processing and marketing of agrobiodiversity products for local, regional or specific markets is that they have access to information and resources on specific

Table 4.1.1 Comparison between global agro-industrial and local livelihood-based food market systems

Characteristics	Global agro-industrial food market system	Local livelihood-based food market system
Scale, location and approach	Global, large, uniformity, location-neutral	Local, regional, small-scale, diversity, location-specific
Goal	Generation of private, corporate or state wealth, ensuring sustainable supply of raw materials	Sustainable livelihoods, and thereby providing several services to society
Source of innovation	Science-based; location-neutral and new; up-scaling	Embedded in tradition and culture; locally-based and the result of long-term processes, matching producers and consumers, science and local innovation based, and up-scaling innovation processes rather than products
Market orientation	Profit: dominance of economic capital	Livelihood: balance between social, economic and environmental capital
Approach to resources	Control of energy resources	Sustainable use of natural and biological resources
Inputs	Defined by global technologies	Based on local technologies
Value chain and scale	Integrated, globalized and long, driven by economies of scale	Specialized and locally adapted, (g)localized and short, driven by economies of consumer and producer relationships
Regimes defining resource use	Private and state property regulated by globally harmonized laws; governance within the chain through monetary means	Common property regimes that are community-based, regulated by customary law; governance by relationships
Power relationships	State and private business control, backed up by bureaucratic and political power; lead firm in charge	Social organization of communities and forms of reciprocal benefits for, and relationships between, stakeholders; led through partnership
Decision-makers in value chains	Economically-tied individuals, shareholder/corporate tied or state decision-makers acting in global lead firm	Tight-knit social organizations, closely interacting with society and acting local actors/partners
Production and conservation	Separation of production and conservation; diversity is avoided	Integration of production and conservation; diversity is used
Approach to diversity	Agricultural uniformity, few crops and varieties; retailers define diversity for consumers	Diversity in local landscapes, crops and varieties define diversity for consumers
Function of farmers	Producers/processors of primary materials in globally organized food systems	Providers of social, economic and ecological services within local food systems; and sustaining agricultural food and production systems

Table 4.1.1 (Continued)

Characteristics	Global agro-industrial food market system	Local livelihood-based food market system
Relationship between consumers and production, and producers	Disconnected from production and producers	Connected to the livelihood of producers
Consumer behaviour	Influenced by marketing strategies	Defined by conscientious understanding of production and appreciation of producers
Values	Price and market influenced behaviour	Religious, spiritual, cultural and social values attached to production and products

Source: Authors; adapted from Borrini-Feyerabend *et al.* (2007).

requirements, such as product quality, packaging standards and other marketing information, and, where appropriate, are capable of seeking the services needed to support them. Equally important is that the farmers have good access to market information, and it is here that conservation and development organizations are often in a position to create direct partnerships with processors, retailers or consumers.

Since the products are agrobiodiversity-based, they should be distinguished from products resulting from the global agro-industrial market system; several differentiation mechanisms are shared in the case studies. Because of their limited volume and embedding in local food cultures, most agrobiodiversity products target local rural or nearby urban markets. Examples can be seen with the healthy and nutritious millet products from the Kolli Hills, which reach rural urban markets as well as the major cities of Tamil Nadu, in India; citrus chuk and squash from Ghanteshwor, which reach rural and nearby major cities in western Nepal; and yerba mate tea, for which the leaves are harvested from natural rather than cultivated populations in the faxinal forests in southern Brazil. Premium markets that may be interesting options include fair trade or organic markets, or those that target products carrying certificates of origin or geographic denomination; each has its own structure and organization for certification and market differentiation. Such markets do, however, place additional requirements on the degree of sophistication of the producer's organization and the value chain. Basically, what these premium markets do is they shape new arrangements between consumers and producers, revitalizing the connection that has been lost in the global agro-industrial food market system.

These new markets are emerging and are definitely interesting for agrobiodiversity products; however, as the cases in this book show, the first option is to target local markets, which is also more practical because of the often-limited production capacity.

Social organization as a market advantage

We have already emphasized that producers, by being small-scale farmers, need to join forces in order to generate volume and continuity in production. This requires the establishment of specific groups, like, for example, the self-help groups in southern Asia, who work together in the joint production, processing and marketing of their products. Such groups must set up some kind of legal entity within the larger community-based organizations (CBOs) for seeking access to technical, financial and marketing information and resources. In conditions like those created by the agrobiodiversity project in Nepal, such self-help groups become eligible for microfinancing, for example through the CBM trust fund (Shrestha *et al.*, Chapter 2.9). In this manner, value addition is promoted as a means to create opportunities for income generation for vulnerable groups in communities. Consequently, this type of entrepreneurship contributes to social and often gender inclusion, and, when undertaken within a larger structure of a CBO, this contributes to social cohesion. Once the CBOs are able to facilitate and support this type of development process, in which they relate the promotion of unique local agrobiodiversity assets with aspects of social inclusion, they are generally at a stage where they know how to access the required information and resources by themselves, and no longer need external facilitation in the CBM process.

One of the case studies from Brazil, the Seed of Passion network, provides an interesting example of how continuity in production can be achieved (Dias, Chapter 2.7). This network of community seed banks is engaged in producing seed and grain of local varieties for the institutional market (i.e. the government food purchase programmes that supply public entities such as schools and hospitals). Through this programme, which is positioned at the heart of the programme of the Federal Government of Brazil to combat hunger and poverty, cooperatives of seed producers of local varieties, and subsequent grain producers, deliver an agreed quantity of food grains to government institutions in a continuous flow, but according to initially favourable market conditions. In this manner, the food purchase programme develops the capabilities of those farming cooperatives for meeting demands of future large-scale business partners for quality and continuity, creating a learning environment for entrepreneurship (Grisa *et al.*, 2011). This example shows that when linking conservation, livelihood and social programmes, win–win situations can be created in which all parties benefit.

Challenges in entering the value chain

With the example of Brazil, we reach the stage where farmers' groups enter into value chains. In order for them to become viable entrepreneurs in those value chains they need the capabilities to collaborate, participate and negotiate with other business entities and service providers, such as quality control entities, government bodies that facilitate marketing or auditing services for cooperatives, or banks or credit programmes that provide financial services. Such involvement in value chains puts significant demands on the capacities of these oftentimes small-scale farmer and community-based organizations. However, the fact that the CBM methodology invests much in CBOs appears to be an advantage for such organizations.

In approaching value chains, the farmers' groups encounter a institutional and legal environment that aims to facilitate the flow of products through the chain. CBOs require the capacity to understand and develop the modalities to respond and adapt to, as well as innovate within, the rules of the game of value chains and the marketing of their products. By focusing on agrobiodiversity products, farmers may face institutional obstacles when operating in what are considered the common and 'normal' value chains. For example, the wheat variety Meunier d'Apt, which is cultivated by a farmers' group in the Lubéron in France, does not meet the standardization requirements for its release and registration, and as such the sale or sharing of its seed is illegal. Through its participation in the Réseau Semences Paysannes, the farmers' group was able to come to an innovative arrangement through a membership scheme that allowed the dissemination of seed of the Meunier d'Apt variety.

It is here that conservation and development organizations who support this type of entrepreneurship need to play a facilitating and catalysing role, providing services to strengthen the capabilities of the farmers' groups not only in production, processing and/or marketing, but also their engagement in value chains, and in contributing to and facilitating innovation when faced by institutional or regulatory obstacles. The niches in which these groups operate are beyond the mainstream globalized agroindustrial food market system and enabling environment, as shown in the case of the

Lubéron. As such, they need to innovate in order to bypass the obstacles that block the marketing of agrobiodiversity products.

It is vital that, from the onset, external organizations aim to secure the autonomy of farmers' groups, thereby ensuring the long-term viability of their entrepreneurship. To achieve this, the organizations need to engage in a business rather than development type of relationship, as can be learned from the food purchase programme in Brazil. What this programme also does is it contributes to the development of short or locally based value chains. Local consumers, such as children in schools, appreciate local products. As such, a significant contribution is being made by the government towards the establishment of future local food systems and short value chains (i.e. where consumers are aware of and appreciate the origin of the products they consume, and farmers know and have a more direct relationship with their customers, which in the end contributes to creating a space for local livelihood-based food market systems) (Table 4.1.1).

The custodian's position in the value chain and market

Most communities approach genetic resources – crops, varieties or species – in their surrounding landscapes, as traditional and therefore common property. By marketing these products or even engaging in further genetic enhancement, the communities may seek protection or exclusivity in the market. Kalajeera is a pioneering example of a farmers' variety that was formally recognized by India's Protection of Plant Varieties and Farmers' Rights (PPVFR) Act, following a process in which the M.S. Swaminathan Research Foundation (MSSRF) played an important role as a catalyst (Bala Ravi, Chapter 6.3). Another example can be seen with Jethobudho Pokhareli, which has been formally released as a variety in Nepal. However, varieties resulting from participatory genetic enhancement in Vietnam do not achieve such recognition or release, as highlighted by Thomas and Anh (Chapter 4.5). Similarly, farmers and scientists in Central America encounter obstacles blocking the release of varieties from participatory plant breeding (Alonzo *et al.*, Chapter 5.8).

Such obstacles do not necessarily impede farmers' access to seed, which they can achieve through the informal seed system, but they do prevent farmers from accessing services such as maintenance breeding and the production of breeders' seed. Santilli (Chapter 6.7) and Kastler (Chapter 6.8) illustrate how seed and variety laws, in Brazil and France respectively, become obstacles to small-scale farmers who, with their organizations, engage in participatory crop improvement and promote the use of local varieties. Consequently, farmers' organizations avoid seed and variety laws, and, in the cases of the Tam Xoan variety in Vietnam and the Meunier d'Apt variety in France, they seek to protect the product of their variety with intellectual property mechanisms that are associated with geographic protection or regional denomination. It is interesting to observe that trade laws provide farmers with protection when seeking market benefits from the genetic resources they maintain. These aspects are related to two of the four elements of farmers' rights included in the International Treaty on Plant Genetic Resources for Food and Agriculture (ITPGRFA) (i.e. the protection of traditional knowledge, and the fair and equitable sharing of benefits).

The PPVFR Act in India is one of the few examples where such a legal space has been created. In addition to plant breeders' rights, variety and seed laws in most countries create obstacles to promoting the use of local varieties for market purposes. NGOs, such as MSSRF in India, or civil society networks, such as the Réseau Semences Paysannes in France, play important roles, first, by providing access to the resources at community level and supporting farmers' actions in the marketing of agrobiodiversity products, and second, through providing advocacy, fostering innovation and creating the legal and institutional space for market systems that value agrobiodiversity.

Lessons learned: livelihoods, agrobiodiversity and markets

Within a CBM process, contributing to conservation and livelihood development, and encouraging and strengthening farming communities to engage in value addition and in the marketing of agrobiodiversity products, seem to be logical steps. In learning lessons from the experiences shared in the book, we realize that value addition or marketing actions do not stand alone; they relate to topics such as market strategies, social organization, value chain development and intellectual property mechanisms (Figure 4.1.1). One important lesson is that from our familiar domain of conservation and livelihood development, we can now relate to a larger discussion that addresses the role and space of small-scale farmers, their livelihoods and agrobiodiversity in different food systems. Here, we meet the dichotomy between the global agro-industrial and the local livelihood-based food market systems (Table 4.1.1). It is clear there is no simple solution or pathway; the two systems are worlds apart. But just the realization that multiple food systems exist and may complement each other is in itself a great achievement. Such a position creates room for incentives and the space for discussing and transforming legal and regulatory frameworks to address dissimilar systems, no longer favouring one system while creating obstacles for the other. In any case, such processes will be very complex, as they involve imbalances in power relationships that are linked to established networks of political and corporate control over food, and which must be addressed in direct relation to the governance of plant genetic resources (De Boef *et al.*, Chapter 7.1).

A number of examples in this section are encouraging, detailing conservation actions, and the construction of new short value chains in which the food producers and consumers reconnect. Producers and consumers attempt to re-establish a local food system, thereby opting to divert from the dominant globalized agro-industrial food market system. In the world of today, civil society organizations in emerging economies, such as Brazil and India, support the development of alternative food systems that are complemented by global movements associated with the Slow Food and Transition Towns. At the same time, we can see that over the past 5–10 years a reconfiguration in the structure of the value chains for global commodities, such as cocoa and coffee, has been motivated by discussions on fair trade and sustainability. Multinational companies, such as Unilever and Mars, are gradually engaging themselves in fair trade and sustainability, motivated by a business strategy that is aimed more at achieving sustainability in the long term, than by short-term price-conscious consumer behaviour. As such, we can see that if we consider the two market systems presented in Table 4.1.1 as the extremes, new arrangements emerge in the grey zone in between the two systems. If we follow our perspective of CBM and livelihood development, we may be able to contribute to the creation of markets that are 'pro-diversity' or 'pro-farmers' livelihood', or in some cases 'pro-poor'.

When value-addition initiatives and the marketing of agrobiodiversity products are embedded within CBM processes, we can strengthen several key assets, such as the awareness and understanding of diversity, and the level of organization in communities. However, in our actions as conservation and development organizations, particularly with regards to facilitating the marketing of products or setting up value chains, it should be clearly understood from the start that the groups of farmer entrepreneurs must remain autonomous. We need to be continuously aware of our exit strategy to avoid the 'dependency syndrome', thereby avoiding the creation of subsidy-based markets, which would as such be unviable in the long term. As with the CBM methodology, our aim is to contribute to strengthening the capacity of the communities to initiate and engage in processes of value addition and marketing of agrobiodiversity products in a collective and purposeful manner, thereby fostering their agency in terms of participating within the economy (De Boef *et al.*, Chapter 7.2).

It is when such a situation is achieved that value addition can be considered as contributing to livelihood development and to the conservation and use of agrobiodiversity. After all, we are referring to farming communities as the custodians of agrobiodiversity; and when they engage themselves in value addition and marketing, they obtain direct benefits from the diversity they maintain. In such a manner, this pathway creates an immediate benefit for the livelihood and the community, from their agrobiodiversity. If we consider this a contribution to benefit-sharing, the result obtained in benefits as shared in this book certainly outnumbers those achieved through formal systems, facilitating access and benefit-sharing over genetic resources. Thus, promoting markets for agrobiodiversity products produced by small-scale farmers in an active manner emerges as a promising and more forceful strategy for contributing not only to the conservation and use of agrobiodiversity, but also to the fulfilment of the third objective of both the Convention on Biological Diversity (CBD) and the ITPGRFA, by ensuring the equitable and fair sharing of benefits arising from the use of genetic resources.

4.2 Value addition and marketing of local citrus products in Nepal

Bharat Bhandari, Bijaya Raj Devkota and Sajal Sthapit

Value addition: sustaining the community management of local species and varieties

Farmers grow a diversity of crops and varieties with different traits, fulfilling multiple needs in relation to their socio-cultural, economic as well as agro-ecological contexts (Rijal *et al.*, 2000). Even so, they only maintain those crops and varieties that they see as having specific domestic or market use values. Community biodiversity management (CBM), through value addition and product promotion, aims to enhance such values, thereby supporting the continued cultivation and use of threatened or rare varieties and species. The CBM project of Local Initiatives for Biodiversity, Research and Development (LI-BIRD) in Nepal, which is described in more detail by Shrestha *et al.* in Chapter 1.2, has been working on value addition as one of the strategies to increase the use value of local citrus species. As part of this project, we aimed to create economic incentives for conservation through use of local citrus species, by processing citrus fruits into less perishable products with a high market demand, and by supporting the marketing of these citrus-derived products. This chapter shares our experiences in Ghanteshwor, in Nepal.

Ghanteshwor, its citrus diversity and the CBM project

Doti district lies in the remote hills of western Nepal (Figure 4.2.1) and is known for its potential for cultivating citrus fruit species, including lime (*Citrus aurantifolia*), lemon (*Citrus limon*), sweet lime (*Citrus limetta*) and mandarin (*Citrus reticulata*). Citrus is an important fruit tree in the home gardens of the mid-hills of the country; its fruit is eaten fresh, processed into juice or eaten as pickle or salad. Many citrus species have traditional, medicinal values, and fruits and leaves are used during festival rituals. In general, the women manage the citrus trees, and together with their children take care of fruit harvesting and selling. Whereas sweet lime and mandarin can easily be sold in the local market, fresh lime and lemon are difficult to sell, fetching low prices in the production season when their fruits are abundant. Huge quantities of lime and lemon are wasted every year. Local processing of citrus fruits into products that can be kept for a longer time, and for which there is a high demand in rural areas as well as in nearby cities, makes good use of such resources that might otherwise be lost. The processing of fruits has the additional benefit of generating local employment

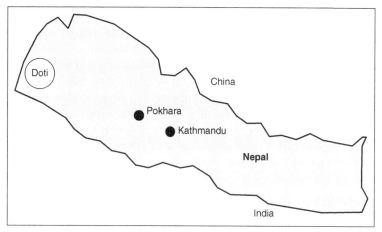

Figure 4.2.1 Map showing the location of Doti district, in Nepal.

and raising the income of those community members involved. As such, farmers in Ghanteshwor, in Doti district, where more than 400 families cultivate citrus species in their homesteads, asked LI-BIRD for support in the establishment of a processing plant for lime and lemon processing.

Citrus value addition and marketing

Step 1: Feasibility study

We carried out a feasibility study in Ghanteshwor in 2009, to assess the area coverage of lime and lemon, the number of plants, and the percentage of plants that bear fruits in each of the nine wards (a 'hamlet' or 'sub-village', the smallest administrative unit in Nepal) that are part of Ghanteshwor. Based on these parameters, we estimated the annual fruit production. In addition, we explored farmers' interest in future citrus cultivation, and studied the potential of marketing lime and lemon fruits, and their various processed products. Based on the study, we calculated that Ghanteshwor produces 5.2 tons of lime and 20 tons of lemon annually. We estimated that 250 ml of juice can be extracted from 1 kg of lime or lemon, and concluded that the establishment of a community processing plant seemed technically feasible and economically viable. This is further supported by the possibility of collecting additional fruits from the surrounding villages of Doti and Dadeldhura districts.

Step 2: Institutional set-up and division of responsibilities

One of the key principles of CBM is to build and strengthen the capacity of local institutions to make their own decisions in relation to the management of biodiversity (Thijssen *et al.*, Chapter 1.1). As such, we supported the establishment of a CBM-oriented community-based organization (CBO) in Ghanteshwor, the Village-level Biodiversity Conservation and Development Committee (VBCDC), which includes representatives of farmers' groups from the wards. We trained the VBCDC members in technical, managerial and financial issues. In addition, we agreed to: support the VBCDC in the establishment of a citrus processing plant; train its members in the management and operation of the plant; link the VBCDC to government agencies; and provide services for promoting and marketing the produce. The VBCDC agreed to promote the proper establishment of fruit orchards, coordinate fruit collection, sustain the processing plant, determine appropriate prices for fruits and processed products, and coordinate the sharing of benefits with participating farmers' groups organized at ward level. A sub-committee of three members within the VBCDC was appointed to coordinate the processing activities. Farmers' groups organized at ward level assumed responsibility for the collection and transportation of fruits.

Step 3: Establishment of the processing plant and identification of products

In November 2009, the VBCDC, together with LI-BIRD and partners, established a small citrus processing plant to produce two products: *chuk*, a dark and viscous, traditional vinegar made from lime that has culinary and antiseptic uses, and lemon squash, a concentrate that is mixed with ice and water to make cold lemonade. The idea was to use lime and lemon by-products, after the extraction of juice, to make pickles. The plant was equipped with two juice-extraction machines for lime and lemon, utensils for boiling and a bottle-sealing machine. In addition, 2000 half-litre bottles and printed labels were provided.

Step 4: Strengthening of farmers' skills in citrus processing

During that same period, we developed a relationship with the Regional Food Quality Control Office, a government authority responsible for controlling these types of businesses. They supported us in carrying out training in citrus fruit processing and in the use of by-products. Ten people identified by the VBCDC were trained in the theory and practice of fruit juice extraction, juice preservation, vinegar and squash making, product quality issues, and labelling, storage and packaging techniques. Farmers were also trained in the techniques of producing pickles from lime and lemon after juice extraction.

Step 5: Enhancement of the capacities of farmers to establish citrus orchards

Citrus plants have the potential to produce fruits for 30–40 years. However, the trees in Ghanteshwor generally die within 8–15 years. This short lifespan is mainly because the seedlings are often planted in soil that is not appropriate for the cultivation of citrus (e.g. white soil, which contains high quantities of calcium, or soil with sedimentary rock below the soil surface, which is very common in Ghanteshwor). We encouraged farmers to use a more sustainable citrus production system by providing training and practical guidance on subjects including orchard layout, good soil management practices, appropriate locations for planting trees, and transplanting and mulching techniques. The VBCDC realized that if the community had its own seedling nursery it would be able to meet local demands for lime and lemon saplings. Community members selected fruits from the best trees and collected seed locally in preparation for the establishment of the nursery. In 2010, the saplings were ready to distribute in the village. In this way, they were able to distribute 3600 citrus seedlings in the area.

Step 6: Initial production and marketing of citrus products

During the first year of operation, 2009, the citrus processing plant produced and sold 200 bottles of chuk and 300 bottles of lemon squash. In addition, the local women's groups produced 265 kg of spicy and sour pickles (Table 4.2.1). The chuk and lemon squash were sold in half-litre bottles, for an average price of US$1.25 and US$ 0.90, respectively. The price set was based on the production costs plus an additional margin of 20%. Marketing the processed products, particularly the pickle, was not as easy as we expected. With our help, 50% was sold in Pokhara; the rest was sold in local shops and through shops in nearby towns. Because of the poor quality of the pickles it was difficult to market them.

Step 7: Adaptation of products and improvement of marketing

In 2010, the VBCDC specifically focused on strengthening the marketing of their products. We assisted in exploring new markets and establishing new market links. VBCDC members visited some of the markets of western Nepal, such as Dhangadi, Attaria and Dadeldhura; they interacted with interested salesmen and shopkeepers at strategic locations (e.g. along highways) and followed up by sending out samples

Table 4.2.1 Production and marketing data for chuk, lemon squash and pickles: Nepal, 2009 and 2010

Production and marketing data	Units	Product and production year				
		Lime and chuk		*Lemon squash*		*Lime and lemon pickles*
		2009	*2010*	*2009*	*2010*	*2009*
Fruit price	US$/kg	0.05	0.10	0.10	0.20	—[b]
Quantity processed	No. of bottles or kg[a]	200	75	300	15	265
Price of processed product	US$/bottle or kg[a]	1.25	2.50	0.90	0.95	0.70
Households involved	No.	200	120	200	15	
Sold through local shops	% of total production	50	100	66	100	25[b]
Sold through wholesalers	% of total production	50	0	34	0	—

Source: Bhandari (2012).

a Amount of chuk and squash is indicated in half-litre bottles; amount of pickles is indicated in kg.
b Because of quality problems with the pickles, 75% remained unsold in 2009; no pickles were produced in 2010.

for promotion. The market assessment showed that the products have a high market potential, but that more effort needed to be made to improve quality, including packaging quality and product information. In 2010, we assisted the VBCDC with improving their product packaging.

In 2010, the VBCDC did not have any difficulty in selling the produce, both at its own selling point and in local shops in Ghanteshwor. Prices were higher, as is illustrated in Table 4.2.1; however, due to a bad fruit season, fruit yields were much lower, which was reflected in the amount of chuk and squash produced. The VBCDC decided not to produce pickles because of the lack of capacity to produce a quality product.

Future steps to expand and sustain markets

To perform well in a growing market and sustain an acquired position, the VBCDC depends on an adequate quantity and quality of fruits, and on a reliable and continuous supply. In the CBM project, we are putting a lot of effort into ensuring the appropriate fruit volumes can be obtained, by reaching out to neighbouring villages, in order to be able to sustain a continuous supply of processed products in the market. Up to now, the small volumes of fruit have limited citrus processing to a seasonal activity.

By maintaining regular contacts with the Regional Food Quality Control Office, the VBCDC is making a continuous effort to improve the quality of its products. A team of VBCDC members have been trained in post-harvest handling, and are now capable of maintaining strict quality control procedures for processing, preserving and packaging. Consumers do appreciate the quality of the chuk and lemon squash. However, the quality of the pickles is still below expectation and needs much improvement. There is also room for improvement in the packaging and labelling of chuck and squash. Consumers, retailers and wholesalers have indicated that besides the half-litre bottles, 250-ml and one-litre bottles should also be available. All bottles should be filled with exactly the same volume, and should provide full information on content. The VBCDC is currently working on these issues.

Through the marketing of chuk and lemon squash in nearby cities, consumers are able to access unique products that are associated with the traditional food culture of Nepal. In addition to this, the marketing of these products in Ghanteshwor enhances the knowledge and skills related to the production of traditional food items. Since they are the only producers of these products in the far western part of Nepal, the VBCDC has a strong market position. To expand and sustain the market for the products, sufficient efforts must be made to raise consumer awareness and promote the products. Product promotion is now being addressed through participation in fairs and festivals as well as through radio advertisements.

Value addition in a context of CBM

Creating economic incentives is generally an effective means to promote the use of specific crops and varieties. Jarvis *et al.* (2011) demonstrate that in many parts of the world market-based actions and incentives are vital for motivating farming communities to continue to use traditional or local crops and varieties, as can be seen in the various practices of our CBM project (Gauchan *et al.*, 1999). Value addition, through the processing of local crops and varieties, and establishing market linkages for the derived products, can provide such market-based incentives. Access to markets for their produce provides rural farmers with opportunities to generate household income and to benefit economically from local agrobiodiversity.

In the specific case of citrus processing and the marketing of its products, the farmers in Ghanteshwor were able to earn some extra cash income. They also benefitted from increased prices for lime and lemon (see Table 4.2.1). Two hundred households, of which 30% are poor and marginalized farmers, directly benefit from access to the local citrus processing plant. The community worked closely together in establishing fruit nurseries and organizing the group collection and marketing of fruits to the VBCDC, which has strengthened social cohesion. The processing and marketing of citrus-derived products has motivated farmers to better manage their orchards and increase the plantation of lime and lemon. The value-addition initiative allows farmer producers to gain an income from, and promote, species that were previously neglected or underutilized. As such, our project has been able to link conservation with livelihoods through increased market access.

To sustain the success of this initiative, the full commitment and ownership of the VBCDC, good collaboration among participating farmer groups and continued support from external organizations will be needed for the coming years. Building the capacity of community-based organizations and their networks in marketing management is vital for ensuring the sustainability of the value-addition efforts. Strong linkages and transparency between producers, consumers and intermediaries in the value chain may increase efficiency and benefits for both producers and consumers.

4.3 Creating an economic stake for conserving the diversity of small millets in the Kolli Hills, India

E.D. Israel Oliver King, S. P. Bala Ravi and Stefano Padulosi

The adaptive potential of neglected and underutilized crops

The general trend in agriculture, particularly concerning grain and cash crops, is an increasing shift towards monoculture, with the focus on a few, high-yielding varieties and hybrids. Thus, agriculture is becoming dependent on just a handful of cereal crops for meeting the demands of the food supply, resulting in a diminishing genetic diversity both on the farm and in dietary habits. According to the Food and Agriculture Organization of the United Nations, three crops – maize, wheat and rice – make up an estimated 87% of all food grain production. This has led to the neglect of a large number of diverse crops that have been contributing to local food security, playing an important role in the livelihoods of local communities in many developing countries. These neglected crops are of particular importance to those living in the semi-arid, mountainous and remote regions of the tropics and subtropics, where crops with a high genetic diversity, suited to stringent ecological niches, have evolved over the years. Most of these regions, besides being hotspots of genetic diversity, are burdened with high populations, poverty and frequent food shortages. Oftentimes, the hardy, adaptive traits of these local crops can save the people from total starvation. Under these conditions, increasing the competitiveness and improving the economic viability of these neglected crops is a must for promoting their effective utilization (Padulosi *et al.*, 2008).

The global initiative on neglected and underutilized species

The M.S. Swaminathan Research Foundation (MSSRF) supports programmes on community biodiversity management to improve the livelihoods of marginalized tribal and small-scale farmers, who are the principal custodians of agrobiodiversity. In this chapter, we introduce those CBM efforts of MSSRF and partners that aim to increase the economic potential of small millets as an important group of neglected, underutilized species (NUS) through value addition and marketing. These activities are part of a global NUS initiative, sponsored by the International Fund for Agricultural Development (IFAD) and coordinated by Bioversity International. The NUS initiative is linked to the MSSRF project 'Integrated Management of Biodiversity Resources in Partnership with People', which is being implemented in tribal areas of India, with support from the Swiss Agency for Development and Cooperation (SDC).

Millets and their value in the Kolli Hills

The Kolli Hills are located in the eastern part of Namakkal district, in the state of Tamil Nadu in southern India (Figure 4.3.1). The area covers about 440 km², is at an altitude of between 1000 and 1350 m above mean sea level, and is inhabited by the tribal group called Malayalis. Within this small area, the Kolli Hills offer diverse agro-climatic conditions in terms of terrain, soil and precipitation. Millets were initially the main food crops of the communities who first settled in the area; they include finger millet (*Eleusine coracana*), little millet (*Panicum sumatrense*) and Italian (or foxtail) millet (*Setaria italica*). The lack of communication between the hill-dwelling communities and the plains increased their dependence on these grains for their food security for a very long time. Over the years, this isolation led to substantial genetic variability in these millet species and the region has become notable for this diversity.

Millets and their nutritional value

The large upland area of the Kolli Hills provides space for the extensive cultivation of millets, and these grains have constituted the principal food grain of the community for many years. The varying distribution and volume of rainfall in the area has resulted in the development of millet varieties of different maturity periods, which have abilities to withstand adverse climatic conditions (King *et al.*, 2009; Bhag Mal

Figure 4.3.1 Map showing the location of Namakkal district, Tamil Nadu.

et al., 2010). Traditionally, the community recognizes their dietary superiority over other cereals such as rice, and those engaged in carrying out hard physical work particularly appreciate them. Nutritional studies have shown that these grains are rich in dietary fibres, calcium and iron, and various essential amino acids that are limited in other major cereal crops. In view of this, Professor M.S. Swaminathan suggested that these grains should be referred to as 'nutritious millets' rather than 'coarse grains' (Bhag Mal *et al.*, 2010). Recently, there has been an increasing awareness among urban consumers of the specific health value of these grains, both in view of their better nutritional composition and their nutraceutical properties.

Decline in the cultivation of millets

The introduction of commercial crops such as cassava, which supplies the regional starch manufacturing industry, and horticultural crops such as pineapple has led to a shift from subsistence to commercial farming in the Kolli Hills (Gruere *et al.*, 2009). These new crops occupied the land that was formerly used for cultivating millets. The ready availability of food grains such as rice and wheat, which were supplied at highly subsidized rates by the government under its anti-poverty programme, also contributed to the decline of millets. The hard work associated with the traditional cultivation and processing of millets also led farmers to choose to cultivate other grains that are easier to process and consume. In 2001, MSSRF began a number of projects to counter these developments and to promote the cultivation and conservation of millets. We share here the various practices that are part of our approach, which can also be referred to as community biodiversity management (CBM).

Practices for creating an economic value for millets cultivation

Step 1: Establishing self-help groups

We organized local farmers – mainly women, as they are more enthusiastic when it comes to the cultivation and consumption of millets – into self-help groups (SHGs). Up to this day, we have been supporting more than 35 SHGs in the Kolli Hills, with over 386 members (of whom 214 are women), in the institutionalization of the various millet operations (Padulosi *et al.*, 2009). Figure 4.3.2 illustrates the linkage between different SHGs in the millets value chain. The various SHGs joined together to form the Kolli Hills Agrobiodiversity Conservers' Federation (KHABCoFED).

Step 2: Strengthening the local millet seed system

Farmers cultivate different local varieties of millets in different zones. After collecting the seed of local varieties from different locations, we were able to identify 21 distinct local varieties of finger millet, little millet and Italian millet. As the seed of some of these varieties was in extreme shortage, and the seed of most varieties is mixed with others during the traditional practice of mixed farming, we trained SHGs in the production of quality seed. We further promoted their safe storage through the establish-

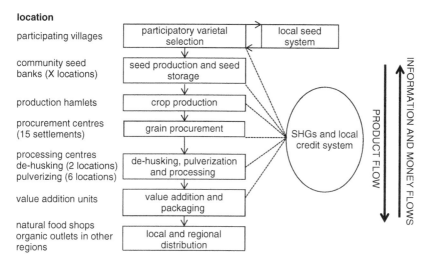

Figure 4.3.2 Schematic representation of the market chain for minor millets, in Kolli Hills, India. SHG, self-help group.

Source: Adapted from Gruère *et al.* (2009).

ment of community seed banks (CSBs), as discussed by Shrestha *et al.* (Chapter 2.8). These CSBs were crucial in the revitalization of the traditional seed storage systems and the promotion of seed exchange practices (Rengalakshmi *et al.*, 2003). Today, SHGs in 12 villages have improved their local seed systems through several practices. The CSBs and their associated practices are important village-managed institutional structures for supporting the conservation of local millet varieties.

Step 3: Increasing yield by improving millet cultivation practices

The availability of quality seed has contributed substantially to the promotion of millet cultivation. Productivity enhancement was essential for the continued cultivation of millets. We supported the SHGs by promoting a number of good agronomic practices, such as planting in rows, using appropriate seed rates, applying farmyard manure and inter-cropping millet with cassava. With the final practice, we aim to increase millet yields and augment the net income from cultivation. The promotion of good agronomic practices resulted in an average yield increase of 39% in finger millet, 37% in little millet and 30% in foxtail millet (Bhag Mal *et al.*, 2010).

Step 4: Carrying out participatory selection for better varieties

We evaluated a wide range of varieties of the three millet species in participatory varietal selection (PVS). We included several hundred accessions from the germplasm bank of the International Crops Research Institute for the Semi-Arid Tropics (ICRISAT) in Hyderabad, and a number of improved cultivars from the national programme of the All India Coordinated Research Programme on Small Millets in Bangalore. We compared these with local varieties during PVS. With the benefit of farmers' knowledge and skills we were able to identify several varieties that have 20–30% higher yield than those local varieties in the production system in the Kolli Hills.

Step 5: Introducing processing technology

Except for finger millet, all millets have a very hard seed coat that requires a highly abrasive force to remove from the grain. The decortication process, which is almost exclusively carried out by women, usually involves a very tedious, physical labour using a pestle and mortar. Our introduction of a small mechanical milling facility signalled a major change in the outlook of women. It substantially contributed to a revival in finger millet cultivation and consumption. An important spin-off from this mechanization was a new interest, shown in particular by the women, in millet value addition and commercialization.

Step 6: Developing new millet products

We used market studies to identify value-added products with good commercial potential. Subsequently, we trained members of the SHGs in value addition at the College of Rural Home Science in Dharwad, a rural campus of the University of Agricultural Science in Bangalore. The SHGs worked collectively to put several products in the production line. We encouraged different SHGs to specialize in different products. Among the most important value-added products that have been developed by these groups are malt, from finger millet; 'rava', the broken de-hulled grains of little millet and Italian millet, which are used in local recipes like *payasam* (sweet gruel), and the various different types of flours of these grains. During the early stages of production of these value-added products, our assistance extended to capacity-building on product quality, packaging, labelling, marketing and account keeping. Currently, five villages in the Kolli Hills are engaged in value-addition activities for millet (Bhag Mal *et al.*, 2010).

Step 7: Marketing millet products

Farmers are experienced in marketing their primary produce, but they often lack capacity in marketing value-added products. We assisted SHGs in marketing through promotional campaigns and awareness-raising activities. We gradually identified members of SHGs with potential and encouraged them to carry out product marketing with local retail outlets. We supported SHGs under the banner of KHABCoFED to establish retail outlets for all Kolli Hills products. Over the last few years, millet products carrying the label 'Kolli Hills Natural Products' have been available in department stores in Chennai, and in several other towns in Tamil Nadu. The profitability of these products is 5–10 times higher than that of grains (Table 4.3.1).

Table 4.3.1 Cost-benefit analysis of the production and marketing of value-added millet products

Product	Production costs (US$/ton)[a]	Total return (US$/ton)[a]	Net return (US$/ton)[a]	Profit as % of costs	Labour days for production of 1 ton
Little millet rava (samai)	0.80	1.20	0.40	50	100
Italian millet rava (thinai)	0.80	1.20	0.40	50	100
Little millet flour (samai)	0.84	1.28	0.44	52	150
Italian millet flour (thinai)	0.84	1.28	0.44	52	150
Finger millet flour	0.36	0.48	0.12	33	40
Finger millet malt	1.12	1.60	0.48	43	400
Samai uppuma mix	1.44	2.40	0.96	67	300
Samai rava dosa mix	1.52	2.40	0.88	58	300
Samai bajji mix	1.52	2.40	0.88	58	300
Thinai payyasa mix	1.52	2.40	0.88	58	300
Thinai laddu	2.67	4.00	1.33	50	300

Source: Adapted from Bhag Mal *et al.* (2010).

a US$1 = Indian Rupee 50.

Step 8: Promoting the products

Product differentiation and branding are important tools for obtaining a competitive market position. Products are nicely packaged and are accompanied by instructions and recipes. We have been able to develop a specific brand name for the products, and present them as locally grown and organic. MSSRF and the SHGs promote millet products during local events and annual temple festivals to increase awareness of the nutritional quality of millet and its products. In addition, we advocate the use of these products in institutional markets such as the Integrated Child Development Services Scheme of the Government of India.

Increased economic benefits that lead to increased cultivation and use

The economic potential of neglected and underutilized crops, including minor millets, is often under-exploited because of a lack of public awareness of the crops' qualities, and also because they are truly neglected in terms of public investment in scientific research (Bhag Mal *et al.*, 2010). Our activities in the Kolli Hills have renewed the interest of tribal farmers in cultivating millets. The value chain interventions have generated employment and income for millet producers and other actors in the value chain.

Small millets have to compete for farmers' interest with high market value crops like cassava, while for consumer interest they must compete with subsidized crops like rice. We consider the key element of success in developing a millet value chain to be the collective process of decision making, in tune with CBM. MSSRF, together with stakeholders, supported SHGs as local community-based organizations to assume ownership over, and be autonomous in, the value chain, pooling many types of resources, understanding economies of scales, and sharing information (Gruere *et al.*, 2009). However, we have not yet reached the point of sustained success. For that, we need to work on the expansion of a sustainable demand for value-added millet products (i.e. to achieve a transformation from a supply- into a demand-driven value chain).

Our current investments focus on the development of robust marketing channels for this, up until now, supply-driven value chain. Difficulties faced include the competition with subsidized commodities like rice, the loss of knowledge concerning the culinary preparation of millet, and changes in consumer behaviour and taste. We realize that it is easier to overcome such problems with consumers inside, rather than outside, the production area. The entry point in our marketing strategy is to understand the past uses of millet, and to develop and promote new uses for millets. We then follow this by raising awareness and exposing consumers to millet and its derived products, including novel foods, during regional events. At present, rice, flour and five value-added products of little and Italian millet, and finger millet malt, reach organic outlets in ten districts. We foresee an increased supply of diversified products to urban consumers in the future.

Providing farmers with an economic incentive to cultivate millets also supported the conservation of local varieties in the Kolli Hills, and the use of associated sustainable, environmentally friendly cultivation practices. With a successful marketing approach and market demand expansion, minor millets have the potential to flourish in their original production environments. As key crops for food and nutritional security, they will continue to support tribal communities, contributing to their resilience in coping with challenges in their rapidly changing environment.

4.4 Value chain development and the regional branding of Kalajeera rice in Jeypore, India

Sushanta Sekhar Chaudhury and Saujanendra Swain

Tribal communities and rice diversity in the Jeypore region

The Jeypore region of Orissa state in India is considered to be a centre of origin of the Aus ecotype of rice, which is known for its early maturing upland varieties. As such, it is home to an enormous diversity in rice landraces (Arunachalam *et al.*, 2006). Figure 4.4.1 shows the location of the Jeypore region in India. Rice has been the staple food of tribal farming communities in Jeypore for generations, and the conservation and use of local rice varieties is central to maintaining their traditional and characteristic lifestyle. Specific varieties are cultivated for use in different occasions, such as festivals, ancestral ceremonies, family functions and rituals. However, the tribal communities are very poor, often suffering from severe malnutrition. This contrast between wealth in diversity on the one hand, and absolute poverty and malnutrition

Figure 4.4.1 Map showing the location of the Jeypore tract, India.

on the other, signals the need for development strategies within a larger framework of community biodiversity management (CBM), for linking conservation with livelihood development. One of the key practices that contribute to poverty alleviation is that of transforming the rich diversity of rice into an economic asset, through a process of value-chain development and market promotion. In this way, communities can obtain economic benefits from their rice diversity.

The M.S. Swaminathan Research Foundation (MSSRF), through its Biju Patnaik Medicinal Plants Garden and Research Centre in Jeypore in the state of Orissa, began working with the tribal communities in 1998. We have been working as agricultural scientists with MSSRF on agrobiodiversity issues for many years, focusing particularly on enhancing food security and livelihoods of local communities. The Swiss Agency for Development and Cooperation (SDC) and the Indian Council of Agricultural Research are supporting MSSRF in this work. MSSRF uses an approach to support tribal communities in the conservation and use of valuable rice diversity in Jeypore, which is quite similar to the approach taken by MSSRF concerning small millets in the Kolli Hills, in the state of Tami Nadu, as described by King *et al.* in Chapter 4.3. In both Jeypore and the Kolli Hills, MSSRF uses a CBM methodology in which the creation of economic incentives is embedded.

CBM practices that contribute to the value chain development of Kalajeera rice

In this chapter we will discuss the main steps and challenges involved in the value chain development of Kalajeera rice. The tribal farming families of the Jeypore region were actively involved in and committed to this process, generously sharing their knowledge and their time. For further details on the process of genetic landrace enhancement and PVS, see Chapter 5.6.

Step 1: Good agronomic practices

In 1998, in order to improve the yields of local rice varieties, we supported training and demonstration exercises in which farmers compared rice plots that were cultivated using good agronomic practices, with plots where farmers applied traditional practices. The farmers observed a 30–70% increase in yield gains. It is important to highlight that the good practices did not include the use of external inputs and did not increase the costs of cultivation (Arunachalam *et al.*, 2006). In this manner, we quickly convinced farmers to adopt many of the good practices.

Step 2: Genetic enhancement of local varieties and participatory varietal selection

As a next step, we worked with farming communities on the genetic enhancement of local rice varieties and participatory varietal selection (PVS). These strategies for participatory crop improvement (PCI), applied within a context of CBM, are further explained in the general overview by De Boef *et al.* in Chapter 5.1. Farmers appreciate the Kalajeera rice variety for its black colour, good aroma, taste and cooking qualities, as well as for its suitability for use in rice snacks such as puffed rice. These qualities, combined with the fact that it fetches a higher market price than conventional rice types, led communities to select this lowland rice variety for commercial rice production (Arunachalam *et al.*, 2006). For further details on the process applied for the genetic enhancement of the Kalajeera variety, see Chapter 5.6.

Step 3: Kalajeera quality seed production

To produce high-quality commercial grain of the Kalajeera variety, we first needed to ensure farmers had access to pure-quality Kalajeera seed. We organized field demonstrations and training workshops on quality seed production and commercial grain production in several villages. We showed farmers the difference between grain and seed, and trained one group of farmers in large-scale quality seed production. This technical know-how was gradually shared with other communities. At the same time, we worked with groups involved in community seed banks (CSBs). We built upon the social and technical capital developed within this community institution, as supported in each of the villages in which we had been operating, in order to promote quality seed production. Shrestha *et al.* compare our MSSRF model for CSBs with several other approaches for CSBs around the globe in Chapter 2.8.

Step 4: Grain production and initial marketing

In 2004, farmers in the Jeypore region produced 31 tons of Kalajeera grain (Table 4.4.1), half of which they hand pounded and sold as unpolished rice for a good price to farmers in surrounding villages. However, local markets, rice mills and traders were still not offering remunerative prices. This prompted MSSRF to approach the National Agricultural Cooperative Marketing Federation of India Ltd (NAFED) and the Government of Orissa. In recognition of the value of the Kalajeera variety and its grain quality, NAFED procured large quantities of Kalajeera grain for highly competitive prices. The prices were similar to those usually offered for Basmati rice in the export markets (Arunachalam *et al.*, 2006). NAFED and the Government of Orissa signed a memorandum of understanding for the procurement of Kalajeera, which

Table 4.4.1 Details on Kalajeera production in the Jeypore region

Information	Year data			
	2004	2006	2008	2010
Cultivated area (hectares)	12	41	50	71
No. of villages	14	27	44	30
No. of households	49	126	159	212
Total production of grain (ton)	31	121	93	146
Seed production(ton)	4	11	41	64
Grain sold to NAFED/ORMAS (ton)	Nil	45	43	67
Household domestic consumption (ton)	28	65	50	79
Price of Kalajeera grain (US$/kg)	0.20	0.30	0.30	0.34

Source: MSSRF (2009).

created economic benefits for the tribal communities in Jeypore. This expanding market encouraged tribal farmers to increase production to 121 tons in 2006, and to 491 tons in 2007. However, in 2007 the Government of Orissa stopped supporting Kalajeera production and marketing, compelling the farmers to look for other, perhaps less government-dependent, marketing channels.

Step 5: Establishment of the Kalajeera cooperative society

With the production of Kalajeera steadily increasing but faced with an insecure marketing system, we realized that the establishment of an institutionally sustainable operational model was required to facilitate the better-coordinated production and marketing of Kalajeera. In 2007, the Kalinga Kalajeera Rice Growers Cooperative Society (KKRGCS) was established under the umbrella of a registered body of tribal farming communities, the Pachabati Gramya Uanayan Samiti. The cooperative society is involved in seed production and distribution, the monitoring of crops, grain procurement and central grain storage. Its executive committee has ten members that the general assembly selects based on their skills and capacities, and these members represent the participating villages. In 2007, KKRGCS commenced with 107 members from 29 villages in the Jeypore region; by 2011, it had reached up to 212 members. It became engaged in organized rice processing in 2009; today, all rice sold is processed. Members of the cooperative society hire local mills to process the rice themselves. All processed grain is sent to a rented central storage facility in Jeypore, and is subsequently packaged.

Step 6: Establishment of seed producer groups

KKRGCS, and a few knowledgeable farmers, maintain a stock of breeders' seed of the Kalajeera variety. New villages and groups of farmers who join the cooperative society in commercial Kalajeera grain production obtain the seed of the Kalajeera variety from this informal seed stock. Each new village commits itself to maintain its own basic seed that originates from this informal breeders' seed. This seed is then multiplied on a yearly basis and maintained in the CSB, itself then serving as informal basic seed stock for the grain producers in the village. In each village, a seed producer group is responsible for the regeneration of the basic seed stock, and the production of quality seed for dissemination to grain producers. Members of the KKGRCS are responsible for ensuring quality control, organizing the seed storage, and transporting grain to a central storage facility. They conduct field inspections to ensure genetic purity, and carry out tests on post-harvest moisture content, seed viability and varietal purity. This quality seed is then used for commercial Kalajeera grain production. Grain producers purchase the quality seed from the CSB at an interest of 50%, to be paid back in kind, in the form of returning seed to the CSB. In 2004, five CSBs in the Jeypore region had a surplus stock of Kalajeera from which they were able to supply quality seed to grain producers. In 2011, 23 villages had Kalajeera seed stocks in their CSBs.

   To ensure exclusivity and benefit-sharing, only members of the cooperative society may produce Kalajeera seed and grain. In order to monitor the quality of the procedures put in place, the quality of 40 samples collected in farmers' fields was compared with MSSRF breeders' seed. The tests showed no variation in purity and homogeneity. We showed that tribal communities can handle Kalajeera grain production on quite a large scale and, with a sophisticated design, using their CSBs and forming specialist groups of farmers, can autonomously manage the process of quality seed production (Chaudhury *et al.*, 2007). This institutional, social and technical capital is an asset for any future practices to support the communities in their contribution to the conservation of rice diversity and other agrobiodiversity, but also for improving their livelihoods.

Step 7: Enhancement of autonomy and creation of a robust marketing strategy

From 2007 to 2009 the cooperative society continued to market a large part of its produce through NAFED (see Table 4.4.1), without, however, the direct involvement of the Government of Orissa. Farming communities prefer to sell to NAFED as it pays a higher price for the Kalajeera grain than the private sector (i.e. US$0.30/kg, see Table 4.4.1). In addition, NAFED pays for the produce in full within two months after harvesting, while the private sector takes more time, often paying in several instalments.

In 2009, MSSRF and the cooperative society began to sell packages of processed Kalajeera rice through the parastatal Orissa Rural Development and Marketing Society (ORMAS). Its major contribution is that of assisting the cooperative society in organizing the marketing channels in Orissa; for example, it links the cooperative society with supermarkets and ensures its participation in pallishree melas, exhibitions for promoting rural, tribal products of India. The exhibitions provide the cooperative society with a platform to access prospective buyers. ORMAS also provides credit, with minimal interest, to support marketing. The price of Kalajeera grain now fetches up to 50% more than that of the high yielding varieties promoted and subsidized by the Government of India. These favourable market conditions have resulted in a significant increase in the area under Kalajeera cultivation, and in the number of households cultivating the variety (Table 4.4.1).

Step 8: Regional branding for protecting the Kalajeera variety

In order to obtain a stable market, it is important to have a unique and high-quality product. MSSRF contacted Tilda Riceland, which is a private company dedicated to processing and exporting rice, and asked them to test the unique quality traits of Kalajeera. Their results showed that Kalajeera from the Jeypore region outperforms the export-quality Basmati rice varieties, in terms of quality traits like relatively low stickiness, short cooking time and softness (Chaudhury *et al.*, 2007). Additional research conducted by MSSRF confirmed that the Kalajeera variety, which is cultivated and marketed by tribal farming communities, has its origin in the Jeypore region (Arunachalam *et al.*, 2006). To protect this unique rice variety we supported the tribal communities in the regional branding of their variety. In 2007, the cooperative society branded the unprocessed Kalajeera rice from their district as Kalinga Kalajeera. Because the Kalajeera production potential in the Jeypore region appeared insufficient, Tilda Riceland's plans to export Kalajeera did not materialize.

Step 9: Registration and protection of Kalajeera as a farmers' variety

In 2009, KKRGCS applied to register Kalajeera as a farmers' variety under the farmers' rights legislation of the Protection of Plant Varieties and Farmers' Rights Authority (PPVFRA). Currently, the cooperative society has fulfilled all requirements, but since it is the first application of its kind the process takes time and official approval is still pending. Through the registration of Kalajeera, the tribal farming communities of Jeypore will be recognized for their role in the development of this rice variety. Furthermore, the registration will help the farmers to protect their variety and ensure their rights under potential future access and benefit-sharing (ABS) mechanisms, which are also further outlined by Bala Ravi (Chapter 6.3).

The sustainability of the Kalajeera rice business and varietal portfolio management

Processed Kalajeera rice from Jeypore is a high-quality unique product with a clear niche market potential. The variety has a number of characteristics that make it interesting for local consumers, elite urban consumers in India, as well as for consumers outside India who appreciate Basmati rice. Through the regional branding and registration of Kalajeera as a farmers' variety, the cooperative society is better able to protect Kalajeera as a unique business venture. The cooperative society is very keen on the fact that only its members have access to Kalajeera seed to produce Kalajeera grain. To ensure this continues, the cooperative society aims to obtain the exclusive right to produce this variety.

The increase in yield achieved by applying good agronomic practices has resulted in an increase in surplus marketable rice at household level. This, together with the higher market price, has encouraged farmers in the Jeypore region to take up Kalajeera cultivation as a business. Farmers can obtain an average net profit of US$500 per hectare of Kalajeera (MSSRF, 2011a). The area cultivated with Kalajeera, its total production and the number of households taking up its commercial cultivation continues to grow (Table 4.4.1). Through this market chain project, Kalajeera rice has been transformed from a locally consumed, largely ignored and threatened local variety into a commodity in the Jeypore region contributing to the conservation and use of local rice varieties. MSSRF and the cooperative society are currently taking on more local varieties in their portfolio, such as Machahakanta and Haladichudi, to demonstrate that the model developed for Kalajeera rice is a successful strategy within a larger CBM framework (MSSRF, 2011b).

Our successful experience with Kalajeera marketing shows that a market, local as well as beyond, exists for the processed rice of local varieties and their products. This allows the tribal farming communities in the Jeypore region to benefit from the efforts made by their ancestors in developing these varieties over the ages, recognizing the value of their rice diversity and encouraging them to maintain this diversity. The path laid over the past 12 years by the CBM process and its practices for promoting the cultivation and marketing of Kalajeera rice has developed into a model for supporting the management of rice diversity, and for ensuring secure and sustainable livelihoods for tribal communities.

4.5 Marketing local rice varieties in Vietnam, supported by their participatory genetic enhancement and intellectual property rights

Frédéric Thomas and Dao The Anh

From bulk production to product differentiation

The Vietnamese are beginning to feel the negative effects of the accelerated economic growth that has been taking place in their country over the last 20 years, and its impact on agricultural prices, the environment and human health (Dao The Anh *et al.*, 2003). Because of this, researchers and policy-makers from the Ministry of Agriculture and Rural Development (MARD) are becoming increasingly interested in establishing quality value chains. There has been a switch from a production economy, which was inherited from the collectivist period, to a quality-based economy with niches based on product differentiation. The identity of the product and its uniqueness and reputation are now valued as much as the formerly more important economies of scale (Allaire, 2002). Using intellectual property concepts, such as trademarks and geographical indications, these local products can be differentiated from bulk production and provide protection for the producers, who are also custodians of the genetic resources on which the product is based.

This is a turning point for rural communities in their role as custodians of plant genetic resources (PGR). Will this development allow farmers to regain their central role in the development and maintenance of varieties, and in seed production, a role that they pretty much lost during the Green Revolution? And how will they benefit from regaining this role? In this chapter, we aim to answer these questions based on lessons learned from various experiences concerning the marketing of local varieties in Vietnam.

Intellectual property rights: creating benefits for custodian farmers

Officially, 22 agricultural products and foodstuffs are registered under the protection of geographical indications and designations of origin in Vietnam. In addition, a large number of other products are protected by trademarks that refer to geographical names. Laws concerning geographical indication strive to promote agricultural activity, help producers to obtain a premium price for their authentic products, and eliminate unfair competition and the misleading of consumers by non-genuine products.

In Vietnam, the initiative for protecting a local product often comes from local authorities at the provincial and district levels, which start mobilizing agronomists, geneticists and economists from research institutes to help small-scale farmers in building up these new channels. Subsequently, the farmers join in this dynamic process. The initiation of the process by local authorities or researchers, rather than by farmers, makes sense since intellectual property rights (IPRs) are complex mechanisms for which an understanding of larger markets, and the design and implementation of regulatory frameworks, is required. In other similar experiences shared in this book, non-governmental organizations have played vital roles in seeking such types of protection, as can be seen in the process for registering Kalajeera rice as a farmers' variety in India, which was facilitated by the M.S. Swaminathan Research Foundation (Chaudhury and Swain, Chapter 4.4); and in the support provided by Local Initiatives for Biodiversity, Research and Development (LI-BIRD) for the formal release of Pokhareli Jethobudho rice in Nepal (Silwal *et al.*, Chapter 5.5).

Nevertheless, our partners in Vietnam are fully aware that the farmers need to be properly involved in the whole process in order to be successful. To do this, they help the farmers to organize themselves into small groups of producers that then gradually expand in size. In the case of Tam Xoan rice from the Red River Delta (Hai Hau), for example, the Hai Hau Association for the Production, Processing and Marketing of Tam Xoan Rice joined 43 small-scale farmers together in 2004, on just a few hectares of land. Today, the association includes 442 farmers, and involves an area of 54 ha. The price for Tam Xoan rice produced in Hai Hau can reach up to 55% more than rice of the same variety produced in another location.

One of the difficulties in constructing value chains for a localized product is that no form of labelling provides a genuine guarantee of origin. Producers may still suffer from unfair practices involving the use of the name of their reputable product for a product that does not offer the same characteristics. It is estimated that 30–70% of Tam Xoan that is sold as originating from the Hai Hau district is in fact a mixture. The large wholesale buyers and retailers are often more concerned with volume than with quality (Binh and Duc, 2007; Moustier *et al.*, 2010).

To ensure that a product is genuine, producers must organize themselves, define their unique production area, and ensure that consumers can distinguish a product by its origin. They should also differentiate their products from other similar products if possible, through packaging and the use of typical characteristics that help the consumers recognize the origin. In Vietnam, if the product is a local variety, researchers and producers invariably enter into a process that is referred to as *phuc trang* ('restoration'), concerning a *giông co truyên* ('traditional variety'). The Vietnamese term *phuc trang* also includes the notion of forcefulness and could be translated as 'reinvigoration', although it also includes a sense of 'homecoming', which refers to the true origin of the variety.

Participatory genetic enhancement of local varieties in support of their marketing

In Vietnam, we use the term 'restoration' for selection that is carried out within local varieties in order to make them uniform and stable, which is seen as an essential prerequisite for strengthening their marketing. We follow the terminology for approaches to participatory crop improvement (PCI) within a context of community biodiversity management (CBM), as outlined by De Boef *et al.* in Chapter 5.1.

Below, we detail the steps involved in the participatory genetic enhancement (PGE) of local rice varieties in Vietnam.

Step 1: Collection of accessions and social reconstruction of the variety

Agronomists from the Centre for Agrarian Systems Research and Development (CASRAD) first gathered accessions from the main zone in which the local variety is grown, and these then formed the basis for its genetic enhancement. At the same time, they collected information from farmers about the characteristics that, in their view, shape the ideal variety. Blind tastings were then held, during which farmers were invited to score accessions for their aromatic quality and taste. These initial activities generated a consensus among farmers, researchers and provincial administrators concerning the identity of the variety. This social reconstruction of the variety was a very important step for organizing farmers into an association of producers that will, in future, respect specifications. During each of the following steps, the agronomists worked to give a more institutionalized form to the producers' association.

Step 2: Establishment of selection criteria

Two or three samples that were considered to be the best were used as the genetic basis for follow-up activities. Ms Pham Thi Huong, a geneticist at the National Centre for Variety Evaluation and Seed Certification (NCVESC) in Hanoi, explains that such accessions are used for what she calls the 'purification and discovery' of the original characteristics of the variety. Ms Pham Thi Huong assisted farmers and agronomists in the PGE of the sticky rice variety Nếp Cai Hoa Vang from Kinh Môn and the aromatic rice variety Tam Xoan from Hai Hau. The location of Hai Hau and Kinh Môn are illustrated in Figure 4.5.1. Based on the farmers' initial descriptions, together they compiled a matrix with some 60 characteristics that they used to guide selection among the accessions. They applied a selection process that follows the protocol to produce pure lines of rice varieties, based on the professional standards for seed production set by MARD in 2006. They used this protocol for the PGE of the Tam Xoan and Nếp Cai Hoa Vang local rice varieties.

Step 3: Purification of the seed

Farmers planted the seed of the two or three accessions that had been identified as being the best during Step 1 in their own fields, in several plots of 200 m^2 with 8000 plants. From the tilling stage and after each crop stage, the farmers gradually eliminated all the plants that did not have the characteristics required. They were then invited to select only 150 plants that corresponded best to the 60 characteristics they wanted to retain. Farmers harvested the panicle from each selected plant individually and Ms Pham Thi Huong, the geneticist, assessed the number of grains per panicle

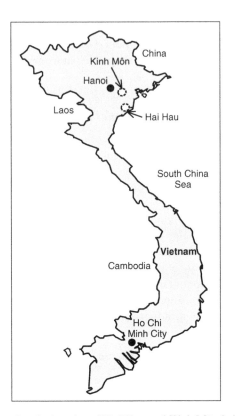

Figure 4.5.1 Map illustrating the location of Hai Hau and Kinh Môn in Vietnam.

and evaluated their quality. The selection pressure during this first year of cultivation in field trials to purify the seed was intense as we only maintained 1% of the plants.

Step 4: Pure line selection

The grains of each selected plant constituted one line and the seed of the lines was sown in small separate plots of equal size in the farmers' trial fields. At this point, the farmers no longer eliminated plots that deviated from the standard variety type; instead, they marked the ones that best matched the unique traits of the variety, applying a process of positive selection. Two days before harvesting, the farmers randomly selected 20 individual plants from each plot that they had marked earlier. Ms Pham Thi Huong evaluated these plants for five quantitative criteria: plant height, number of ears per plant, ear length, number of grains per ear, and weight per thousand grains. She calculated the average values of selected plots using these criteria, and when the average values for a selected line deviated from an allowed variance, the line was removed, as it was considered to be too heterogeneous. The seed of those plots that passed the uniformity tests was considered breeders' seed of the restored variety.

Implications of the selection process

Farmers originally cultivated broad ranges of slightly heterogeneous varietal populations that together constituted a local variety. However, the current process of genetically enhancing local varieties results in a significant reduction of their allelic polymorphism (Fukuoka *et al.*, 2006). The originally diverse nature of local varieties has been reduced to meet the stringent formal requirements of the product's commercial development. At the end of the process of PGE, only a few lines that are close to the pure-lineage template are retained. This product of the selection process serves as a type of breeders' seed, which is then multiplied to obtain seed classes similar to basic and certified seed for dissemination to producers. In this way, the PGE supports the commercial exploitation and marketing of local varieties, while at the same time contributing to a reduction in diversity, narrowing the genetic basis of crops and varieties used by farmers, and reducing their potential adaptive capacity. There are those who consider this to be the price that must be paid for variety commercialization, but we now know that with a genetic paradigm less in favour of pure-line varieties, it would be possible to apply a softer selection pattern.

Protection of commercial interests through IPRs

Within the genetically enhanced local variety, each plant cultivated has an identical genotype and therefore a stable and uniform phenotype, just like formally released modern varieties. As such, their characteristics are less and less dependent on interactions between the genome and the environment. Consequently, it has become easy and possible for any external economic agent to cultivate this enhanced local variety outside the area in which it was originally cultivated, and thereby compete with the original local producers, as can be seen with the sticky rice Nêp Cai Hoa Vang from Kinh Môn. The PGE process of Nêp Cai Hoa Vang had barely been completed when provincial authorities began a programme to extend the area of cultivation from a hundred hectares or so up to several thousand hectares. The provincial authorities are quite easily able to ignore the rights of those farmers who have not only conserved the variety but have also contributed to its enhancement. It is therefore vital for farming communities to protect their genetically enhanced local varieties through IPRs by applying collective trademarks or geographical indications.

The balance between meeting market requirements and encouraging the use of diversity

The PGE process promotes the use of local varieties, lessening the threat of their disappearance from competing with modern varieties. However, the PGE and adaptation of diverse local varieties to standardization requirements reduce rather than increase their diversity, and as such diminish rather than reinforce the role of farmers in PGR conservation. With this intervention in the existing system of community management of local varieties important questions emerge: how can we correct the direction of these practices, which are currently used for market-based upgrades of local varieties? How can we insert socio-environmental goals into practices of market orientation? Can PGE and market development focus on the farmers' involvement and encourage rather than deter the use of diversity when targeting the market? The new PGR professionalism, as elaborated on by De Boef *et al.* in Chapter 7.1, may guide us in finding the approaches to positively answer these questions.

Normal PGR professionalism in Vietnam led us towards standardization, restricting room for manoeuvre, and constraining us to fix the identity of a variety to a uniform genetic basis. If we want to reverse such a process, we must accept that we will have to move away from standardization. To do this, we should use selection schemes that are better suited to protecting local varieties with their diverse genetic make-up. Our challenge will be to find a balance between the market requirements for standardized quality, and retain the complex relationship that exists between the diversity in the varieties cultivated, their place of origin and their custodian farmers. This can only be achieved when researchers accept that they have to move beyond normal professional standards. We need to adapt our breeding methods to such professionalism, and divert from those used for the production of modern varieties. In Vietnam, we propose the use of population-based breeding methods (Ceccarelli *et al.*, 1994; Bonneuil and Thomas, 2009). Such methods would make it possible to meet requirements for marketing the variety in terms of the social construction of a commercial identity of the variety, and maintain consistent quality; moreover, they would facilitate interactions between farmers, genetic resources and their environment.

4.6 The dynamic management and regional marketing of a local wheat variety by farmers in the Lubéron, France

Guy Kastler and Anne-Charlotte Moÿ

Local varieties, farm management and the environment

Each local variety develops specific characteristics through interactions between farmers, with their own traditional knowledge, culinary and spiritual traditions, and the environment in which it grows. Local varieties can only be maintained if they are managed and cultivated by farmers year after year. Farmers may share and exchange the seed of those varieties with others who will then cultivate it in other places, leading to the adaptation of the varieties to other production environments and management practices. Local varieties are often used as a basis for traditional regional dishes, and as such they are part of the local culture. To promote the maintenance of local varieties it is essential to attach economic value to the products derived from them (Demeulenaere, 2008).

The environment, the soil and the farm management practices applied give local varieties their own particular taste. The use of significant amounts of inputs and intensive mechanization makes the environment more homogeneous but also artificial, and it weakens the relationship between the varieties and the environment in which they developed their distinctive properties. The result may be that specific characteristics of local varieties are no longer expressed or appreciated. In the current chapter, we share the experiences of a group of small-scale farmers, referred to in France as *paysans*, in promoting the use of a local wheat variety in the Lubéron region in France, and their search for the right legislative framework to support their on-farm management practices.

The Lubéron region is a dry, mountainous region in south-eastern France. Farming is still often a small, family-run business. Much of the produce is processed on the farm itself and sold in local and regional markets. These paysans cultivate their wheat using minimal inputs, and as such the wheat is often organic. Their agricultural system promotes an expression of the characteristics of local wheat varieties that have evolved in this region.

Meunier d'Apt: a local wheat variety in the Lubéron

About 20 years ago, a group of paysans revived the cultivation of an old, local wheat variety, Meunier d'Apt, with the aim of using it for the preparation of a local flour and special bread. The variety had almost completely disappeared. One farmer obtained some seed from a neighbour and began to grow and propagate it, distributing its seed to other paysans in the region. The flour of this bread wheat variety, when mixed with other local varieties, gives the bread a special taste that is much appreciated. Today, paysans have joined forces with artisan millers and bakers to make the most of Meunier d'Apt wheat. Some of the paysans themselves are also millers and/or bakers. It is this social process that promoted the maintenance and renewed cultivation of the variety in its original environment, adapting it over the course of time to changes in agricultural practices, and to the eating habits and fashions of modern life. Consequently, traditional aspects were incorporated into modern lifestyles, such as the appreciation of regional and organic products. This adaptive process is restricted to farming and processing methods that do not alter the intrinsic qualities of the variety or the products made from it, and rejects those methods that could have a negative effect on its distinct properties (Zaharia, 2008).

Trademarks for products unique to the Lubéron region

Today, the group of paysans, artisan millers and bakers are not alone in their venture to promote the cultivation of Meunier d'Apt wheat. Their initiative is being followed by local officials, who want to promote the conservation of local heritage associated with the Lubéron Regional Nature Park, through the registration of the trademark Produit du Parc Naturel Régional du Lubéron ('Product of the Lubéron Regional Nature Park'). The paysans, millers and bakers have drafted a set of regulations to ensure the quality and authenticity of Meunier d'Apt flour and bread from the Lubéron region. We further refer to this set of regulations as the Meunier d'Apt certificate of quality and origin.

The Meunier d'Apt trademark helps promote the product to consumers; moreover, it aims to recognize its importance and usefulness. The trademark aims to encourage consumers and farmers to value genetic heritage, agricultural products and the knowledge of paysans. The Meunier d'Apt certificate of quality and origin clearly defines the links between local varieties, farm management practices, farmer seed multiplication, and farmer-to-farmer seed exchange or trade. The certificate states that those farms that intend to produce flour or bread bearing the trademark must be located in one of the municipalities in the Lubéron Regional Nature Park area. Furthermore, it requires that the grain produced must come from family-run, organic farming operations (as far as possible). Seed used should originate from a farm within that production zone. The conditions defined in the certificate support on-farm management of agrobiodiversity as a means to contribute to the *in situ* conservation of local plant genetic resources (PGR), and aim to promote agricultural production systems that are adapted to their environment. As such, the certificate aligns well with the objectives of a regional nature park, in that it contributes to sustainability and conservation.

Intellectual property rights for protecting Meunier d'Apt

Option 1: The conservation catalogue

The Lubéron Regional Nature Park looked at the possibility of listing the variety in the conservation catalogue. However, this would involve the application of a number of constraints to the production system and to the farmers' management of the local variety. These constraints relate in particular to the homogeneity and stability of the variety, limitations in quantity of seed to be produced and disseminated, quotas for the production and marketing of grains and derived products, and general bureaucracy in relation to seed production activities. These requirements are inapplicable to the situation in the Lubéron, where farmers save seed from their grain production using common practices that link seed production to grain production (i.e. they don't sustain the variety just for its conservation). Kastler and Moÿ in Chapter 6.6 further describe the barriers created for farmers by regulations on the conservation of local varieties as developed by the European Union Commission.

Option 2: The seed catalogue

Meunier d'Apt cannot fulfil the standards set out in the formal seed catalogue concerning homogeneity and stability for the simple reason that it has been developed from a population and not just one or few lines. Its heterogeneity, which is vital for the ability of the variety to adapt to the difficult terrains in which it is cultivated, is highly valued by the paysans in the tough farming conditions of the Lubéron region. The variety has low yields but produces better than other varieties in the Lubéron. It would be unable to pass the required cultivation and use trials if it were tested in an experimental station under good conditions. In addition, its limited cultivation area does not allow for the production of a sufficient volume of seed stock to cover the costs of registration and to ensure its continued inclusion in the seed catalogue. Neither the certificate that associates its production with the Lubéron Regional Nature Park, nor the collective quality trademark for Meunier d'Apt flour and bread from the Lubéron, allow for the circumvention of the rigid formal seed regulatory framework. Kastler (Chapter 6.8) further discusses the implications of the seed and variety laws in France on the management of local varieties by farmers.

Option 3: An experimentation agreement for variety maintenance and exchange

The group of paysans from the Lubéron have found an alternative that is compatible with both their management practices of the local variety, and the certificate. They continue to use and exchange traditional cereal varieties in the Lubéron region by way of an experimentation agreement. The farmers' association and paysans sign agreements for experimentation in the production of seed and grain of the local wheat variety, Meunier d'Apt. Through these agreements, the association commits to providing the farmer with (experimental) seed of the variety, monitoring crop production, organizing crop demonstrations, and documenting results obtained. In return, the paysan agrees to sow the seed, maintain a test plot, and record observations before harvesting. The experimentation agreement specifies that once the farmer has harvested the crop and met all the obligations, the farmer becomes the sole and complete owner of the variety of that crop. In this way, the farmers gain the right to freely use and sell the grains of this crop. The farmers may not, however, exchange or sell their seed with other paysans. Kendall and Gras (Chapter 1.7) describe how the Maison de la Semence Paysanne (seed bank) in the Périgord region of France shares the seed of local maize varieties with farmers across the country using a similar mechanism. Kastler (Chapter 6.8) further elaborates on how farmer organizations that are members of the Réseau Semences Paysannes (Farmers Seed Network, RSP) are compelled to use such mechanisms to circumvent rigid seed and variety regulations in order to achieve on-farm management of agrobiodiversity, and engage paysans in practices for contributing to community biodiversity management.

Recognition of the need for the dynamic management of local varieties

Current operations in seed production and the exchange of local varieties in the Lubéron and in other regions of France are based on experimentation agreements. We realize that we are innovative in avoiding rigid seed and variety regulations. What would be more appropriate is a conservation framework that supports the dynamic management of local varieties, and which provides clear room for farmers to produce, exchange and sell seed of those local varieties. The practices that are used by the farmers in the Lubéron, where they use their own seed, and exchange or sell the seed of local varieties, have long been considered only relevant to small-scale farmers in developing countries. Even the on-farm management of PGR as a strategy for contributing to their *in situ* conservation was considered only relevant to developing countries and not to a country with a highly developed and industrial agricultural system such as France. However, our experience in the Lubéron with Meunier d'Apt, and experiences in other French regions (Demeulenaere *et al.*, 2008) and locations, as described by Kendall and Gras (Chapter 1.7), show the contrary.

Towards the end of 2009, the *in situ* conservation of local varieties was recognized by the Fondation pour la Recherche sur la Biodiversité (Foundation for Biodiversity Research; FRB) as a relevant practice in France. In this context, we should explain that the FRB was officially launched in 2008 by the ministries in charge of research and ecology, and includes eight public research organizations that work with biodiversity, including those organizations responsible for PGR conservation. FRB is a point of convergence between the different actors involved in science and society that relate to biodiversity. We hope that their recognition of *in situ* conservation will open the way for the development of frameworks that support rather than limit the dynamic management of local varieties. Such frameworks would allow the variability and heterogeneity of varietal populations in a country where static conservation appears to be the only officially sanctioned conservation strategy.

Although the geographically based quality trademark seems quite appropriate for the protection of local varieties, it still necessitates a change in the dominant scientific approach in France. This current approach views the seed as mere software for reproduction, following the logic of economies of scales, with one seed allowing the production of an infinite number of copies of the same variety. The existing legal framework in France allows considerable freedom for the exchange of seed for conservation or for research. The reason for this is simply because such varieties are not included in the catalogues, which either regulate seed sales or the conservation of local varieties. Experience shows that the advanced concepts of population genetics do not match the dominant logic that guides the catalogues. The concept of a variety should be modified in order to combine the science of population genetics, with the desire of groups of paysans and their associated millers, bakers and even consumers to use and appreciate locally adapted varieties. In such a new approach, varieties should be considered as living organisms, endowed with an ample capacity to adapt, favouring their co-evolution with the farmers who cultivate them, and with the environment in which they have gained their distinctive properties.

Part V

Participatory crop improvement in a context of community biodiversity management

5.1 Participatory crop improvement in a context of community biodiversity management

Introduction

Walter Simon de Boef, Marja Thijssen and Bhuwon Sthapit

Synopsis

Participatory crop improvement (PCI) is based on a series of methods in which farmers and scientists collaborate in plant breeding with the basic objective of more effectively addressing the needs of farmers in the marginal areas of developing countries (Almekinders and Elings, 2001; Morris and Bellon, 2004; Ceccarelli *et al.*, 2009). A second and often considered secondary objective is that PCI contributes to the *in situ* conservation or rather the on-farm management of plant genetic resources (PGR). This objective is based on the assumption that the diversity of farmers' preferences, and the environments in which they cultivate their crops and varieties, results in the local selection and use of a diversity of materials that is wider than when one or two broadly adapted varieties are introduced and disseminated over large areas (Witcombe *et al.*, 2001; Jarvis *et al.*, 2011). A third objective for PCI, which is expressed by non-governmental organizations (NGOs) in particular, is the potential to contribute to the empowerment of farmers in agricultural research and development, and agrobiodiversity management (Almekinders *et al.*, 2006). This third objective relates to policy goals and to the creation of mechanisms through which farmers can express their rights (Andersen, Chapter 6.2) and can share benefits from the use of genetic resources in a fair and equitable manner (Vernooy and Ruiz, Chapter 6.4).

As indicated by the term 'participatory', PCI implies a link between farmers, plant breeders and other scientists involved in crop improvement research. However, within the context of the current book we have to look past that link and approach PCI as a methodology that leads to collective action that is integrated into a wider setting of community biodiversity management (CBM) (Jarvis *et al.*, 2011). Through this action, the collective capabilities of communities are strengthened so that they can engage themselves purposefully in the dynamic management of agrobiodiversity (Shrestha *et al.*, Chapter 1.3; De Boef *et al.*, Chapter 7.2).

The chapters in Part V aim to explore the potential of the PCI methodology and examine the ways in which it can contribute to CBM. The section includes case studies covering a wide diversity of PCI methods. In Chapter 5.2 Sthapit *et al.* introduce grassroots breeding as a method in which breeders strengthen the skills of farmers in the selection, seed production and marketing of neglected and underutilized crops (i.e. crops and species for which no or very limited capacity exists within formal breeding programmes). The authors present five case studies from Nepal that share a common methodology but which vary in types of species. For perennial and often semi-domesticated species, such as tropical fruits, collaboration between scientists and farmers may result in a process that we refer to as participatory domestication. Santos *et al.* (Chapter 5.3) share the experience of such collaboration for the domestication of feijoa (*Acca sellowiana*) in Brazil. Lassaigne and Kendall (Chapter 5.4) share the experience of using a wide variety of PCI methods in France to strengthen farmers' dynamic management of local maize varieties. Two cases from south Asia illustrate variations of participatory genetic enhancement (PGE) of local rice varieties that were known for their quality (culinary) traits, but which at the same time were on the verge of disappearance. Silwal *et al.* (Chapter 5.5) describe an experience from Nepal involving the local rice variety, Jethobudho, which was at risk of genetic erosion. A similar experience from India involving the local rice variety Kalajeera is shared by Chaudhury and Swain (Chapter 5.6). Where the above two PGE experiences targeted rice as an autogamous crop species, Ogliari *et al.* (Chapter 5.7) share an experience with maize from Brazil that results in a very different PGE process because of the allogamous nature of maize, but which has basically the same goal of genetically enhancing a local variety through the collaboration of breeders and farmers.

In Chapter 5.8 Alonzo *et al.* describe the experiences of the Collaborative Programme on Participatory Plant Breeding in Mesoamerica (PPBMA), which uses a diversity of PCI methods, including PGE, participatory varietal selection (PVS) and participatory plant breeding (PPB). They share case studies from Honduras, Guatemala and Nicaragua, where PCI methods were applied to beans, maize and sorghum. Mohammed *et al.* (Chapter 5.9) approach PVS from another angle, aiming to enhance the capabilities of seed producer cooperatives in Ethiopia so that they can manage their crop and varietal portfolio in a more autonomous and entrepreneurial manner.

A different aspect related to access, which is illustrated in the concluding chapter in this section, shows how PVS and PPB are instrumental in enhancing the access of Samoan farmers to the global gene pool of taro germplasm. In Chapter 5.10 Iosefa *et al.* describe the importance of being able to access healthy and resistant or tolerant varieties of a vegetatively propagated crop such as taro. This chapter indicates clearly that even though PCI may be instrumental in supporting farmers' dynamic management of agrobiodiversity at local level, a global interdependence in PGR is crucial for sustaining farmers' livelihoods and even their production systems. This taro case study provides encouraging, albeit alarming, examples of where global and local efforts need to meet in order to ensure that farmers can access good taro varieties and secure a healthy, productive and sustainable crop, thereby safeguarding their livelihoods.

Participatory crop improvement: a diversity of methods

PCI constitutes a diversity of methods on which farmers and scientists collaborate. Table 5.1.1 provides an overview of the experiences shared in this book. The manner in which farmers and scientists work together depends on the crop, its reproduction system and the major objectives that the farmers and breeders want to achieve by participating in crop improvement. Below, we briefly describe the methods involved in the PCI process.

Table 5.1.1 Overview of the crops and methods used in the participatory crop improvement case studies included in the book

Country	Chapter	Author	Crop(s)	FrB	GrB	PGE	PVS	PD	SFB	PPB
Nepal	5.2	Sthapit *et al.*	NUS[a], rice		X		X			
Brazil	5.3	Santos *et al.*	Feijoa					X		
France	5.4	Lassaigne and Kendall	Maize	X		X	X		X	X
Nepal	5.5	Silwal *et al.*	Rice			X	X			
India	5.6	Chaudhury and Swain	Rice			X	X		X	
Brazil	5.7	Ogliari *et al.*	Maize	X		X	X			
Nicaragua	5.8	Alonzo *et al.*	Beans			X	X			X
Guatemala	5.8	Alonzo *et al.*	Maize			X	X			
Honduras	5.8	Alonzo *et al.*	Sorghum				X			X
Ethiopia[b]	5.9	Mohammed *et al.*	Wheat, beans, barley				X			X
Samoa	5.10	Iosefa *et al.*	Taro				X			X
Thailand	1.6	Doctor	Rice	X			X		X	X
Bhutan	1.6	Doctor	Rice, maize				X			
Vietnam	1.6	Doctor	Rice	X			X		X	X
India	3.7	Ramesh *et al.*	Garcinia			X		X		
Vietnam	4.5	Thomas and Anh	Rice				X			X

F, farmer; S, scientist; FrB, farmer breeding; GrB, grassroots breeding; PGE, participatory genetic enhancement; PD, participatory domestication; SFB, scientist-supported farmer breeding; PPB, participatory plant breeding,

a Neglected and underutilized species, which in this chapter include sponge gourd, taro, cauliflower, rice bean and chilli pepper.

b The chapter of Mohammed *et al.* addresses in particular the use of PVS in local seed business development, but refers in two of its examples to cases where the products result from PPB.

When we refer to farmer breeding in the context of PCI, basically the farmers are engaged in all the steps of crop improvement. The dynamic management of local varieties is not isolated from any scientific inputs. The most prominent examples in the book are those that illustrate the farmers' management of local varieties in France and Brazil (Lassaigne and Kendall, Chapter 5.4; Ogliari *et al.*, Chapter 5.7), involving the selection and development of farmers' or local maize varieties. The examples apply to maize because, owing to its allogamous nature, it responds to and encourages farmers' individual selection and management. Doctor (Chapter 1.6) describe experiences in which rice breeding by farmers in both Thailand and Vietnam has resulted in the development of popular varieties.

In grassroots breeding, the collaboration between farmers and scientists focuses on the use of scientific approaches in assessing and selecting within existing diversity. As illustrated by experiences from Nepal (Sthapit *et al.*, Chapter 5.2), rapid results can be achieved in the improvement of neglected and underutilized species, including fruits and vegetables. Advances with simple selection within existing genetic variation; investment in the establishment of seed production or other structured forms of multiplication (such as nurseries) and seed dissemination can have an impact on promoting the continued use of local crops or varieties by farmers. Varieties are not released but are distributed through an informal seed system, building on the logic that if the resources or interest for breeding are lacking, they will also be unavailable for more formal or commercial structures of seed production and marketing.

Participatory genetic enhancement (PGE) aims to select within existing variation, and in that manner it is similar to grassroots breeding. However, PGE continues the process of selection up to the development of more stable and uniform varieties. The variation in diversity within which PGE selection is carried out depends on the crop's reproduction system; examples provided in this section include PGE for maize (Brazil, France and Guatemala), rice (India, Nepal) and beans (Nicaragua). In all these examples, the interactions between breeders and farmers, and their different combinations of knowledge and skills, result in varieties with varying degrees of particularity and refinement. These varieties can be released and seed production may have a more formal set-up, or the varieties may return to more informal seed systems and follow paths of seed dissemination similar to those of local varieties or the products of grassroots breeding.

A number of chapters in this section describe situations in which scientists support farmer breeding. This method distinguishes itself from the other PCI methods in that the farmers themselves participate in making crosses. This is considered a crucial method for contributing to the empowerment of farmers. The example from India (Chaudhury and Swain, Chapter 5.6) shows a long process of scientist-supported farmer breeding, in which farmers were engaged in making crosses and selection in early generation materials, and in developing advanced rice populations. Unfortunately these were then lost as a result of natural disasters. Although such natural events are common and frequently jeopardize PCI programmes, they are simply not often reported. It was, however, regrettable that no back-up had been established. Fortunately, a PGE process was set in motion simultaneously in the same location, resulting in a popular, improved local variety.

Another example of scientist-supported farmer breeding is provided by Lassaigne and Kendall (Chapter 5.4), who describe how a group of farmers in France joined together with a scientific breeder to form a collective as part of a sophisticated scheme to develop a hybrid maize variety. Here again its original set-up was strongly motivated by empowerment, with possible political connotations, illustrating that the farmers themselves are capable of developing hybrid maize varieties. This programme was discontinued when participating farmers concluded that the resulting hybrids would create dependencies among farmers in terms of sustaining the variety (maintaining inbred lines), and would interfere with the crucial dynamic ways in which the farmers want to maintain and further develop their own local maize varieties.

The two case studies from India and France were driven by a process aimed at facilitating the farmers' appropriation of scientific principles of plant breeding, but in reality resulted in impractical situations. When farmers incorporate some aspects of scientific breeding into their farmer breeding they may become more successful. Several case studies described in this section share experiences where varieties resulting from farmer breeding, with the limited involvement of breeders, have subsequently been released (e.g. the cases of rice varieties in Thailand and Vietnam as described in Chapter 1.6) or have become popular, such as the local maize variety, MPA1 in southern Brazil, which was not released but was disseminated through the informal system (Ogliari *et al.*, Chapter 5.7).

The example of scientist-supported farmer breeding from India does not differ greatly from participatory plant breeding (PPB), in which farmers and breeders collaborate in basically all the steps. In PPB, however, a more effective division of responsibility is sought to define what the farmer and breeder should contribute. This section presents several case studies of PPB from Ethiopia, France, Honduras and Nicaragua, Samoa, Thailand and Vietnam, which cover all reproduction systems (i.e. barley, beans, maize, rice, sorghum and taro).

All the case studies involving PGE and PPB include participatory varietal selection (PVS) as a practice for varietal evaluation. PVS, however, can be used more in isolation when applied as a means to evaluate (re-)introduced gene bank accessions or local varieties. It approaches the CBM practice of diversity blocks by complementing the original objectives of the practice (i.e. to demonstrate and characterize local diversity). Public PGR programmes engaged in CBM, such as those being implemented in Brazil (Dias *et al.*, Chapter 2.5), use PVS as a tool to (re-)introduce and enhance farmers' access to gene bank accessions as a means to achieve on-farm management.

Participatory domestication is a variation of PPB that is applied to perennial species such as fruit trees. As illustrated in the cases of feijoa in Brazil (Santos *et al.*, Chapter 5.3), and garcinia in India (Ramesh *et al.*, Chapter 3.7), scientists and farmers work together in most steps of the breeding. The process is complex; on the one hand it facilitates quick wins, by building upon farmers' practices, such as the identification of good genotypes and the best ways to propagate them clonally, as with garcinia in India. On the other hand, the work with feijoa in Brazil emphasizes first and foremost the need to gain an understanding of existing genetic variation in order to identify good parent plants that can be used for the development of superior varieties. In this way, scientists and farmers collaborate together in a long process for the domestication of a species. Scientists play a major role in the complex design and implementation of such a process.

PCI methods and their contribution to CBM

We return to the focus of this section, which is the potential role that PCI methods can play in achieving a situation in which farmers conserve and utilize agrobiodiversity in a collective and purposeful manner. We use key characteristics of the CBM methodology to identify topics, and complement them with those that are relevant for assessing various strategies for seed sector development in their potential contribution to PGR conservation (De Boef *et al.*, 2010). In our synthesis of the case studies, we identified the following topics: conservation and use of agrobiodiversity; farmers' access to superior varieties; efficient use of breeders' and farmers' capabilities; diversity and farmers' autonomy in the structure of seed value chains; and empowerment of farming communities in their agrobiodiversity management. Figure 5.1.1 illustrates these five topics, and in the following paragraphs we will discuss them in more detail, linking each to several experiences of PCI in the book.

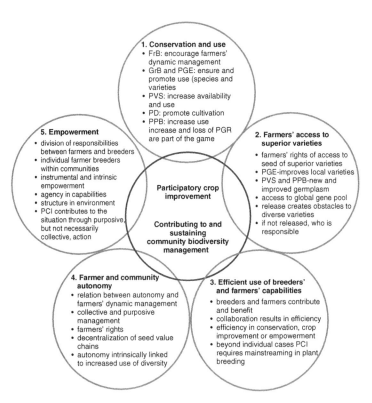

Figure 5.1.1 Participatory crop improvement contributing to community biodiversity management, as characterized by five topics. F, farmer; S, scientist; FrB, farmer breeding; GrB, grassroots breeding; PGE, participatory genetic enhancement; PD, participatory domestication; SFB, scientist-supported farmer breeding; PPB, participatory plant breeding.

Conservation and use of agrobiodiversity

The various PCI methods contribute in dissimilar ways to the on-farm management of agrobiodiversity. Promoting farmer breeding in essence strengthens farmers' capabilities in the dynamic management of local diversity, as illustrated by Lassaigne and Kendall in Chapter 5.4. However, the example from Vietnam, described by Doctor (Chapter 1.6), shows that when farmers, through their farmer breeding, are successful in the development of varieties, the effect on on-farm management is similar to that of formal breeding (i.e. it may be considered either a threat, by replacing local varieties, or an asset, by making diversity available).

Grassroots breeding develops adapted or enhanced varieties of neglected and underutilized species and makes them available, thereby encouraging their continued cultivation. The effect of this is an increase in the farmers' use of a diversity of species rather than varieties. Similarly, but more at variety level, PGE contributes to on-farm management. The case studies of Jethobudho and Kalajeera rice illustrate how local varieties that become rare following a decline in their quality can be consolidated as varieties through PGE, thereby regaining their popularity with farmers and consumers and becoming once again abundant (Silwal *et al.*, Chapter 5.5; Chaudhury and Swain, Chapter 5.6).

The case study from Ethiopia concerning PVS, described by Mohammed *et al.* (Chapter 5.9), illustrates that when the method is appropriated by farmers' groups it is instrumental for demonstrating and disseminating varieties, whether these are local or improved. PVS has the potential to increase the number of varieties cultivated in a community and thereby strengthens informal seed systems that indirectly contribute to on-farm management.

Through the domestication of forest species, scientists and farmers can contribute to their gradual development as crops, accompanied by the improvement of production and multiplication systems. The aim is to reduce the pressure on natural stands of species, such as feijoa in Brazil and garcinia in India. However, as shown by Peroni *et al.* (Chapter 3.4), when the association between the people, species and landscape is very strong, then strengthening the dynamic management of the species and its associated landscape, rather than domesticating its species, contributes to *in situ* conservation. Consequently, understanding when to apply domestication with the aim of conservation depends greatly on the relationship between people, species and landscape. Based on this, we consider that domestication may have an adverse effect on the *in situ* conservation of genetic diversity.

When PPB promotes the wide use of diversity and incorporates local varieties in breeding, it encourages the use of local diversity, but if the resulting varieties are very successful, for example, those resulting from crop improvement, they may replace existing local diversity. In association with various PCI rice projects that have been implemented over the last 15 years in Nepal, scientists have been monitoring the impacts of PCI products and have learned that although PCI does not significantly reduce richness and evenness measures of diversity, the level of community divergence, and of diversity used, does in fact decrease. Farmers are interested in a set of traits rather than the variety itself; if a more complete set of traits are available in a different variety they will change varieties. This is part of the dynamic process within which the interface between farmers and breeders in PCI advances the evolutionary processes.

When farmers manage agrobiodiversity, there is both a gain and a loss of diversity; these should be considered part of the dynamics we are contributing to with PCI methods in the context of CBM. This illustrates that when PCI methods are used within CBM processes, communities need to monitor the effects on local diversity, and those conservation and development organizations that are involved need to approach on-farm management with PCI methods in such a manner that it can be complemented if necessary by *ex situ* conservation.

Farmers' access to superior varieties

Any action in PCI results in a gain but also potentially risks the loss of diversity, which links us to what S. P. Bala Ravi (personal communication) indicates is one farmers' right (i.e. the right of access to quality seed of superior varieties, independent of whether they result from conventional breeding or PCI or are local varieties). In terms of access, the case studies on Jethobudho and Kalajeera rice in Nepal and India, respectively, illustrate that PGE can make a significant contribution to farmers' access to local varieties, simply because the variety is better and the seed value chain is organized and integrated in a product chain in a structured manner. This can be both informal and formal, though its more professional organization ensures the availability of quality seed of the improved local variety. Similar arguments can be used for PPB and especially PVS, where PVS is an effective method for promoting farmers' access to new materials. The issue of access does not only apply to improved or local varieties, but may also apply to PGR in general, as shown by the case study on taro from Samoa. This example shows that access to the global gene pool became, and continues to be, vital for sustaining the farmers' access to viable varieties and therefore their crops.

In the event that varieties resulting from PCI prove superior, the farmers do not necessarily have easy access to such varieties; several case studies show that legal obstacles hinder the farmers' access to those varieties. Experiences from Vietnam, described by Thomas and Anh (Chapter 4.5), and from Central America, as detailed by Alonzo *et al.* (Chapter 5.8), show that the release of PCI varieties was not possible because the materials were neither stable nor uniform. However, release of such varieties may not be that important as they can be disseminated in any case through the informal seed systems. The case study of Jethobudho in Nepal shows how the breeders were able to create uniformity in the variety in order to secure its release. With the case studies involving PCI in Brazil, Central America and France, farmers and scientists opted not to release their PCI-improved maize or sorghum varieties, as they would not have passed the formal procedures. However, even if they could have released them it would have proven incompatible with the farmers' dynamic management of those open-pollinated varieties. Consequently, in addition to the legal obstruction, these PCI products are not released by choice, but they are still able to find their way to farmers through the informal system. One important technical and institutional issue that is of relevance in the cases of rice and maize is the identification of who is responsible for the maintenance, breeding and production of breeders' and basic seed.

Once a PCI variety becomes available, a structure should be put in place to sustain the continuous flow of quality seed of the variety to farmers, or among farmers. Within the formal structure of the seed value chain associated with release and registration, breeders have such obligations. Similarly, institutions that develop and register varieties in Nepal have the responsibility to maintain breeders' seed and provide basic seed to seed producers. Consequently, the NGO Local Initiatives for Biodiversity, Research and Development (LI-BIRD) has taken on responsibility for maintaining and producing basic seed of Jethobudho Pokhareli and other varieties. This is in contrast to the maintenance of local varieties; few community seed banks produce diversity catalogues (with farmer descriptors for distinguishing varieties) that serve as a reference. However, sustaining this structure of maintenance, breeding and basic seed production, either formally or informally, is a continuous struggle for breeders in institutions, NGOs, community-based organizations and farmers.

If a variety is only made available through the informal system it evolves further and no such institutional back-up exists. In such a situation, a partnership could be explored between farmers' groups engaged in commercial seed production and marketing, and public institutions, as illustrated in the case of Ethiopia. Public–private (i.e. farmers) partnerships that are based on agreements that include clear divisions of responsibilities and resource allocation may create innovative arrangements to ensure that farmers can access quality seed of those varieties.

Efficient use of breeders' and farmers' capabilities

Both farmers and breeders participate in PCI processes and contribute their time and resources, but they also gain capabilities, information and resources. The balance is defined by their objectives for joining in such collaborative action. Where PCI seeks to rationalize the breeding process, the scientist takes the lead in PPB and considers the farmers' involvement in pre-breeding or making crosses less important. If empowerment motivates joint action, then another balance may be found, for example, in scientist-supported farmer breeding or farmer involvement in all the stages of PPB. Table 5.1.2 illustrates the division of responsibilities throughout the steps of the PCI process.

With the relatively short processes of grassroots breeding and PGE, both scientists and farmers seek fast outputs that result in the farmers' use of species or varieties. A more sophisticated design of PGE (e.g. MPA1, Brazil) and of participatory domestication (e.g. fruit trees, both in Brazil and India) is possible in cases where a research project is hosted by a university; PCI is then the subject of investigation within a larger research and development process. NGOs promote conservation and empowerment work with scientist-supported farmer breeding, PGE and PPB. Key for these PCI processes, whether they are supported by universities or NGOs, is to find ways to incorporate such PCI action within mainstream programmes. The development of more effective plant breeding that also contributes to conservation and use has a lasting impact, as can be seen when following the release of Jethobudho and Kalajeera, or the release of a first variety for garcinia or feijoa. The individual experiences with PCI that are shared in this book together form an innovative methodology, where farmers and breeders collaborate on crop improvement, resulting in more efficient processes and structures in crop improvement, and in the conservation and use of genetic resources.

Table 5.1.2 Involvement of farmers and/or scientists in the process of participatory crop improvement: comparison of methods

PCI steps	FrB	GrB	PGE	PVS	PD[a]	SFB[b]	PPB
Diversity assessment	F	F/S	F/S	F/S	F/S	F/S	F/S
Setting breeding goals	F	F/S	F/S	F/S	F	F/S	F/S
Selection of parents	F				S	F/S	F/S
Making crosses	F		F/S		F/S	F/S	S
Early selection	F		F/S		S	F/S	F/S
Advanced selection/trials	F	F/S	F/S	F/S	F/S	F/S	F/S
Release [c]	F	F/S/-			F	F/S	F/S
Seed production	F	F	F	F	F	F	F
Dissemination	F	F	F	F	F	F	F

Source: adapted by the authors from Morris and Bellon (2004).

F – farmer; S – scientist; FrB – farmer breeding; GrB – grassroots breeding; PGE – participatory genetic enhancement; PDo – participatory domestication; SFB – scientist-supported farmer breeding; PPB – participatory plant breeding.

a Participatory domestication of feijoa, as described by Santos et al (Chapter 5.3), has so far been implemented up until the step 5, i.e. development of a breeding strategy; and as such the final steps of mass selection (step 6) and on-farm evaluation (step 7) are still in the process of being carried out .

b Scientist-supported farmer breeding, as described by Chaudhury and Swain (Chapter 5.6) was carried out up until the final step – evaluation and advancement up to F5 populations (step 7), which was interrupted owing to bad weather conditions and the absence of a back-up.

c Release is conducted in some cases by scientists, others by farmers, and in some cases release is not possible because the PCI varieties do not pass owing to lack of stability and uniformity

Farmer and community autonomy in the management of varieties

If we follow the logic of the CBM methodology, which results in a situation where communities manage their crops and varieties, thereby contributing to *in situ* conservation, we can construct a similar rationale for PCI. It is crucial that communities are supported in crop improvement in a way that strengthens their capabilities to manage their crops and varieties in a sustainable manner. The example from France provided by Lassaigne and Kendall (Chapter 5.4) is most explicit, as PCI is carried out by farmers within a context of strengthening their autonomy and dynamic management. Other experiences, such those of the South-East Asia Regional Initiatives for Community Empowerment programme (SEARICE) in Thailand and Vietnam (Doctor, Chapter 1.6), and of the Collaborative Programme on Participatory Plant Breeding in Central America (PPBMA) in Central America (Alonzo *et al.*, Chapter 5.8), aim towards the enhancement of farmers' autonomy in a more indirect manner. Autonomy is vital for strengthening the capabilities of farmers to manage their varieties in a collective and purposeful manner. PCI methods offer ways for farmers to exercise their rights (e.g. through PGE or grassroots breeding). If such motives are explicit, these PCI experiences (such as farmer breeding, grassroots breeding and PGE) can be distinguished from those where breeding efficiency and food security, rather than autonomy and conservation, are the leading objectives. Although breeders play major roles in such applications of PVS and PPB, farmers' organizations play an increasingly important role in the seed value chain, thereby further appropriating their space and increasing their autonomy, as can be seen for example in the case study in Vietnam, in which SEARICE partners collaborate with seed producer clubs (Doctor, Chapter 1.6).

If farmer organizations, such as seed producer cooperatives, begin to strengthen their position in seed value chains, and the PCI methods in breeding programmes are up-scaled, the most logical result will be that farmers will have more access to diversity. Small, organized and local market-oriented seed producers will be able to produce varieties that are better adapted and respond to specific demands. Gradually, such local seed producers may even take on the responsibility of producing basic seed, and of establishing the necessary relationship with public breeding for new varieties. Since more farmers are now directly engaged, more varieties are available, and breeders are attending to the demands of a larger and more diverse group of farmers' and production systems; collaboration between breeders and farmers can be considered more effective than working through the linear model of conventional breeding and formal seed production. Public breeding is no longer solely embedded in the model of economies of scale, which works towards the wide adaptation and release of just a few varieties to cover large areas, with the result that farmers become dependent on distant breeders and seed providers. By applying participatory approaches to crop improvement, accompanied by the restructuring of responsibilities in the seed value chains, the principles of new professionalism are embraced (De Boef *et al.*, Chapter 7.1), increasing rather than reducing diversity and enhancing farmers' autonomy. Although such a situation is far from mainstream, if we consider that participatory approaches in breeding have become common over the course of the past two decades, then the next logical step towards mainstreaming PCI is to restructure the seed value chains, which does not appear to be too ambitious.

Empowerment of farming communities

Since their development in the 1990s and further consolidation over the past decade, PVS and PPB have proved to be effective in developing farmers' capacities in various aspects of breeding and selection, as well as in raising the confidence of farmers to make their own choices in crop improvement (Almekinders *et al.*, 2006). As illustrated by the diversity of cases shared in this book, PCI methods have become aligned with CBM in a larger process aimed at enhancing the empowerment of communities in their management of agrobiodiversity. In analysing the diversity of experiences, we need to address the following two questions: (i) who is being empowered – participating breeder or farmers, or both; and (ii) within the communities, which farmers are being empowered? To respond to these two questions, we use a typology of participation based on Pretty (1994), for identifying the responsibilities in PCI methods, as elaborated in Table 5.1.2. The results of the assessment are shown in Table 5.1.3. Since most chapters in this book share advanced experiences with PCI, most collaboration can be characterized as interactive; those contributing to farmer breeding are clear examples of self-mobilization. Because of the close interaction between farmers and breeders, all the examples have elements that enhance farmers' capabilities in the breeding process, thereby contributing to the active and purposeful participation of farmers in breeding, and promoting their management of agrobiodiversity. However, the degree to which such a process results in collective action, which is a prerequisite for establishing and sustaining CBM, is not very clear in the examples.

The analysis provided in Table 5.1.3 does not detail whether the interaction is between breeders and individual farmers, or between breeders and communities of farmers. From our knowledge of the case studies, examples involving PVS may aim to develop the variety portfolio management of seed producer cooperatives (e.g. in Ethiopia), and achieve farmers' dynamic management of local varieties (e.g. in France). Although the typology of participation in the examples from Ethiopia and France (rows I and IV of Table 5.1.3) may be different, they stand out in the collective structure of decision-making within the communities. Many of the cases, however, focus on individual farmers collaborating with plant breeders, who only join together as a collective when reaching the stage of variety testing, seed production and marketing. This also seems to be logical in the farmer management of agrobiodiversity, where some farmers become specialists and others do not. PCI may contribute to community empowerment when individuals are identified within community organizations as having the skills and knowledge to participate in crop improvement, and are associated with groups of seed producers and other farmers. The case of the farmer who developed a particular rice variety in Vietnam, as described by Doctor (Chapter 1.6), is a clear illustration of this; seed of the variety was disseminated through a highly structured, though still informal, system of seed production and marketing.

When the major goal for engaging in PCI is to increase efficiency in breeding, the process results in instrumental empowerment; empowerment then equals participation (De Boef *et al.*, Chapter 7.1). This situation is applicable to most of the cases in rows I and II in Table 5.1.3, which are characterized as passive, functional and interactive participation. Collective and purposeful action in the management of agrobiodiversity (i.e. collective agency) can only be achieved within PCI following

Table 5.1.3 Modes of participation in methods of participatory crop improvement

Interaction between breeders and farmers	Case studies[a]	PCI methods[c]	Role of farmer	Type of participation[b]
I Farmers are given finished varieties, developed by plant breeders	Dias et al. (various, Brazil) Mohammed et al. (various, Ethiopia)	PVS	Decide to adopt or not	Passive
II Farmers evaluate relatively large quantities of unfinished materials (lines, families, populations, large numbers of clones, local varieties) on-station or in breeder managed on-farm trials, and select materials for further improvement	Silwal et al. (rice, Nepal) Ogliari et al. (maize, Brazil) Santos et al. (feijoa, Brazil) Mohammed et al. (barley, Ethiopia)	PGE PPB PD	Identify interesting materials used for further selection Identify breeding objectives and traits	Functional to interactive
III Farmers conduct evaluation trials of germplasm (lines, families, populations, large numbers of clones, local varieties) in their own fields and use their own management practices and experimental design	Sthapit et al. (various, Nepal) Chaudhury and Swain (India) Alonzo et al. (various, Central America) Iosefa et al. (taro, Samoa) Ramesh et al. (garcinia, India) Thomas and Dao The Anh (rice, Vietnam)	GrB PGE PPB SFB PD	Contribute in all selection and testing steps Identify breeding objectives, traits and materials	Interactive to self-mobilization
IV Farmers use scientific breeding methods, including making crosses, and are further autonomous in their breeding programmes; they release or disseminate their varieties autonomously	Sthapit et al. (various, Nepal) Lassaigne and Kendall (maize, France) Doctor (rice, Thailand and Vietnam)	GrB FrB SFB	Define breeding objectives, traits and materials Decide in all selection and testing steps	Self-mobilization

Source: Adapted by the authors from Morris and Bellon (2004).

a References are made to specific chapters, although some chapters may include several experiences that consequently correspond to dissimilar types of participation.
b Types of participation are based on Pretty (1994).
c FrB, farmer breeding; GrB, grassroots breeding; PGE, participatory genetic enhancement; PD, participatory domestication; SFB, scientist-supported farmer breeding; PPB, participatory plant breeding; PCI, participatory crop improvement.

processes with intrinsic empowerment, where farmers and farming communities are autonomous in their decision-making and, if required, seek information, resources and germplasm from breeders. This situation is achieved when the participation is typified as self-mobilization, as illustrated in row IV of Table 5.1.3. The interactive type of participation, as exemplified in row III, takes the more intermediary position of contributing to instrumental empowerment (i.e. participation), though gradually resulting in intrinsic empowerment (De Boef *et al.*, Chapter 7.2).

It can therefore be concluded that PCI contributes to the empowerment of farming communities in CBM when reaching collective agency, and can therefore be typified as interactive and self-mobilized. Given the nature of PCI, however, most of the methods do not result in self-mobilization. Grassroots breeding, PGE and PPB, which encompass most of the examples in the book, have an intermediary position between interactive participation and self-mobilization. This variation is inherent to the process of interactions between breeders and farmers that make up PCI. What is vital is that more of such methods are used in plant breeding, thereby resulting in changes in structure, which, in an indirect way, contribute to farmers' empowerment.

Lessons learned

When PCI is associated with CBM, we learned that it contributes to the dynamic management of agrobiodiversity, promoting farmers' use rather than conservation. With these dynamics, successful PCI actions result in farmers adopting new, improved and local materials, which inevitably leads to changes of use. When we consider those dynamics as part of the process in which farmers manage their agrobiodiversity in a collective and purposeful manner, PCI illustrates how CBM, in contributing to on-farm management, is more about dynamic approaches to conservation rather than static approaches, and needs to be complemented by *ex situ* conservation.

Participation in crop improvement can be approached as a process with a means to an end. It is clear that when the major goal is to enhance efficiency in breeding, participation is primarily the means, and farmers' access to improved varieties is the end (Morris and Bellon, 2004). However, this perspective changes when PCI aims to contribute to a situation referred to as CBM; then the participatory processes are the means that contribute to the end, which can be translated into the farmers' capacities to manage and improve agrobiodiversity in a collective and purposeful manner. When efficiency is the primary goal, the breeders drive the process, whereas in the case of CBM, breeders take on the role of facilitator, contributing to and strengthening farmers' capabilities. In this way, PCI in a context of CBM addresses key elements of farmers' access to varieties and their autonomy, and relations between breeders and farming communities are transformed to a new professionalism in crop improvement (De Boef *et al.*, Chapter 7.1). When PCI targets efficiency, farmers are consulted on and participate in crop improvement, thereby enhancing the empowerment of breeders for achieving a higher efficiency. However, where PCI contributes to empowerment in a context of CBM, this situation is reversed. This illustrates how the division of responsibilities and interactions between breeders and farmers in PCI are defined by the orientation of the CBM methodology (i.e. towards the empowerment of farmers for managing their agrobiodiversity).

5.2 Grassroots breeding of local crops and varieties in support of community biodiversity management and resilience in Nepal

Bhuwon Sthapit, Kamal Khadka, Pitambar Shrestha, Shreeram Subedi and Indra Prasad Poudel

Participatory crop improvement and agrobiodiversity

Most of the poor live in rural areas and depend on agriculture for their livelihoods and well-being. A key livelihood asset that they maintain is their agrobiodiversity, which is made up of their crops, varieties and forest species. Improving the use of agrobiodiversity through enhanced understanding and better management offers opportunities to advance the well-being of small-scale farmers by providing more sustainable production systems, better nutrition and enhanced income opportunities (Jarvis *et al.*, 2011). In the past, the world invested many resources in collecting and conserving farmers' varieties of the major food crops for future use by plant breeders, rather than developing mechanisms to make local biodiversity accessible to poor and needy people. Many small-scale farmers in marginal areas have not yet benefited from the investments and great successes in modern plant breeding (Bellon, 2006). Various methods that together shape participatory crop improvement (PCI) have been developed in recent decades with the aim of meeting those farmers' needs (Sthapit *et al.*, 1996; Witcombe *et al.*, 1996; Ceccarelli *et al.*, 2009). However, most PCI methods do not assess the usefulness of plant genetic resources (PGR) that are easily available to farmers in the form of local varieties or semi-domesticated species before launching conventional breeding schemes with elements of farmer participation.

Breeding methods for targeting a wide range of locally important species

Farmers' varieties are the fruits of many generations of observations, selection, exchange and breeding by farmers and communities. There is a growing recognition of the importance of local varieties as sources of genetic variation for modern plant breeding (Park *et al.*, 2005; Teklu and Hammer, 2006). Based on traditional knowledge concerning factors essential for local crop development, farmers consciously maintain their highly diverse local varieties (Teshome *et al.*, 1999; Jarvis *et al.*, 2011). This local capacity can be further fine-tuned and/or strengthened for the efficient and effective use of genetic resources. This fine-tuning can be achieved through the use of a new method that is part of wider group of PCI methodologies; we call this method 'grassroots breeding'.

There will never be enough scientific plant breeders with sufficient institutional support to carry out plant breeding for all crops in all production environments. Global public resources and, increasingly, private funds are invested in the public breeding of major food security crops. The private sector is responsible for, and successful in, carrying out plant breeding for crops in which investments in breeding can be earned back through the sale of quality seed of improved commercial, and often hybrid, varieties. Besides the major staple crops that are addressed by the public sector, and crops that are addressed by commercial companies (mainly maize and vegetables), the main sources of food, nutrition and livelihoods of many millions of small-scale farmers originate from agrobiodiversity that is extant in their direct farm or surrounding area. Given this scenario, we identified the following strategies to increase food security using this local agrobiodiversity:

- increasing crop yields through PCI methods;
- strengthening local seed systems;
- diversifying sources of food and nutrition by promoting local and wild edible species;
- enhancing benefits through market incentives for such species;
- supporting the development of policies for promoting diverse agricultural products.

In this chapter we focus on the development of the grassroots breeding method, since it has the potential to incorporate all these strategies. This method strengthens the capabilities of farmers and community-based organizations (CBOs) to apply scientific principles for the selection and maintenance of local PGR. We share a range of examples applying grassroots breeding to a wide range of crops and species in Nepal and draw some lessons on the comparative advantage of applying grassroots breeding in a context of community biodiversity management (CBM).

Conceptual framework

Definition of grassroots breeding and its principal steps

Grassroots breeding is a simple, effective and efficient method of plant breeding that strengthens farmers' skills in seed production and marketing (Sthapit and Rao, 2009). It focuses first on pre-breeding through locating, assessing, multiplying and making genetic diversity available with the objective of immediate use; second, it addresses the enhancement of germplasm through simple selection techniques and the production of quality seed. Grassroots breeding is a simple PCI method that enhances the capability of the CBO and farmers to search existing genetic diversity, select niche-specific plant material, multiply and produce sufficient quality seed and distribute it within the community. The method involves more than just crop improvement; it is a selection method, embedded in a much wider approach for associating conservation goals with livelihood development. The method involves the following practices, all of which are carried out by CBOs, with the assistance of an agricultural scientist, especially in the first four practices:

- participatory rural appraisal of local PGR;
- assessment of farmer-preferred traits to assess available variability;
- selection of preferred traits by farmer-breeders and scientific-breeders;
- selection of varieties or species samples from the target environment;
- community-based seed multiplication of selected varieties/plants;
- distribution of the seed/planting material;
- monitoring the use in terms of households, communities and area.

Grassroots breeding and other PCI methods

Grassroots breeding can be applied in situations where the production systems are rich in genetic diversity but many individual farmers have problems in terms of its access, availability and quality. In those areas where modern varieties have already replaced local varieties, grassroots breeding is not an option, since its basis is to work with local and existing diversity in crops and varieties. In such cases, participatory varietal selection (PVS) and participatory plant breeding (PPB) might then be more appropriate to enhance access to varietal diversity. It is relatively easy to out-scale grassroots breeding as it requires fewer resources than other PCI methods. The grassroots breeding model is developed expressly to focus on the constraints of accessing and using local genetic materials, and on enhancing the knowledge and capacities of CBOs to improve those materials through selection. As such, the role of breeders or scientists is much more limited than in PPB.

Grassroots breeding experiences in Nepal

We have been involved in the development of grassroots breeding of major food crops and neglected, underutilized species in particular in Nepal for many years. Here, we outline six case studies.

Basaune ghiraula variety of sponge gourd

In 1998, a diversity fair was organized in Begnas district, one of the sites in the agro-biodiversity project described by Subedi *et al.* (Chapter 1.2). Mrs Durga Maya Lama, a farmer from Jaman Kuna, displayed the unique and rare sponge gourd variety (*Luffa cylindrical*) called Basaune ghiraula. This variety is appreciated because of its aroma and sweet taste, and because the texture of the fruit does not become fibrous as quickly as with other varieties. Fellow farmers who participated in the fair were not familiar with this local variety. Local Initiatives for Biodiversity, Research and Development (LI-BIRD) assisted a local women's group in its multiplication in a diversity block; the group distributed 70 diversity kit packets with seed of the variety to farmers. In 2005, more than 200 households cultivated the variety. The Horticulture Research Station in Malepatan, Pokhara, is now maintaining 'breeder seed' to ensure wider access to this variety, and because of the growing market for this variety agro-dealers have started to sell the seed they collected from the women's group. This experience illustrates the initial stage of grassroots breeding, where local diversity is identified and, through a series of CBM practices (as described by Shrestha *et al.*

in Chapter 2.2), turns into a valuable resource within the community and beyond (Sthapit *et al.*, 2006).

Panchamukhe pidalu variety of taro

Also in Begnas, Mr Khim Bahadur Gurung maintained a special variety of taro (*Colocasia esculenta*) known as Panchamukhe pidalu. Compared with other local varieties of this crop, the corm of this variety is appreciated for its low acridity, and excellent cooking quality and taste; when cooked, its corms combine very well with fish and meat. It is also used for the production of masura, which are nuggets made of dried taro petioles and black gram flour. Five members from the Majhthar women's group multiplied seed of this variety in 2003 as a part of a collective income generation activity. In 2005, six out of 25 households in Bhangara village cultivated this variety; while in 2010, more than 12 households were cultivating it. In addition, the variety spread to many other villages in Begnas. The identification of this unique variety and recognition of its traits, followed by simple multiplication by a group of women in the initial steps of grassroots breeding resulted in this rare variety becoming abundant.

Kalokauli variety of cauliflower

This variety of perennial *Brassica* spp., found in Gulmi, produces a dark green-coloured cauliflower. A favourable trait of this variety is its continued production during the dry season, which is impossible with the improved varieties. While carrying out a participatory rural appraisal for a home garden project, we found that only a small number of households (about 10% per village) were cultivating Kalokauli plants, with fewer than ten plants per household. One major reason why this variety became rare was the limited access to good planting materials. As part of our project, we carried out simple selection among the accessions available in the district and found that farmers maintained white and green Kalokauli types. Since the taste of the green type was preferred, we started to promote its vegetative propagation through the selection of suckers from plants with compact curds. Within a few years, more than 70% of the farming households in Gulmi were cultivating at least 20 Kalokauli plants in their home gardens. This case illustrates how, through grassroots breeding, a simple procedure can be used to select and promote a vegetatively reproduced crop that would never be addressed by larger participatory or conventional plant breeding programmes.

Rice bean: simple selection among local varieties

Rice bean (*Vigna umbellate*) is a legume that is grown by intercropping with maize, or as a sole crop under rain-fed conditions. Floral morphology suggests rice bean to be a self-pollinated crop but some level of out-crossing has been reported. In 2006, LI-BIRD targeted this particular crop in a project, collecting 156 rice bean accessions that we tested in observation nurseries. Subsequently, we evaluated them in mother trials for different agro-morphological and quality traits, selecting four accessions based on yield, seed size and seed colour. In 2010, LI-BIRD staff created uniform populations for the four varieties. We produced more than 70 kg of seed of each variety, which was distributed to farmers. The experience with rice bean illustrates that in grassroots breeding we conduct only selection within existing variation and disseminate the results of such selections through informal seed systems.

Bhathi rice variety

During a diversity assessment conducted in Kachorwa (one of the villages of the agro-biodiversity project), we found that one farmer was maintaining a local rice variety known as Bhathi, which he cultivates in deep-water rice fields in association with fish farming. Only this particular variety is adapted to this type of production system, which is locally referred to as *ghol*, meaning 'deep, waterlogged area'. Other ghol areas were identified in neighbouring districts and far distant villages, where farmers were suffering because the modern varieties that had been recommended to them by extension services did not survive. We identified similar micro-niches in different project sites, in collaboration with a local non-governmental organization (NGO). With the help of our partner, farmers produced small packets or kits of Bhathi rice seed (1–2 kg) that were distributed among farmers in ghol areas. Following the initial promotion of the variety, farmers and the community seed bank in Bara became involved in its commercial seed production and marketing. This example shows that grassroots breeding can be a very simple process of identifying unique diversity and promoting its dissemination or marketing through the use of the CBM practice of the diversity kit.

Akabare khursani chilli

In the eastern mid-hills of Nepal farmers cultivate the native chilli variety they refer to as Akabare khursani. It fetches a high price in domestic and international markets. We found that in the market Akabare chilli show a wide variation in shape and size, as well as in taste and pungency. Because of this variability, consumers have lost their trust in this chilli. The key reason behind the variation in quality is its partial out-crossing nature. LI-BIRD, together with the Nepal Agricultural Research Council and Anmolbiu Seed Company, were able to improve Akabare chilli with simple population improvement that was followed by the establishment of a proper system of seed production and marketing. Our work with Akabare chilli shows that grassroots breeding is not only valid for dealing with subsistence food or neglected crops; it can also be used within a commercial setting and in response to market demands.

Synthesis: the major potential and comparative advantages of grassroots breeding

In grassroots breeding, each variation in method can deal with a specific situation, thereby providing an opportunity to work with neglected and underutilized species, for which there are usually no resources available to run fully fledged conventional or PCI programmes. As has been illustrated for several cases from Nepal, grassroots breeding is a PCI method that can be embedded in CBM processes to support the management and use of local varieties. It is a simple methodology for breeding and dissemination, which addresses crops and varieties adapted to rather specific agro-ecologies that will never be addressed by larger programmes.

Applying grassroots breeding in the context of CBM has a series of advantages over other methods of PCI and conventional plant breeding. It contributes to the empowerment of individual farmers, CBOs and communities in PGR conservation and use, as was illustrated in the cases of the sponge guard and taro. It is an effective PCI tool for encouraging farmers and their CBOs to assert their farmers' rights over their PGR, as highlighted by the engagement of the CSB in Bara, in the commercial seed production and marketing of the Bhathi deep-water rice variety. In all the cases we can see that grassroots breeding generates immediate benefits to communities by increasing access to rare and unique germplasm and associated traditional knowledge, and by contributing to the on-farm management of PGR. In most cases described in this chapter, the grassroots breeding process started with a participatory rural appraisal to identify whether a variety or crop found in a community was abundant or rare. With grassroots breeding we were able to lift varieties and crops from scarcity to abundance. Grassroots breeding addresses the weaknesses of local varieties or their production system, as shown in the cases of rice bean and chilli. Moreover, it generates a better understanding of PGR for local and specific adaptation, along with a preference for culturally valued traits, as we saw with the local varieties of cauliflower.

At institutional level, grassroots breeding enhances the understanding of formal breeders and other scientists about the needs of small-scale farmers. Through simple identification of locally valued genetic resources, as used in all the cases, or simple selection, as applied in the cases of rice bean and chilli, significant advances can be made without the use of complex conventional or participatory selection procedures. However, CBOs and their partners need to carefully monitor the changes in use of local PGR, and be aware of the genetic erosion that may logically result when valuable varieties are identified, selected and disseminated through grassroots breeding. But this risk also applies to all other PCI or conventional breeding methods.

Grassroots breeding and the enhancement of community resilience

Millions of small-scale farmers who live in regions of uncertain rainfall with marginal production potential have yet to benefit from grassroots breeding as a simple and community-oriented crop improvement method. Local varieties are unable to cope with evolving and adapting to climate change at the currently reported speed. Farmers may not have the ability and facilities to predict seasonal climatic variability. However, they can use a set of crop varieties or select traits from local PGR through grassroots breeding to increase the options to protect against unpredictable changes. In this context, grassroots breeding has a potential to enhance community resilience because of its focus on the use of local PGR, and strengthen the capabilities of communities themselves to use such resources in a more strategic manner. Through grassroots breeding, farmers and researchers may seek to understand the value of local PGR before resorting to the use of improved germplasm in the form of modern varieties. In a context of enhancing community resilience, grassroots breeding may be a rather successful strategy when compared with PVS and PPB, which primarily focus on selection and dissemination of improved germplasm.

Challenges for the further development of the grassroots breeding method

The idea of grassroots breeding has merits that should be evaluated in a wider geographical, institutional and socio-economic setting. In order for grassroots breeding to work, formal breeding and research institutions must enhance the capabilities of interested farmers and CBOs, and as such they should be encouraged to allocate resources to strengthen innovations at community levels. As illustrated by the variation of rather simple examples from Nepal, a range of initial experiences has been gained, but much remains to be learned from using grassroots breeding in a context of CBM and the enhancement of community resilience. Such a synthesis of experiences could guide global investments in crop research to be better targeted towards poor, small-scale farmers in marginal production environments. The work of both international and national public breeding programmes in using variations of grassroots breeding within larger PCI frameworks has the potential to increase their efficiency in responding to the changing demands of those farmers.

5.3 Participatory domestication of the fruit tree species feijoa (*Acca sellowiana*) in Brazil

*Karine dos Santos, Nivaldo Peroni,
Raymond P. Guries, Joel Donazzolo and
Rubens Onofre Nodari*

Feijoa: a native species in the process of domestication

Brazil is home to hundreds of fruit tree species. Despite their economic potential, several of these species are being threatened by changing landscapes. A number of efforts are being made to domesticate them, by contributing to their conservation. Feijoa, or pineapple guava (*Acca sellowiana*, synonymous with *Feijoa sellowiana*), is such a fruit tree species. Feijoa is native to southern Brazil and northern Uruguay. Since the 1950s, the potential of feijoa has been explored in several countries. The most successful breeding programme resulted in wide-scale commercial production in New Zealand, as well as in Colombia, albeit on a smaller scale. In Brazil, feijoa is cultivated on just a small scale at local level, even though the fruits have a unique flavour and aroma, and are attributed with having a number of medicinal values (Vuotto *et al.*, 2000; Bontempo *et al.*, 2007). Although feijoa is well adapted, its use remains limited. Since it has the potential to contribute to forest conservation and to ensure sustainable livelihoods for small-scale farmers in rural areas through its use as a non-timber forest product, several activities are being carried out in Brazil in order to promote its use (Ducroquet *et al.*, 2000).

Research programme on feijoa domestication

In this chapter we share our experiences as professors and doctoral students associated with the Federal University of Santa Catarina (UFSC) in using a participatory approach for the domestication of feijoa. Domestication can be achieved when the selection, propagation and improvement processes result in phenotypic and genotypic changes. Those plants that possess the desired traits can be used to promote commercial cultivation and are therefore selected (Clement, 1999). Since we are only at the initial stages of developing feijoa as a crop, we refer to the process as domestication, but the participatory structure of the process has several similarities with grassroots breeding, as described by Sthapit *et al.* (Chapter 5.2). We focus on participatory domestication in this chapter and do not address aspects of conservation and marketing, though we recognize that one cannot be separated from the other. We share our experiences working with participatory feijoa domestication in the states of Santa Catarina and Rio Grande do Sul, in southern Brazil.

Feijoa in Santa Catarina and Rio Grande do Sul

Genetic diversity of *A. sellowiana* can be found in the Atlantic Forest, situated in the states of Santa Catarina and Rio Grande do Sul in southern Brazil. The historic development of livestock and monoculture cropping systems in this region has limited the marketing of feijoa products. In addition to being dominated by livestock, commercial apple orchards, pine plantations and cereals are important economic activities in the region; these result in monoculture systems that are predominantly embedded in a matrix of forest vegetation on steep slopes. As such, little room exists in this agricultural landscape for the commercial use of indigenous tree species like feijoa. Recently, development has begun to focus on rural tourism and organic agriculture, creating more room for alternative production systems in which feijoa may become a driver for more 'biodiversity'-friendly fruit production, with value chain development that supports small-scale producers. In Santa Catarina, UFSC and Santa Catarina State Enterprise for Agricultural Research and Rural Extension (Epagri) are working together with several groups of farmers engaged in feijoa cultivation and management.

In the municipalities of Ipê and Antonio Prado, in Rio Grande do Sul (Figure 5.3.1), wild populations of *A. sellowiana* exhibit substantial genetic and phenotypic variation. The Centre of Ecology (CE), a local non-governmental organization, has been promoting organic, agroforestry-based production systems for the past 20 years. CE emphasizes the use of native fruit species, including feijoa; disseminates seedlings and cuttings of indigenous species; and provides assistance in developing value chains for marketing

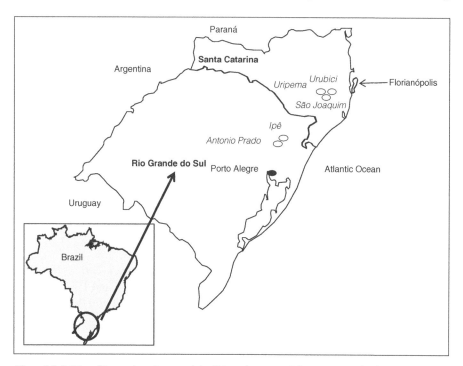

Figure 5.3.1 Map illustrating the municipalities where participatory genetic domestication of feijoa is being carried out.

fruits and fruit products. CE uses participatory methods in its developmental work and relies on the availability of genetic diversity maintained on farm and in native forests. In 2008, CE and UFSC formed a partnership, in which Econativa Cooperative, the State Foundation of Agricultural Research (FEPAGRO) and other institutions joined, to further structure the participatory domestication of feijoa.

Steps towards the participatory domestication of feijoa

We identified the following steps as being necessary for the participatory domestication of feijoa based on our experiences in carrying out research on feijoa genetic resources in Santa Catarina and Rio Grande do Sul. We identify seven steps, but have so far we have only completed up to step 6 of this programme, since we are dealing with a perennial species and the process takes time.

Step 1: Development of a collaborative framework

To begin with we identified and mobilized partners, especially farmers, as well as representatives of local development and research organizations. An agreement was then drafted, in order to define everyone's responsibilities and rights. In Brazil, such an agreement, concerning genetic resources and associated traditional knowledge, is governed by federal legislation related to access and benefit-sharing (ABS).

Step 2: Formation of key farmer working groups

For the selection process, we found it was crucial to work with groups of farmers only. This made it possible to avoid having to deal with any individual farmer's demands that were not in line with conservation, for example. The diverse nature of these small groups enabled us to encounter a variety of demands and, as we made progress in crop improvement, to become more efficient in disseminating selected materials.

In Santa Catarina, the characterization of four types of feijoa users facilitated the identification of farmers willing to take part in the participatory domestication. In Rio Grande do Sul, CE actively supported the establishment of a group of farmers who meet on a regular basis to plan and direct the actions of the feijoa participatory breeding programme.

Step 3: Assessment of production system

We conducted an inventory of traditional knowledge associated with *A. sellowiana* in Santa Catarina in the municipalities of São Joaquim, Urubici and Urupema (see Figure 5.3.1). Both qualitative and quantitative ethnobotanical methods were used; 54 informants were included in this assessment. The inventory used the following typology of informants:

- *custodians*, who simply use the plants for home consumption and derive little or no income from feijoa;

- *managers*, who collect and propagate selected plants and, to some degree, manage their feijoa plantations; they use the fruits in domestic value addition activities (such as making jams, juices and medicinal infusions) from which they are able to generate some income;
- *cultivators*, who actively establish feijoa orchards and market fruits, have strong links with research and extension agents, and use released, improved genotypes;
- *users*, who utilize the fruit, for example, in cooking, reforestation, and in the production and sale of seedlings, but who do not actually cultivate the species.

We found no significant difference in knowledge between the four groups concerning the use of feijoa, implying that conservation and use activities can involve everyone. However, we did find a significant difference in knowledge concerning the management of feijoa by managers and cultivators, and as such any activities to promote the management and production of feijoa, in particular through participatory feijoa domestication, should start by addressing these differences (Santos *et al.*, 2009).

In Rio Grande do Sul, a number of working groups were set up by CE, with the aim of promoting the use and conservation of feijoa, using participatory methods to obtain higher yield genotypes and resistance to pests and diseases; and to enhance the community's capacity in marketing feijoa products. In addition, we interviewed 39 farming households, with a total of 70 informants, in order to systematize the traditional knowledge associated with the use and management of the species. Through the assessment we were able to identify 11 different management practices.

Step 4: Participatory and market-oriented identification of key traits

We are currently in the process of incorporating the demands of farmers and consumers into the selection criteria, and have found that priorities can vary among groups and consumers. With regards to the group of farmers that was formed in Rio Grande do Sul, we used participatory methods to establish the selection criteria. In this manner, farmers contributed to the selection of promising genotypes.

Step 5: Development of a breeding strategy

Since *A. sellowiana* is an out-crossing species – the simplest and least expensive breeding strategy is that of 'mass selection', using open-pollinated seed collected from selected (maternal) plants from each generation. For many genetically diverse species, this method is an adequate way of obtaining improvements in each generation, and of contributing to the maintenance of genetic diversity. The vegetative propagation (grafting) of unusually productive individuals can 'secure' or 'genetically fix' desired traits and allow for multiplying certain selections (Leakey and Akinnifesi, 2008). However, this method cannot be considered breeding, since it does not create new genotypes for selection. If we followed this method, we would minimize our contribution to conservation. Vegetative propagation carries the risk that only one or a few preferred selections will become widespread, thereby reducing the level of diversity overall. Among the various breeding strategies, mass selection offers the best opportunity to balance cost, technical feasibility and genetic improvement, while also contributing to conservation.

In Rio Grande do Sul, before beginning its collaboration with UFSC, CE supported the establishment of more than a dozen small orchards, using cuttings originating from seed collected from approximately 15 selected trees. The selection criteria used to choose the parental plants were fruit size and quality, skin texture and distinct ripening time. As expected, the populations generated from the seed of selected trees showed wide segregation, especially for fruit size. Currently, UFSC and CE are engaged in Rio Grande do Sul in identifying potential parents by segregating populations for further selection. In addition, the group of farmers selected ten promising genotypes to be utilized in crosses. Using their own selection criteria, farmers selected from over 300 genotypes. Eight of the farmers decided to test 500 plants from the progenies obtained.

Step 6: Cycles of mass selection

We need to know which genetic parameters to use when selecting traits, in order to predict to what degree improvement would be possible. Those heritability estimates that are available for some traits, especially for fruit weight and the ratio of fruit length/diameter, suggest that mass selection would probably be effective in selecting among wild plants of *A. sellowiana* (Santos, 2005). The process of domestication takes several cycles and years with a perennial like feijoa, but in the long run we aim to have completed a number of selection cycles to balance genetic improvement concerns with the maintenance of genetic variation related to plant adaptation (Leakey and Akinnifesi, 2008). Ten field tests have already been carried out in the southern region of the country, with distinct progenies from selected parents.

In Rio Grande do Sul we are currently conducting such genetic studies on population structures and testing progenies for future evaluation. In Santa Catarina, these studies have already identified promising genotypes (Santos, 2009). The results of the studies from both Rio Grande do Sul and Santa Catarina will help us to identify parents to be used in the cycles of mass selection that we are planning for the future. Furthermore, the genetic studies will also be useful for defining strategies that contribute to the conservation of natural populations. We are currently providing guidance to our local partners in both locations related to the identification of priority areas for conservation, the genetic characterization of natural and managed populations, and the characterization of fruit traits of selected trees.

Step 7: On-farm evaluation

Members of the farmer working groups carry out trials in order to facilitate the identification of desirable plants in dissimilar environments and production conditions. Farmers evaluate distinct progenies from selected parental materials in ten field tests located in both states. These trials allow for the evaluation of agronomic performance under grower conditions, while providing estimates of production costs and the efficiency of agronomic treatments under realistic conditions.

Development of capabilities to maintain interest and conduct follow-up actions

For each of the steps of the participatory domestication process, we organize workshops to manage and discuss data and results and also to maintain the interest of all the partners. As such, we aim to be efficient in the selection process and provide an opportunity to share identified and selected materials at any one time. The workshops enhance the capacities of farmers' groups, as well as research and development partners. They also encourage participants to think beyond the breeding process, to contribute to the development of value chains, and to promote the creation of local and sustainable markets for feijoa fruit and other feijoa products.

The research group and its role in participatory domestication

Since we work as a university group on topics related to plant genetic resources, we are able to conduct genetic and ethnobotanical studies that can contribute in a vital manner to designing a long-term process, such as the participatory domestication of a fruit tree species like feijoa in southern Brazil. In this way, we are able to contribute to the grassroots breeding of such a perennial species, which can be considered a reference species for the use of indigenous species within larger domesticated landscapes. Since we are dealing with a time-consuming process, the collaboration between universities, farmers' groups and their organizations is vital, as other breeders may not be able to engage themselves in such a long process in a similar manner. It is crucial that we are able to link up through our studies and encourage our partners, so that we do not lose the context of conservation during such domestication processes. At the same time, we can foster value chain development, creating benefits for small-scale farmers for treasuring this potentially valuable genetic resource. It is our long-term goal that through the development of such studies, we will be able to balance conservation, crop improvement and economic benefits in promoting the use of feijoa in the region.

5.4 Farmer and participatory maize breeding

Increasing farmers' autonomy and promoting the use of diversity in France

Bertrand Lassaigne and Jennifer Kendall

Farmer and participatory breeding: increasing the autonomy of paysans

Small-scale farmers (*paysans*) who are engaged in organic agriculture and regional production in France are increasingly avoiding the use of commercially developed hybrid maize varieties. They are also increasingly rejecting the use of – or are prohibited from using – genetically modified maize varieties. Instead of purchasing seed each year, they are maintaining local maize varieties. Since 2001, the Association for the Development of Organic Agriculture in Périgord (AgroBio Périgord) has been collecting and testing local maize varieties from France and other countries in its diversity platform. AgroBio Périgord has established a seed bank, Maison de la Semence Paysanne, for maintaining and providing farmers with access to local varieties. The Maison de la Semence Paysanne forms the basis for work related to the community management of biodiversity in France; it is described in more detail by Kendall and Gras in Chapter 1.7.

One of the key motivations for farmers to use local maize varieties is their desire to rebuild their autonomy. They manage their maize varieties in a dynamic manner and even conduct some breeding and selection activities, facilitating cross-pollination that is inherent to maize. The diversity platform has been instrumental in motivating farmers to get involved in breeding activities, and to adapt scientific breeding techniques to their own capabilities and resources. In this way, farmers contribute to the management of maize diversity in France, while increasing their autonomy by using their own varieties. In the current chapter, we share some of our experiences working with farmers on participatory maize breeding.

Seed multiplication for the Maison de la Semence Paysanne

Farmer members of the Maison de la Semence Paysanne are vital for maintaining and disseminating seed; their contribution is described by Kendall and Gras in Figure 1.7.4. Farmers produce seed of acquired varieties in on-farm multiplication blocks, using appropriate isolation in space and time, and an agreed selection protocol. These blocks cover an area of about 0.1–0.2 hectares. The seed of varieties of interest are maintained for use, with surplus seed being sent to any new farmers who want to participate the following year, or being returned for storage. When we receive interesting varieties that are not well adapted to our production environments, we regenerate them through on-farm conservation blocks. However, since the purpose of regeneration is to conserve varieties and prevent their loss, the plots are small in size. Only a few farmers are willing to host such conservation blocks because they require isolation, monitoring and special care. The varieties are regenerated through controlled pollination in conservation blocks in the diversity platform. This is a procedure that we would prefer to avoid, but we are constrained for want of a better alternative.

Dynamic management, farmer breeding and participatory maize breeding

AgroBio Périgord and its members are currently engaged in several variations of farmer breeding and the participatory breeding of maize. Farmer breeding involves dynamic management in which farmers maintain and improve their local maize varieties. In this chapter, we look at participatory maize breeding activities that are being implemented in collaboration with Guy Thiebaut, who is a professional breeder. All the activities are conducted by farmers, in farmers' fields. An important motivation for our breeding activities is that we want to show that farmers can become independent from the commercial breeding companies who offer only limited choices of hybrid and genetically modified maize varieties. Such varieties are not well adapted to our (organic) production system, and genetically modified organisms (GMOs) are simply not allowed and not appreciated in organic agriculture. We focus our dynamic management and breeding activities on practicality and on an increase in diversity and autonomy. Here, we summarize our experience with the four methods used for dynamic management, farmer breeding and participatory maize breeding.

Method 1: Farmers' dynamic management of local varieties

Farmers use several variations of mass selection in the maintenance and selection of their varieties. Most commonly, they use positive mass selection, thus selecting better performing plants for seed production. In some cases, they use negative mass selection, thus detasselling plants before flowering. Each farmer conducts selection using his or her own perception of what the local variety constitutes.

Method 2: Farmers' management of composite varieties

Some farmers take an additional step in selection; they create composite varieties from a mixture of different, open-pollinated populations that often include several local varieties, but which may also include improved varieties. The idea is to increase genetic diversity and to combine different characteristics of the chosen varieties. These composite varieties can be the result of crossing just two different varieties, or they can be the product of crosses between more than 15 varieties. Farmers create their own composite varieties according to their own personal objectives and preferences. One farmer, Bernard Andissac, for example, created the composite variety Andico, by crossing the early-maturing French local white maize variety, Blanc de

Bresse, with a late-maturing South American blue maize variety. His goal was to make the blue variety mature earlier by crossing it with the white variety. In another example, farmer Bertrand Lassaigne created the variety called Lavergne. He carried out crosses between more than 15 different varieties chosen to create a new composite variety that would be well adapted to the conditions at his farm.

Every year the farmers continue to use mass selection in their composite varieties for seed production. Several varieties used today by farmers associated with AgroBio Périgord have been grown in this manner for up to ten years. Experiments carried out at the diversity platform and in farmers' field conditions have shown that these composite varieties have valuable characteristics.

In dynamic management, as applicable to methods 1 and 2, composite varieties originating from the same source and accessed through the Maison de la Semence Paysanne become distinct composites that are adapted to different locations and farmers' preferences. As illustrated in Figure 1.7.4, these varieties may be returned to the Maison de la Semence Paysanne and then included in the diversity platform for comparison. If they prove interesting, they can be included on a more permanent basis in the Maison de la Semence Paysanne for wider distribution. The network of farmers, in combination with the diversity platform and the Maison de la Semence Paysanne, contributes to the dynamic management of local maize varieties at a national level beyond that of individual farmers.

Method 3: 'Brazilian selection method'

Since 2008, we have been introducing members of our network to a particular method for developing composite maize varieties. A Brazilian agricultural technician, Adriano Canci, inspired us with the method during an exchange visit, which is why we refer to it as the 'Brazilian selection method'. Canci *et al.* (Chapter 1.5) and Ogliari *et al.* (Chapter 5.7) provide further details on some of the work in Brazil that inspired us to apply their methodology to the participatory enhancement of a local maize variety.

The purpose of the method is to develop a composite variety in a relatively short time. We learned that the Brazilian method takes 3–4 years to create a homogeneous population. An important advantage is that it is relatively straightforward and can be easily implemented by individual farmers to develop their own well-adapted composites. The method is particularly interesting for AgroBio Périgord because it can be used to develop composites adapted to organic production.

The method can be summarized as follows. In the first year, we select, for example, ten populations. From each, we take 300 seeds, using 150 as 'mother' plants and 150 as 'father plants', providing pollen. During harvesting, we collect cobs from the 24 best female plants of each of the populations. In the second year, 13 seeds from each of the 24 cobs are planted to maintain the composite. One hundred and fifty seeds from each of the selected 24 cobs are planted to be mother plants, while another 150 seeds are planted to contribute pollen for the pollination mixture. During the following harvest, 24 plants are selected and subsequently multiplied for another two years using the same procedure. In the fifth year, we start to apply cycles for stratified mass selection to reduce the variation for some traits in the newly created composite.

Bertrand Lassaigne began to apply the Brazilian selection method in 2010, using some of his own composites that he had been developing since 2001, and including some productive but late-maturing Brazilian varieties. For each late-maturing Brazilian variety, he included two early-maturing local ones. He further added varieties with an intermediary cycle that served as a 'genetic bridge', bringing the total number to 25 varieties. Another farmer from Périgord, Didier Margouti, also began to create his own variety in 2010, developing 12 different varieties with this method. Guy Thiebaut began a similar breeding programme in 2009 at his farm, including 12 varieties, of which ten were local varieties from Périgord. Today, several farmers are interested in this method. In 2012, three new farmers in other regions of France began to use the Brazilian selection method. Testing of the first new composite varieties created with this method will start in 2013 in farmers' fields and at the diversity platform.

Method 4: Development of hybrids based on local varieties

AgroBio Périgord has been implementing a breeding programme, in collaboration with Guy Thiebaut, since 2001, aimed at developing maize hybrids based on local French varieties. The idea was to combine their distinct qualities, and explore how to increase their yield potential, while aiming to adapt them to organic production systems and the agro-climatic conditions of south-western France. In Figure 5.4.1, we visualize this complex procedure for the development of the hybrid maize variety as implemented by AgroBio Périgord and its member farmers.

In 2001, we selected 12 individual plants from five local French varieties, planting the seed of each selected cob in individual rows in two sites. The five best half-sib families of each variety were selected based on yield potential, plant vigour, disease resistance and drought tolerance. Instead of self-pollination, we conducted pollination between half-sib plants, developing what we refer to as inbred families. We selected the 11 most vigorous plants within each inbred family, planting the seed of five plants for self-pollination and six for pollination between full-sib plants. In the subsequent year, we only used self-pollination to create the full-sib families that were included in the following year in tests for 'combining ability', for which we used isolated areas to avoid contamination. Four farmers agreed to set up such genetic islands. Each block contained six inbred families that were used as male parental plants, and 12 inbred families used as female parental plants. The female inbred family plants were pollinated with mixed pollen from several plants of one male inbred family; this made one cross. We were able to make 200 crosses between the inbred families, which gave 200 F_1 progenies. We planted these progenies in an 8 m row without repetitions, and used commercial hybrids as control. An initial group of progenies were eliminated based on their visual appearance (e.g. limited vigour, agronomic performance). We selected 50% of the remaining F_1 progenies based on visual observations of their production potential. The selected progenies were included in another trial with two repetitions for extensive agronomic evaluation. A progeny with the code B53, originating from a cross between inbred families, based on Grand Cachalut × Ruffec, proved to be outstanding. In 2007, seed for the hybrid B53 was produced, while the remaining seed was used in more trials to evaluate their production potential. In 2008, we evaluated a number of variants of B53 and they confirmed the potential of this hybrid.

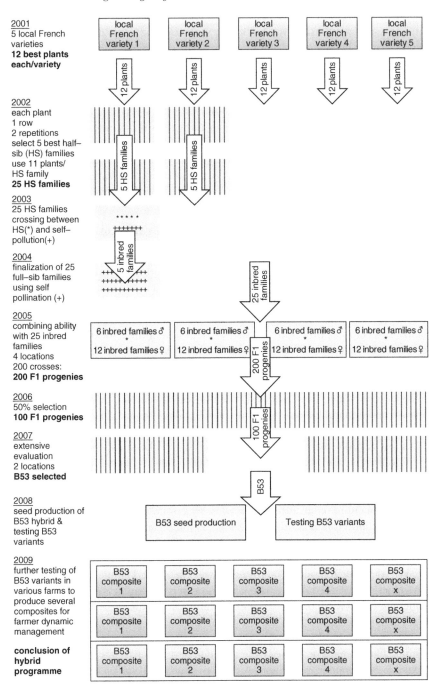

Figure 5.4.1 Procedure as used by AgroBio Périgord and its member farmers for developing a hybrid maize variety based on local varieties.

Source: Based on Bio d'Aquitaine/AgroBio Périgord (2011).

Since 2009, the hybrid programme has taken another direction, following the demands of farmers. We began to test the descendants of B53 in order to evaluate how they evolve through dynamic farmer management.

Reflections on different farmer and participatory breeding methods

In 2009, the farmers who were engaged in hybrid breeding mobilized themselves to finalize the development of the hybrid B53. We were ready to establish a group of farmers who would collectively maintain inbred families, produce B53 seed on a large scale and continue with testing other hybrids. However, parallel to our work in producing the hybrid, the Brazilian farmer selection method had begun to bear fruit. More and more farmers have been adopting this method to improve their local maize varieties as it is less complex, allows individual farmers more freedom, and promotes autonomy in varietal development and seed production. It became clear that our hybrids would require a more sophisticated organization of farmers, and technically complex procedures for maintenance and multiplication. Although we were prepared to engage in the development of hybrids, we realized that this would lead to an increase in dependency among farmers, instead of the envisaged autonomy. In addition, results concerning the production capacity of the B53 hybrid were not convincing enough when we evaluated them. We also considered that the seed production of this hybrid variety would mean engaging in a technically complex process. Moreover, farmers and the network would have to invest much time and resources in producing seed of this hybrid variety.

The Brazilian method allows for a wider use of diversity, which is one of our key goals. Farmers lost their enthusiasm for continuing with the hybrids and we decided to continue trying to develop a composite local population using the proven interesting combination of Grand Cachalut and Ruffec. By focusing on the development of a composite local population, we could tackle the issue of dependency associated with the hybrid approach.

Farmer breeding, maize diversity and autonomy

We have learned many lessons from our work in farmer and participatory maize breeding. We realized that there is room for farmers to use their own creativity in managing local maize varieties, and in maintaining and developing their own varieties. As in many other parts of the world, where small-scale farmers are interested in keeping their autonomy and in operating independently from large-scale seed companies, many small-scale farmers in France maintain and develop their own varieties. Small-scale farmer colleagues in Brazil, Guatemala and in many other countries inspired us to follow similar pathways. We have seen that an increasing number of farmers who produce maize in organic farming and for regional production, but also those who opt to use their local varieties within conventional farming, engage in the dynamic management of local maize varieties.

The diversity platform and Maison de la Semence Paysanne have been instrumental in sharing knowledge and materials, and in motivating farmers to experiment with several breeding methods, varying from simple mass selection to more sophisticated methods of developing hybrid varieties. What we have learned is that in farmer and participatory maize breeding we promote the use of diversity, strengthen farmers' capabilities in dynamic management, and guarantee their autonomy in management and selection. We may seek sophistication in, for example, developing hybrids, but the process of producing seed of hybrid varieties bypasses one of the key pillars of the farmers' management of maize diversity (i.e. the farmers' autonomy in managing their diversity within the specific conditions of their farms and according to their specific preferences). A key lesson to be shared with others in similar situations is that we need to cherish autonomy and, more importantly, the skills and enthusiasm of farmers for maintaining and using diversity.

Note

The editors would like to thank Juliana Bernardi Ogliari for her contributions to this chapter.

5.5 Participatory genetic enhancement of the Jethobudho rice variety in Nepal

Sachesh Silwal, Sanjaya Gyawali,
Bharat Bhandari and Bhuwon Sthapit

Participatory genetic enhancement and community biodiversity management

Farmers have their own individual preferences and needs and this is reflected in the diversity of crops and varieties that they grow. Farmers' preferences depend upon the specific characteristics of their agro-ecosystem, as well as on socio-cultural, religious or nutritional use values. Traditional local varieties, which have developed over a long process of domestication, selection and interaction with the environment, often show variation for adaptive or qualitative traits. That is why small-scale farmers in many production systems still continue to cultivate traditional varieties and why the introduction of improved varieties did not result in the complete disappearance of locally adapted, traditional crop species and varieties. Participatory crop improvement (PCI) aims to further support the use of those local varieties within their production systems (Jarvis *et al.*, 2011).

In the current chapter we share our experience of the participatory genetic enhancement (PGE) of a local rice variety in Nepal. Our PGE work contributes to the conservation and utilization of a famous traditional rice population known as Jethobudho, in Pokhara Valley, Nepal. Our experiences are based on an eight-year process that formed part of an agrobiodiversity project described in more detail by Subedi in Chapter 1.2. The objective of our project, implemented by Local Initiatives for Biodiversity, Research and Development (LI-BIRD) together with the Nepal Agricultural Research Council (NARC) and Bioversity International, was to strengthen the scientific basis for *in situ* and on-farm conservation of agricultural biodiversity. We identified PGE as one of the strategies for contributing to on-farm management of agrobiodiversity.

Jethobudho: the identity of Pokhara Valley

Jethobudho is an aromatic local rice variety from Pokhara Valley, in the middle hills of Nepal. It is valued for its excellent cooking qualities, such as its softness, taste, aroma and volume expansion after cooking, and for its superior milling recovery. In Pokhara Valley, Jethobudho plays a special role in feasts and festivals. With Pokhara being a major tourist hub, the variety is in great demand in hotels and restaurants, which again creates a specific market. Visitors to Pokhara take home the sweet memory of Jethobudho bhat, a traditional Nepali rice dish. The variety has carved out its own identity and it has become an intrinsic part of the culture of Pokhara Valley.

Deterioration of Jethobudho

Jethobudho is a fundamental part of Nepali daily life and is a dominant variety in terms of local area coverage. Market surveys have shown that consumers are ready to pay a good price for quality Jethobudho rice (Poudel and Johnsen, 2009), although variability in quality in the 1990s was a major concern for consumers and those involved in marketing the rice. Limited access to seed, in terms of quantity, quality and also genetic identity, were some of the problems experienced by farmers interested in producing the variety on a commercial basis (Gyawali *et al.*, 2010).

The availability of Jethobudho seed was limited because populations of local varieties were not able to comply with formal release procedures; a population of a local variety like Jethobudho would not pass variety release. Without official release, Jethobudho could not be included in formal seed production programmes. In addition, a number of diseases emerged in the late 1990s; neck blast, for example, was identified as a limitation for the production of Jethobudho. The genetic purity of Jethobudho also deteriorated, contributing to its genetic erosion. These problems diminished the competitive ability of Jethobudho rice compared with other high-quality rice varieties. Because of these issues, in addition to the fact that it would also contribute to its conservation, we selected Jethobudho for population improvement. With the CBM approach, in which the PGE work was embedded, we aimed to link conservation with economic development by integrating traditional varieties into commercial markets. Our goal was to enable a larger number of farmers to benefit economically from the large-scale production of high-quality Jethobudho (Jarvis *et al.*, 2008).

The rescue of Jethobudho through participatory genetic enhancement

With the objective of conserving the true Jethobudho rice, LI-BIRD, in collaboration with NARC and Bioversity, began the PGE of Jethobudho, supported and accompanied by market-oriented value-addition activities. Quality traits of post-harvest grain are important factors for marketing rice varieties. Jethobudho is considered to be a high-quality local rice variety but, until this activity, its post-harvest qualities were poorly understood. Here, we share a series of seven steps that we used for the PGE of Jethobudho.

Step 1: Appraisals and collection

We began the Jethobudho project in 1998, holding extensive participatory rural appraisals and focus group discussions in order to identify quality traits that could then be used as selection criteria for identifying and collecting Jethobudho accessions. In the following year, 338 accessions of Jethobudho were collected from mature Jethobudho populations, each from a different farm in Pokhara Valley, where the local variety was being cultivated in large areas. Farms were chosen because the community considers the farmers to be custodians of Jethobudho, and as such they were known for the quality of their seed. Five panicles were selected and the bulked seed from those panicles was marked as an accession (or Jethobudho sub-population).

Step 2: Assessment of Jethobudho accessions

The 338 accessions were screened on-farm and on-station in subsequent years. During the first visual assessment we selected 183 accessions, focusing on traits such as field tolerance to leaf and neck blast, lodging tolerance and yield components. Over the course of intensive screening, which was carried out in three different sites, we targeted shorter plant height, higher number of effective tillers per plant, longer panicle length, higher grain yield, and, most importantly, good performance in post-harvest grain quality traits. Using these criteria, we selected 143 accessions. The main criterion used in the agronomic screening of Jethobudho was that of field level tolerance to blast and lodging.

Step 3: Market survey and screening

A household-level survey on post-harvest quality traits indicated important characteristics, such as milling and head rice recovery percentage, milled rice colour, and the softness, taste, flakiness and aroma of the rice after cooking. Using these characteristics we screened the 143 Jethobudho accessions for milling and organoleptic traits, and post-harvest physiochemical properties. Milling and head rice recovery were more than 70%, which is an attractive trait for rice millers and consumers. To be selected, the organoleptic score of selected Jethobudho accessions should always be higher than the average of the 143 accessions. We also evaluated the 143 accessions for leaf and neck blast, as well as for lodging, plant length, yield components and maturity. Combining the laboratory and field data, we selected 46 accessions. Furthermore, we identified amylose content (23–24%), intermediate gelatinization temperature, and higher volume expansion and water absorption as future selection criteria. These attributes make the local varieties competitive for national as well as international markets for high-quality, Basmati-type rice (Gyawali *et al.*, 2010).

Step 4: Final screening for post-harvest traits and PVS

We selected under conditions that also allowed us to select for blast resistance. In addition, we selected relatively early maturing accessions, which, combined with higher scores for post-harvest quality traits such as rice colour, the softness, flakiness and aroma of cooked rice, overall cooking quality, and other preferences, contributed to a genetic gain for improved Jethobudho populations. Six accessions were selected for scaling-up on the basis of various post-harvest quality traits, and were later bulked as a multi-line variety for release as 'Pokhareli Jethobudho'.

Step 5: PVS to re-insert the variety into the seed system

Farmers' preference for, and acceptance of, enhanced accessions of Jethobudho was important for the dissemination of Pokhareli Jethobudho. In 2004, 12 farmers compared their pipeline variety with their own local Jethobudho variety, for both agronomic and post-harvest aspects. Few farmers could participate since the quantity of seed available was still too low. In the following year, 130 farmers participated in an on-farm evaluation of Pokhareli Jethobudho and provided their feedback on the performance and the quality of the rice. Each farmer received a small quantity of seed. We asked the farmers to grow Pokhareli Jethobudho alongside their original Jethobudho. We distributed seed over the course of three consecutive years to 10, 33 and 44 farmers, respectively. We received their feedback through a survey on both agronomic and post-harvest criteria. Table 5.5.1 shows the results of the survey of the final PVS trial, which included 44 farmers. PVS proved very effective in promoting the variety, as reflected by the fact that after its implementation many farmers approached LI-BIRD for Pokhareli Jethobudho seed (Gyawali *et al.*, 2010).

Step 6: Release of Pokhareli Jethobudho

In order to make quality seed of Pokhareli Jethobudho available to farmers, we approached six custodian farmers who had provided us with the Jethobudho accessions that were finally selected. Together with these farmers, we took the initiative to release Jethobudho formally. In 2005, we invited the national committee responsible for variety release for a field visit and for assessments, both in farmers' fields and in on-station trials. In the release proposal we included comprehensive data of field screening, laboratory screening, field trials, consumer surveys, PVS and molecular characterization for aroma. The formal release of the enhanced Pokhareli Jethobudho variety in 2006 paved the way for the community-based seed producers to market the variety.

Step 7: Seed production of Pokhareli Jethobudho

Foreseeing the great demand for seed, we supported the farmers in organizing themselves into the Fewa Seed Producers Group. We provided the group with foundation

Table 5.5.1 Farmers' perceptions on preferred agronomic and post-harvest traits of Pokhareli Jethobudho, as compared by 44 farmers with their own local varieties in PVS, 2005

Agronomic traits	Pokhareli Jethobudho, in comparison with farmers' own local varieties (%)		
	Better	*Same*	*Worse*
Plant height	57	17	26
Neck blast tolerance	85	15	0
Lodging tolerance	91	2	3
Grain yield	60	26	14
Post-harvest traits			
Milling recovery	88	10	2
Softness	85	13	2
Aroma	95	3	2
Flakiness	68	30	2

Source: Adapted from Gyawali *et al.* (2010).

seed for seed multiplication. The Jethobudho custodian farmers became involved in producing quality source seed (breeders' seed), while other seed-producer farmers became engaged in producing seed that is truthfully labelled, in order to disseminate quality seed within and outside the valley. We tried to link Jethobudho custodian farmers, rice millers and rice traders to the seed producers group for the rapid dissemination of the enhanced variety. The District Agricultural Development Office (DADO), a government agricultural agency, is supporting the seed producers group, and has been instrumental in the dissemination of Pokhareli Jethobudho.

Current status of Pokhareli Jethobudho

The PGE work resulted in an increase in monetary benefits for custodian farmers of Jethobudho, and seed and grain producers, with the price of Jethobudho seed and grain in 2011 double what it was in 2005 (Table 5.5.2). The area under improved Jethobudho cultivation has been expanded to 11 districts in Nepal, from the original and sole area of cultivation in Pokhara Valley. The increase in area of cultivation has provided a broader group of farmers with better access to the new variety and greater monetary benefits through the higher price obtained for their rice grain crop. Expanding the area of Pokhareli Jethobudho cultivation has also resulted in non-monetary benefits for consumers, by ensuring the availability of the rice in the market, as well as its high quality and stable aroma.

Market linkage for promoting the use of Pokhareli Jethobudho

The experience of PGE of Jethobudho has shown that the farmer-preferred economic traits of populations of local varieties can be enhanced and can add economic value to a local variety, thereby contributing to its conservation and use. In the case of Jethobudho, what was crucial was the selection of both agronomic (blast resistance) and market-related traits. The establishment of a seed production scheme parallel to the PGE process proved vital as a means to disseminate the variety, creating benefits for custodian farmers, and for the farming communities involved in grain production.

Table 5.5.2 Status of Jethobudho rice marketing before and after the variety release

	2005 *(Jethobudho)*	*2010* *(Pokhareli Jethobudho)*
Production data	Original and diverse local varieties	Genetically enhanced local varieties
Total quantity of seed sold (kg)	1382	4788
Price per kg of seed (US$)	0.50	1.10
Farm-gate price of unprocessed grain per kg (US$)	0.40	0.90
Number of farmers involved in seed production	5	10
Area covered by Pokhareli Jethobudho (hectares)	N/A	120

Source: LI-BIRD Seed Unit (2011).

The experience of Jethobudho is the very first case in Nepal to demonstrate the usefulness of PGE for the benefit of the farming community through enhancing the values of local crop varieties, which has ultimately contributed to its conservation through on-farm management. By embedding PGE within a larger community organization, which is required for seed production and its dissemination, it has contributed to raising awareness of agrobiodiversity, establishing the social capital for making community biodiversity management work and creating direct and indirect benefits from conservation and use, as further elaborated on by De Boef *et al.* in Chapter 4.1.

Furthermore, PGE has created possibilities for a more regional branding of Pokhareli Jethobudho rice from Pokhara Valley. Even though geographic indication does not exist in Nepal, specific value chain development, as in the case of Vietnam, described by Thomas and Anh (Chapter 4.5), has been promoted in an informal manner through the PGE process. The experience has also had a positive impact on researchers and policy-makers, in appreciating the genetic enhancement, release and registration of multi-line populations based on selection within local varieties, despite their strong opposition in the past as described by Louwaars *et al.* (Chapter 6.1). Our experiences with PGE, and consecutive efforts to release Pokhareli Jethobudho, are similar to the experiences of the M.S. Swaminathan Research Foundation (MSSRF) with Kalajeera rice in India, as described by Chaudhury and Swain (Chapter 5.6). Our work in Nepal, and that of MSSRF in India, has contributed to creating seed policy environments in both countries that are more conducive to, and compatible with, agrobiodiversity conservation and use.

5.6 Providing scientific support to farmers using local rice diversity in Jeypore, India

Sushanta Sekhar Chaudhury and Saujanendra Swain

Introduction

The concept of the 'evergreen revolution' involves the enhancement of agricultural production and productivity without causing any ecological or social harm (Swaminathan, 2003). For plant breeding, this implies the conservation of genetic diversity and the development of varieties that are adapted to the local agro-climatic and agro-economic conditions, and that respond to the specific demands of the farming communities. The M.S. Swaminathan Research Foundation (MSSRF) has developed a methodology referred to as scientist-supported farmer breeding (SFB), which aims to develop adapted varieties while contributing to the maintenance of local genetic diversity. Linking the conservation and use of local diversity to the improvement of farmers' livelihoods leads to sustainable pathways of poverty reduction (Arunachalam, 2000), as envisaged in the evergreen revolution. The current chapter shares our experience working with this method in Jeypore, India.

The Jeypore Tract

The Jeypore Tract, located in the Koraput district of Orissa state, in India (see Figure 4.4.1), is recognized as a secondary centre of the origin of rice (Arunachalam *et al.*, 2006). It has a vast tribal and cultural diversity. Twenty-nine tribal groups reside in the Jeypore Tract, nine of which have a population of more than 100 000. The area is characterized by highly uneven and undulating land with varying slope gradients. Farmland is classified as uplands, midlands and lowlands. Midlands and lowlands are mainly used to cultivate rice.

Local rice varieties under threat

Most rice is grown in the rainy season. Tribal households in Jeypore cultivate local varieties based on their traditions. Farm-saved seed is the major source of seed. Specific local varieties are maintained for ceremonies and festivals. Since local varieties are of different maturity groups, they contribute to household food security throughout the year. The strategy of using farm-saved seed guarantees the continued use and conservation of a diversity of local rice varieties.

In the period 1955–1959, the Central Rice Research Institute in Orissa collected 1745 germplasm accessions of cultivated rice and 150 accessions of wild rice from the Jeypore Tract (Govindaswami *et al.*, 1966). From 1995 to 1996, MSSRF re-surveyed the area and collected 256 accessions (Tripathy *et al.*, 2005). In 1998, MSSRF re-visited previously explored areas and were able to collect only 98 accessions. The worrying extent to which the rice diversity had eroded led MSSRF to engage in direct action to revitalize the management of this unique genetic heritage (Arunachalam *et al.*, 2006).

Rationale behind scientist-supported farmer breeding

With the aforementioned objective, we at MSSRF aimed to revitalize and enhance farmers' management of traditional rice varieties. We had to embrace a new paradigm in crop development, promoting the collaboration of scientists with farmers as breeders, and the use of diversity as a principle, without, however, rejecting advances in scientific plant breeding. We developed SFB to utilize genetic resources and support farmers in their crop management, working together with tribal communities and farmers as equal partners. Each initiative and activity was discussed in a free and friendly way, offering constructive criticism. Only those steps that we all agreed upon were implemented.

Scientist-supported farmer breeding: the steps

The SFB process was implemented by MSSRF with the tribal communities of the Jeypore Tract, between 1998 and 2007. It included the following steps.

Step 1: Diversity blocks and travelling seminars

We collected local varieties from custodian farmers in eight villages in four different agro-climatic locations. In the following year, those varieties were included in a diversity block in the four agro-ecological zones. To evaluate the performance of individual varieties we provided farming families with a small quantity of seed of a single variety of their choice. We cultivated all varieties as a back-up on-station. We organized a travelling seminar to the four sites, where custodian families and MSSRF scientists evaluated the varieties. Together with the farmers, we discussed the agronomic data, using participatory rural appraisal (PRA) tools to facilitate discussion and encourage collective decision-making. At the end of this first step, we decided to continue with 26 varieties from the 98 collected materials for further selection in different rice ecologies. We subsequently evaluated the 26 varieties.

Step 2: First participatory varietal selection

We planned an experimentation scheme that included trial and demonstration plots for the 26 accessions. Varieties were tested in their appropriate lowland, midland or upland rice ecosystem, across four locations. We helped the farmers to prepare land, layout fields and plant using an experimental layout. One trial included three replications with individual plots of 30 m². We guided the farmers to use refined management practices. In an adjacent trial, farmers laid out single replication plots of 90 m² each using traditional management practices. We evaluated the varieties based on seven agronomic traits and calculated performance scores for each variety, based on those traits. We discussed the results with farmers and it was interesting to see that for each ecosystem, the farmers chose the same two varieties as those that outperformed the others using scientific varietal evaluation. Farmers selected nine varieties out of 26 varieties grown. Those varieties that were not selected, owing to poor performance, were replaced by an additional 22 varieties.

Step 3: Second participatory varietal selection

In the subsequent year, we continued with PVS in the same locations, evaluating the nine selected and 22 additional local varieties for a series of quantitative traits. Farmers learned to distinguish quantitative traits related to yield performance. We collected data on those traits, which were analysed statistically. For each rice ecosystem we identified the two best-performing varieties, based on two years of PVS. It was heartening to observe during the discussion of the outcome of these PVS trials that once again our selection, which was based on performance, matched the selection made by farmers. The decision was made to select one upland variety, two midland varieties and one lowland variety, and to produce and disseminate their seed for promotion on a wider scale. In the same seasons as those during which the PVS trials were conducted, farmers learned to identify and remove off-types as part of a special training on seed production. Over three seasons, farmers acquired the knowledge needed for quality seed production.

Step 4: The involvement of farmers in making crosses

During a capacity development activity, we explained to farmers the difference between pure line and multi-line varieties. We demonstrated how crosses are made, showing emasculation and pollination techniques using rice plants cultivated in pots. This activity raised curiosity and interest. Many farmers volunteered to emasculate and pollinate plants in a hybridization plot, which was laid out in a village (i.e. farmers were eager and curious to make crosses among rice varieties). We further explained the need for, and the process of, the staggered planting of varieties so that the emasculation–pollination process can be completed over a period of time.

Based on their initial performance in the PVS trials, seven varieties were selected as parents. Eight women and seven men from three villages volunteered to assist in carrying out the crosses. Some of the volunteers were old farmers with an increased interest in agriculture, while others were boys and girls. From among the 15, four were selected to be responsible for taking care of the crossed plants. Over a period of four weeks, the group made more than 10 000 crosses. However, due to lack of expertise in making crosses, many of the mother plants did not set seed, leaving only 593 seeds for crosses that involved five parent varieties; the success of pollination was only 6%.

Step 5: Early generation multiplication

F_1 seeds were sown in single plots for each cross. However, due to heavy intermittent rain during crop growth, some of the plots were completely washed out. The rains also caused the plants to mix with plants from adjacent plots, leaving only mixed F_1 plants from five crosses. From each plot allotted to F_1 from one cross, 200 g of F_2 seed were collected. The 200-g samples were planted in five lines. In the following season, populations from each of the five F_2 lines were cultivated in 8-m rows.

Step 6: Early selection and formation of F_3 populations

Two composite samples compiled from the seed of three plants were planted within each row. We measured plant height, number of tillers, number of panicles, panicle length, number of filled grains per panicle, and grain weight. Grain filling percentage was computed from the data. Statistical analysis of variance was conducted for each character. We grouped potential lines using the performance scores we had computed. Multivariate means testing for all traits assisted with the selection of F_3 lines that showed significant superiority over the parents. One F_3 line selected had outperformed four parents (S1); two lines that had outperformed three parents were joined in an S2 population; and two lines that had outperformed one parent constituted S3. Twenty-two other F_3 lines that had performed similarly to the parents formed S4. The remaining lines were discarded.

Step 7: Evaluation and advancement up to F_5 populations

We evaluated the four composed populations and advanced them to F_5 populations over the two subsequent seasons. Field observations showed synchrony of maturity in the populations S2, S3 and S4. S1 was still segregating at F_5, although it flowered early in comparison to the other populations. When compared with the parents, the four populations showed superiority for most traits. During F_6, we started to homogenize the populations. We involved farmers in all the steps, sharing the process of plant breeding with them. Farmers' groups began to produce seed of those populations and lay out demonstration plots to promote their use.

   During the final stage of the breeding programme we lost the seed of the four S1 to S4 populations. This was due to bad weather conditions in the grain-setting stage; heavy rainfall and a cyclone destroyed our work. Unfortunately, we did not have a back-up at that point in time. As a result, the SFB experiment with farmers making crosses with scientists in a collaborative manner, finalizing and releasing varieties was abruptly ended. This disaster concluded the experiment of MSSRF to design a methodology for SFB with tribal communities. Fortunately, we had begun another PCI programme in parallel, which focused on the participatory genetic enhancement of a local variety.

Participatory genetic enhancement of Kalajeera

The 12 years we spent collaborating with the farmers on the scientist-supported farmer breeding programme inspired us to begin working on the development of the improved traditional variety, Kalajeera. However, improved Kalajeera was not the result of farmers making crosses; it was only purified through a process of participatory genetic enhancement (PGE), a participatory crop improvement (PCI) method that is also described by Silwal *et al.* (Chapter 5.5). During the selection process, Kalajeera emerged as a better variety than other traditional varieties that are grown in lowland areas. Following the genetic enhancement of this variety, we began to work on preparing good-quality, improved Kalajeera seed, and to promote its commercialization, as we further describe in Chapter 4.4. We observed that improved Kalajeera has a very good market demand and farmers are interested in growing it on a large scale. The yield per hectare of this improved Kalajeera is substantially higher than that of the traditional Kalajeera varieties.

Registration of Kalajeera

Together with four other varieties, we have applied to register Kalajeera as a farmers' variety with the Protection of Plant Varieties and Farmers' Rights Authority. The SFB methodology has been instrumental in putting this globally innovative legislative and regulatory framework into practice. The formal recognition of Kalajeera as a farmers' variety will provide the farmers with the benefits of improved traditional varieties. Since improved traditional varieties are widely appreciated for their characteristics, a significant demand for Kalajeera seed has emerged. Those farming communities in the Jeypore Tract that are involved in breeding are now being rewarded by becoming commercial seed producers.

Impact on farmers' income

There has been a clear-cut 20% increase in economic benefits in cultivating Kalajeera, compared to the cultivation of other rice varieties. Kalajeera seed is sold at a 40% higher price. Its grain fetches a 50% higher premium because of its aroma, supported by the fact that it is produced in an organic (traditional) production system. The average net profit obtained by tribal households through Kalajeera cultivation, in 2010, was US$500 per hectare. Profits are higher if the farmer sells processed rice, which increases the economic benefits a further 40%. The average household income of farmers involved in the cultivation of Kalajeera rice is 20% higher than those who cultivate other traditional rice varieties.

Lessons learned

The communities participating in SFB expressed excitement at the novelty of the entire process. They also shared with us the fact that their participation in carrying out plant crosses and selection had demystified the notion that rice breeding is for scientists only. Almekinders *et al.* (2006) compared various cases of participatory plant breeding in south-east Asia, and included examples of farmers making crosses. In the design that was developed by MSSRF, SFB, coupled with CBM practices, has strengthened communities, who, as a result, assume responsibilities that lead to the continued use of traditional rice varieties (Arunachalam, 2007). By incorporating SFB into our approach for implementing CBM, we have been able to revitalize the management and promote the use of local rice varieties in the Jeypore Tract, thus strengthening genetic and ecological, as well as social, cultural and economic dynamism, which is crucial to achieving on-farm management, thereby supporting the perspective described by De Boef and Thijssen in Chapter 1.8. We realize that our work has contributed to the social empowerment of farming communities in terms of organization, for example, of experimentation and seed production. Our work has also contributed to the legal empowerment of the farming communities in the Jeypore Tract, in terms of their recognition of their custodianship over unique genetic resources and their formal release of Kalajeera as a farmers' variety. Furthermore, it has contributed to the economic empowerment of these communities, by generating income through the commercial seed production of Kalajeera. An outcome of the process is that with SFB we have been able to create community awareness of the value of, and ownership over, the local genetic resources, guiding a transition from the replacement of local diversity to the conscious use of diversity.

5.7 The participatory genetic enhancement of a local maize variety in Brazil

Juliana Bernardi Ogliari, Volmir Kist and Adriano Canci

Small-scale farmers and their local maize varieties in Santa Catarina

The agricultural sector in the state of Santa Catarina in southern Brazil is character-ized by small-scale farming, with an average farm size of about 20 hectares. Farmers face a number of problems, including changes in rainfall patterns, frequent droughts, low soil fertility, acidic soils, and outbreaks of pests and diseases. These variations in biotic and abiotic stress factors limit crop production and productivity in the small-scale farming sector. Maize is a very important crop as it is associated with the poultry and pork processing industry, which dominates the local rural economy (Ogliari and Alves, 2007). Less significant in size, but no less important, is the amount of maize consumed domestically, in the form of food for the household, animal feed and arti-san wares (for example, hats and purses) that are made with maize straw and grains. About 90% of maize produced is used within the farm (Vogt, 2005).

Autonomy in seed production and the use of maize varieties

The municipality of Anchieta is located in the western part of Santa Catarina. Social movements, such as the Small-scale Farmers' Movement (MPA) and the Women Farmers' Movement (MMC), as well as the non-governmental organization Support Centre for Small-scale Farmers (CAPA) and the Small-Scale Farmers' Trade Union (SINTRAF), are critical of the industrial agricultural model as supported by agri-cultural research and extension services. In order to turn this criticism into practical action, these organizations are aiming to enhance agro-ecology, the farmers' auton-omy in maize seed production and their capacity to maintain local maize varieties. A community-based organization, the Association of Local Maize Variety Farmers and Processors (ASSO), supported by SINTRAF, has been promoting the use of local or *crioulo* (creole) maize varieties (Ogliari and Alves, 2007). Milho crioulo is the term used by farmers for maize varieties that are maintained and multiplied by small-scale farmers, independently of their origin and cultivation period on-farm. They include local varieties, local composite populations, and seed derived from old, important, improved cultivars. In the late 1990s, crioulo maize varieties became a symbol for agro-ecology, autonomy and self-reliance.

NEABio's involvement in research and development

The activities of farmers' organizations in Anchieta attracted the attention of the Nucleus for Agrobiodiversity Studies (NEABio), which was formed by a group of professors and students of the Federal University of Santa Catarina (UFSC) in 2002. NEABio has been working on several research projects with farmers' organizations in support of crop improvement and on-farm management of maize and other crops. Over the years, an informal partnership has been developed for investigating and supporting on-farm management and the conservation and use of crioulo varieties, using maize as a model for other crops and other municipalities.

Motivation for the participatory genetic enhancement of a crioulo maize variety

Although numerous hybrid maize varieties with high yield potential are available in the market, small-scale farmers in the western part of Santa Catarina are reluctant to purchase and use them. Such varieties have specific requirements, such as the use of fertilizers and pesticides, and need efficient machinery and equipment, in order to explore their high genetic potential. The high cost of seed of hybrid maize varieties has been another limitation. The limited adaptation of the maize hybrids to their production system, in addition to the aforementioned economic reasons, motivated some small-scale farmers, after years of using hybrid maize, to return to open-pollinated local varieties. Open-pollinated local maize varieties (milho crioulo) are an interesting alternative for these farmers, as they require fewer external inputs. The varieties are also less vulnerable to biotic and abiotic stress factors compared to hybrid cultivars (Ogliari and Alves, 2007). In this context, NEABio began to carry out research in 2003, looking at the yield potential and other specific characteristics of these crioulo maize varieties. Together, we explored ways of how we could genetically enhance the performance of these varieties, using a collaborative approach, with the aim of strengthening the autonomy of the farming community in managing and using maize varieties.

Supported farmer breeding of MPA1

Before NEABio began its participatory plant breeding activities in Anchieta, farmers were already involved in farmer breeding, with the support of local organizations. The variety that we worked with in the participatory genetic enhancement process is the product of scientific farmer breeding. The development of MPA1 was initiated by a local farmer, Névio Alceu Folgiarini, with technical support from SINTRAF. The first step in its formation was the inter-crossing of 25 varieties, represented by 18 advanced generations of commercial hybrids; four local varieties, maintained by small-scale farmers at Anchieta; and three varieties derived from selection, which was carried out by local farmers from within the crioulo variety group referred to as Pixurum. In order to assemble the base composite, a sample of 50 ears (300 kernels per ear) was taken from each parental variety. In the following season, five other local varieties of unknown origin were incorporated into the base composite. These five additional varieties were recombined with the first 25 varieties over 1999 and

2000, using the 'topcross' procedure. Three cycles of stratified mass selection were applied to establish the population. The farmers that had been engaged in the process decided to name the variety MPA1, in honour of the social movement that supported and inspired them in their work.

Participatory genetic enhancement of MPA1

Step 1: On-station and on-farm evaluation of crioulo maize varieties

Since many agricultural communities maintain maize varieties on-farm, the first step in our research was to assess their yield potential. In 2002/2003, we evaluated 23 varieties at the research station of the Santa Caterina State Enterprise for Agricultural Research and Rural Extension (Epagri), in Canoinhas. The results showed a variation in grain yield for the crioulo varieties, with the variety that is referred to as MPA1 showing the best performance. In a subsequent year, we carried out an evaluation at the same station, separating varieties with a plant height above 3.0 m. These taller varieties showed a grain yield of between 2.1 and 7.2 tonnes per hectare (t/ha); the shorter varieties (less than 3.0 m) had grain yields of between 5.1 and 8.0 t/ha. Again, MPA1 proved to be outstanding. What we considered important was the fact that when compared with the average maize yield in Santa Catarina over the two respective years, the crioulo varieties turned out to be producing, on average, significantly more (62% and 59%). Table 5.7.1 shows the results of the on-station evaluation of the varieties.

In consecutive years, further participatory varietal selection was carried out on-farm, by evaluating crioulo maize varieties using different experimental designs (Canci, 2006; Kist, 2006). In subsequent research, the superior performance of MPA1 was confirmed when it was compared to improved and other crioulo varieties for agronomic traits (Kist *et al.*, 2010). MPA1 also showed superior values for kernel nutritional content (Kuhnen *et al.*, 2011), and for resistance to *Exserohilum turcicum*. At the request of farmers associated with ASSO and SINTRAF, we started a process of participatory genetic enhancement (PGE) of MPA1 in 2003, as illustrated in Figure 5.7.1.

Step 2: Stratified mass selection of MPA1

We continued the PGE programme using a modified convergent–divergent scheme, since it caters well with the demands of improving an open-pollinated variety for sustainable agriculture (Kist *et al.*, 2010). In this design, half-sib selections are tested under low-input farm crop conditions (i.e. in varying rather than uniform testing environments). Initially, we used MPA1 that had originated from the farmer Névio Alceu Folgiarini, over three cycles of stratified mass selection, applying a defined set of criteria during pre- and post-harvest selection.

Step 3: Selection of MPA1 half-sib families

Adriano Canci and Névio Alceu Folgiarini took a sample of 220 ears from the population in an isolated block, located in a uniform field, based on visual selection in the

Table 5.7.1   Yield potential of maize varieties from Anchieta, in experimental trials in 2002/2003 and 2003/2004, Canoinhas, Santa Catarina (SC), Brazil

2002/2003		2003/2004		2003/2004	
Variety	Grain yield (t/ha⁻¹)	Variety (below 3.0 m)	Grain yield (t/ha⁻¹)	Variety (above 3.0 m)	Grain yield (t/ha⁻¹)
MMPA1	6.90	Pixurum 06	7.98	Pixurum 1	7.16
Pixurum 06	6.44	Pixurum 07	7.50	Amarelão 03	6.87
Amarelão 03	6.14	MPA 01	7.45	MPA13	6.81
MPA 02	5.94	MPA 02	7.45	Palha Roxa	6.75
Pixurum 04	5.80	Cateto Vermelho	7.35	Rosado	6.65
Composto S. L.	5.74	Pixurum 05	7.13	Composto S. L.	6.64
Amarelão 02	5.69	Cateto	7.04	Amarelão 02	6.56
Branco	5.68	Mato Grosso 11	6.30	Amarelão 01	6.53
Palha Roxa	5.51	Pixurum 04	6.26	Cunha 01	6.47
Pixurum 05	5.47	Língua de P.	5.80	Pires	6.45
Cateto	5.46	MG-Palha Roxa	5.64	Mato Grosso 39	6.19
Mato Grosso 39	5.37	Cunha 02	5.49	Branco	5.89
Rosado	5.22	Rajado 8 C.	5.45	Roxo Valdecir	5.82
Cunha 01	5.22			Palha Roxa	2.10
Mato Grosso	5.14				
Língua Papagaio	4.94				
MG-Palha Roxa	4.83				
Roxo	4.73				
Asteca	4.65				
Moroti	3.12				
Commercial checks					
BR 106	6.23			BR 106	6.23
Sol da Manhã	5.43	Sol da Manhã	5.06		
average (trial)	5.42		6.62		6.20
Average of Santa Catarina state[a]					
2002/2003					
2003/2004	4.99		4.16		4.16

Source: Adapted from Alves *et al.* (2004).

a   ICEPA (2005).

field. After harvesting, the sample was evaluated at UFSC for several post-harvest traits and we took 186 ears to represent half-sib families.

Step 4: Evaluation and selection of half-sib families

The 186 half-sib families were divided into three experiments, with 62 families in each (three independent samples of half-sib families) plus two checks (the original MPA1 and a recommended, improved open-pollinated variety). We conducted the experiments in three farms, using an 8 × 8 partially balanced triple lattice design (including three replications). Each plot consisted of one row, 5.0 m long with 1.0 m between rows. We evaluated the half-sib families for grain yield and plant height and analysed the data according to a completely randomized block design. Kist *et al.*

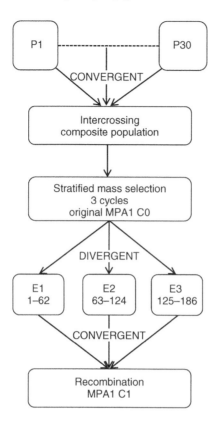

Figure 5.7.1 Convergent–divergent selection scheme used for participatory genetic enhancement of MPA1, Anchieta, SC, Brazil. P1 to P30 are 30 sub-populations of MPA1 (C0), used for the formation of the composite population (convergent); 62 half-sib families (divergent) were evaluated in different environments (E1: São Domingos, E2: São Roque, and E3: Café Filho); C1 is the result of recombination in an isolated block (convergent), represented by samples of constant size of the selected families.

Source: Kist *et al.* (2010).

(2010) further elaborate on the scientific evidence for the effectiveness of the selection for grain yield and plant height in this selection scheme.

Those selected half-sib families that resulted in a decrease in plant height of 5.4% in the three experiments (with 25% selection intensity) did not completely correspond to those outstanding families that had resulted in an increase in grain productivity of around 5%, compared to the average of the original base population. To achieve a positive correlation between yield and plant height, simultaneous selection for both traits should be weighed adequately.

We further selected samples of all superior half-sib families, uniform in size, for the recombination block that associated high grain productivity with low plant height. In this way, we were able to genetically enhance MPA1 for the traits that were preferred by small-scale farmers in Anchieta and its surrounding area. Kist *et al.* (2010) argue that the gain is small when compared with other experiments and selection schemes, but we draw attention to the fact that the current selection was conducted in farms, in contrast to the common experience of plant breeders, who work in experimental farms. Farm-level selection environmental variation takes its share in selection efficiency for such quantitative traits.

Step 5: Recombination of the selected half-sib families and final selection

Selected half-sib families were cultivated in a recombination field, in which we included mild selection for traits other than those for yield and plant height. In this selection, we had to be careful since changes in one character can result in other (non-desired) changes in correlated traits (Kist *et al.*, 2010). Over several subsequent cycles, the newly composed, genetically enhanced MPA1 was further improved in farmers' fields in various locations in the region, and we consolidated the variety as enhanced MPA1. It is currently being prepared for return to the farmers. We have been supported in this process by the same organizations in Anchieta, by Epagri in the municipality of Guaraciaba, by new local government partners in the municipality of Novo Horizonte, and by small farmers' groups from four municipalities in the western part of Santa Catarina.

Step 6: Seed production and maintenance of enhanced MPA1

In the current and concluding phase, seed production and supply, and the maintenance of genetic seed of enhanced MPA1 are our challenges. We need to develop and institutionalize collaboration between researchers and small-scale farmers. In this respect, we must be careful to prevent any contamination from migrant pollen, seed mixing, mutations and seed-borne diseases, which would result in the loss of the genetic gain achieved by the enhanced MPA1.

Crioulo maize and genetically modified maize varieties: posing new challenges

An important challenge that we currently face in working with milho crioulo is posed by the genetically modified maize varieties that are currently being introduced. NEA-Bio and partners in the area are engaged in a project to design a strategy to avoid the contamination of crioulo maize varieties. In 2007, the Brazilian Biosafety Commission fixed the minimal distance between genetically modified and non-genetically modified maize fields to 100 m to ensure there is no contamination or transgression. In contrast, the Ministry of Agriculture recommends between 200 and 400 m as the minimum distance of isolation for maize seed production fields. Contamination would be dramatic in the western part of Santa Catarina because small-scale farmers use and maintain many crioulo varieties. Here, the isolation of maize fields is neither a common nor viable practice, which is mainly due to topography or the density of farms. Avoiding gene flow between neighbours' fields seems almost impossible

(Cordeiro *et al.*, 2008). In this respect, NEABio is currently paying more attention to the participatory mapping of areas with a high density of milho crioulo. We aim to identify safe areas for the conservation of milho crioulo, and promote seed saving and organic maize production. The collaboration of NEABio with its partners has evolved from building the scientific basis for accepting the adaptation potential of crioulo varieties, to developing and adapting a breeding methodology aimed at the genetic enhancement of MPA1. At present, our partnership aims to strengthen and support farming communities in maintaining their crioulo varieties in times of advanced introduction of genetically modified maize.

5.8 Participatory crop improvement in Central America

Encouraging farmers to use local varieties

Sergio Romeo Alonzo Recinos, Mario Roberto Fuentes López, Juan Carlos Rosas Sotomayor, Silvio Aguirre Acuña and Rolando Herrera Torres

The Collaborative Programme on Participatory Plant Breeding in Mesoamerica

The Collaborative Programme on Participatory Plant Breeding in Mesoamerica (Programa Colaborativo de Fitomejoramiento Participativo en Mesoamérica, PPBMA) is a pioneer in its area, and has been operating in the region for the last ten years. It promotes the participation of farmers in decision-making through the various steps of plant breeding, and increases farmers' knowledge on ways to improve the performance of their local varieties. PPBMA is funded by the Development Fund (Norway), which fosters the establishment of relationships between governmental institutions, non-governmental organizations, and national and international research centres in Costa Rica, El Salvador, Guatemala, Honduras and Nicaragua. Through PPBMA, we have been working with small-scale farmers on a number of national projects to conserve, characterize and improve varieties of maize, beans, sorghum and other crops. In the current chapter, we describe the methods used in participatory crop improvement (PCI) in Guatemala, Honduras and Nicaragua. The locations are indicated in Figure 2.6.1 (p. 100).

Participatory crop improvement methods: matching crops, local situations and capacities

Key participants in PPBMA are community-based organizations (CBOs) that include cooperatives, associations and local agricultural research committees (CIALs). We aim to promote capacity development at local level, particularly among male and female farmer leaders of the CBOs, and foster farmer participation in breeding, research and the coordination of activities.

During the process of implementing breeding activities, strategic alliances are established between key stakeholders. Although these alliances are initially formed at local level, they may later cross national borders, thereby legitimizing the regional structure of the programme. Since the professional and institutional capacity for plant breeding in the various Central American countries is in general rather limited, we aim to develop those capacities through PPBMA's regional structure, and to enhance the efficiency of the regional breeding programmes, through PCI.

The programme has identified a range of PCI methods, based on a number of factors, including type of crops and crop reproduction systems involved, the plant genetic resource (PGR) base, the key problems faced, and the capacity of farmers to contribute to, and appropriate, the breeding process. In the current chapter we examine the PCI approaches that we use in the programme, following the terminology outlined by De Boef *et al.* in Chapter 5.1.

Strategy 1: Conservation of local varieties

By supporting farmers and researchers in their activities to rescue local maize, beans and sorghum varieties, the programme aims to contribute to their conservation. Farmers are organized in groups to register, characterize and document their local varieties, and this information is used for the design of both *in situ* and *ex situ* conservation strategies. Farmers display and exchange seed of local varieties through diversity fairs, thereby contributing to on-farm management.

Strategy 2: Use of diversity in PCI

The heart of our programme is to encourage farmers to use the diversity of local crops and to participate in their improvement. Initially, the major problems that farmers want to solve for a specific crop are identified, in order to choose the most appropriate breeding and selection methods to be used. The gene pool of local existing varieties is used as much as possible to solve the problems identified. It is crucial that farmers participate in the identification of criteria to be used during selection, and that the various PCI steps are mainly conducted in farmers' fields.

Strategy 3: Quality seed production of PCI varieties

The third strategy involves the local production and marketing of quality seed of the varieties produced. We aim to increase the availability of high-quality seed to farming communities through the implementation of activities related to seed production and processing. Farmers' capacities in seed production and processing are improved through intensive training, and support is provided for the establishment of farmers' groups for engaging in collective action in seed production and marketing.

General framework: food security, income generation and out-scaling

In addition to the three technical strategies, the programme is also guided by a more general approach that has three aims:

1   to promote access to and availability of food crops, focusing on food and nutritional security;
2   to support income generation, where the aim is to develop higher yielding varieties, for which a demand exists in local, regional, national and international markets;
3   to seek ways for out-scaling the PCI methodologies to other organizations.

We encourage the involvement of other stakeholders in the PCI processes, including farmers, communities, government institutions and non-governmental organizations, and aim to promote a wider use of PCI methodologies, thereby increasing the efficiency of plant breeding in responding to farmers' demands.

Practical experiences

Participatory varietal selection and plant breeding of common beans in Nicaragua

In 2001, the programme began to support a PCI process in the municipalities of Pueblo Nuevo and Condega, in the Esteli Department, in Nicaragua (see Figure 2.6.1, 100). The aim was to generate new common bean varieties by crossing local varieties with elite breeding lines, which served as donors of useful genes. Our partners use the participatory plant breeding (PPB) method, in which farmers and breeders carry out selection within segregating populations of common beans in farmers' fields. PPB is complemented with participatory varietal selection (PVS) to evaluate and select among advanced breeding lines that carry combinations of traits. For several cycles, our partners selected for a broad set of agronomic, commercial and consumer characteristics. They concluded the PPB process with the release by the CBO of outstanding bean varieties. Subsequently, the programme strengthened the capacity of the CBO in the production, distribution and marketing of good quality seed. Through the participatory bean improvement programme, farmers and breeders in Nicaragua have become able to improve the tolerance of local varieties to drought and their resistance to bean golden yellow mosaic virus. To this date, the PPBMA partners in Nicaragua have released six bean varieties.

Participatory improvement of common beans and maize in Honduras

The Zamorano Pan-American Agricultural School (in short, Zamorano School) is engaged in participatory bean improvement in Honduras (Hocdé *et al.*, 2010). It collaborates with non-governmental organizations (NGOs) that include the Foundation for Participatory Research with Honduran Farmers (FIPAH) and the Rural Reconstruction Programme (PRR). CBO partners consist of farmer research groups associated with CIALs. As with our partners in Nicaragua, Zamorano School and its local partners combine PPB and PVS methods, with strategies for producing and disseminating seed of improved varieties to rural families.

Farmers from the participating CIALs received training in PCI methods during the implementation phase of the programme, using a learning-by-doing approach at each stage of PVS or PPB; a process that can take several seasons and years. The training included recognition of reproductive mechanisms and pollination in common beans and maize, collection and characterization of local and introduced germplasm, hybridization techniques, management of field trials, and evaluation of breeding nurseries for physiological traits, major diseases and pests, commercial qualities concerning yield and grain, and other traits to be selected during the PCI process. Farmers also learned about crop management practices, seed production and post-harvest technologies, which are important for the management of breeding nurseries and for improving the productivity of their commercial fields.

Since 2000, our partners have released 13 common bean varieties and four maize varieties that were developed using PPB and PVS methods (Rosas, 2001; Rosas *et al.*, 2003, 2006). These varieties are now being used in the regions where they were released, contributing to an increase in the productivity of maize and bean crops. Some of those PCI varieties were developed using local varieties as parents in crosses with varieties that are resistant to certain diseases, or which possess other desirable agronomic traits, such as better plant architecture and yield in common beans, and shorter plant stature and better lodging resistance in maize. The PCI varieties are particularly valued for being adapted to specific agro-ecological niches, where they perform better than the conventionally improved varieties.

Participatory genetic enhancement of local maize varieties in Guatemala

Most maize in Guatemala is produced by small-scale, poor, subsistence farmers, who generally have limited access to good quality maize seed. If these farmers do have access, it is only to the seed of conventionally improved varieties that are generally not well adapted to their agro-ecological and production systems. Participatory genetic enhancement (PGE) of local varieties creates a possibility to achieve significant improvements by selecting from within existing local diverse maize varieties. The local maize varieties and the diversity that they harbour are a product of the structure and functioning of the informal seed systems. In 2000, partners of the programme began the maize PGE programme in the highlands of Huehuetenango (see Figure 2.6.1, p.100), the highest areas of which reach up to 2200 m above sea level, by collecting and characterizing varieties that represented the local maize diversity.

Subsequently, our partners identified the most promising materials through PVS, taking into account the criteria established for their enhancement. They used recurrent reciprocal selection of half-sib families, and applied a method that allowed us to select within existing variation in preferred local varieties. Ogliari *et al.* (Chapter 5.7) describe in detail a variation to this method, which they used for the PGE of a local maize variety favoured by small-scale farmers in southern Brazil. The method used in Guatemala allowed our partners to evaluate half-sib families in different environmental conditions, and select those half-sib families that scored well for the selected traits. Only selected half-sib families were used for the formation of genetically enhanced local varieties. Since this breeding procedure requires some understanding of genetics, the programme organized community training workshops to improve farmers' skills in selection, and enhance their understanding of plant breeding. Farmers proved to be well able to conduct selection among half-sib families. In fact, they subsequently carried out several cycles of mass selection to create some uniformity within the composite assembled, based on the selected half-sib families. Moreover, with their improved skills in selection, farmers became more capable of maintaining the elite materials. Such enhancement of farmers' skills and capabilities is vital for ensuring that the varieties can be maintained in various agro-ecological niches, *in situ*, through their management by farmers. Our partners in Guatemala have developed ten maize varieties under this scheme for the PGE of local maize varieties, which were subsequently multiplied and marketed.

Participatory varietal selection and plant breeding of sorghum in Nicaragua

A study carried out in four locations in Nicaragua in 2002 showed that local varieties of *tortillero* (short-cycle) and *millón* (day-length sensitive) sorghum did not perform well. They produced poor-quality fodder, were late maturing, and were vulnerable to diseases and pests during storage. We initiated our participatory sorghum improvement programme in 2002, with the evaluation of advanced sorghum lines of African origin, in collaboration with the International Centre for Tropical Agriculture (CIAT, Colombia) and the Agricultural Research Centre for International Development (CIRAD, France). In 2002, our partners trained farmers and local field technicians in the municipalities of Totogalapa and Somoto, in the department of Madriz (see Figure 2.6.1, p. 100), in the use of participatory varietal selection (PVS). They subsequently carried out PVS on local varieties and materials introduced by CIAT and CIRAD. In 2003, more farmers and communities became involved and new materials were introduced.

During the PVS process, one farmer, Orlando Gómez, said that he favoured the black tow variety because of its local adaptation, but that he would like to reduce the plant height, and improve grain and fodder quality. This statement motivated PPBMA's partners in Nicaragua to move from PVS to PPB in their sorghum breeding programme. Farmers began to participate in selecting among segregating populations. Once they had stabilized a new population, they included the materials in the yearly series of PVS trials. In 2004, farmers in more municipalities joined in the development of new composites of both tortillero and millón sorghum varieties. Since 2005, local NGO partners have been ensuring the continuity of the participatory sorghum improvement programme. In 2007, our Nicaraguan partners registered a new white tortillero sorghum variety, which was followed in 2008 by the local distribution of three tortillero, and four millón, sorghum varieties.

The formal release of a variety is a lengthy procedure that involves standardization tests, which the sorghum PCI varieties are unable to pass. This does not mean that participating farmers and their organizations refrain from using and disseminating the PCI varieties. The sorghum varieties are becoming popular, resulting in their production and marketing through informal seed markets. However, they are not included in formal seed production, simply because they have not passed, and will not pass, the formal release procedures. Until these procedures change, such sorghum PCI varieties will remain in the informal seed sector by default.

The use of local diversity and collective action for enhancing farmers' autonomy

The Central American region is a strategic reservoir of biodiversity for maize, beans and other food crops. The PCI methodology, with its variations in methods ranging from PVS to PGE and PPB, as applied in the PPBMA, constitutes a pathway for contributing to the management, conservation and utilization of biodiversity at community, national and regional levels. PPBMA brings researchers, technicians and farmers together. We use quite simple procedures that help both farmers and breeders to improve and select within local varieties. The results can be identified easily, since farmers are increasingly using local, genetically enhanced, or PCI varieties.

We consider the transformation of farmers into plant breeders as another important achievement of PPBMA. These farmer-breeders improve their own varieties and consequently market the seed of such unique varieties, which provides them with the opportunity to increase their income and share the seed of their breeding work with others. In our network, the seed of most PCI varieties is distributed through informal markets.

To date, our regional network has released 31 bean, 18 maize and 10 sorghum varieties, which is only a fraction of the material produced and informally distributed and used by farmers. Another, more indirect result that concerns food security and income generation is that farmers who have access to our varieties have been able to increase the productivity of maize, beans and sorghum up to 200%. More than 5000 farmers have become affiliated with CBOs that are our partners. About 450 farmers and 30 technicians have improved their capacity to manage agrobiodiversity, thereby enhancing the overall capacity in the region in plant breeding and the *in situ* conservation of local plant genetic resources. Farmers' organizations in Costa Rica, Guatemala, Nicaragua and Honduras have increased their share in the market for seed and grain production. These organizations have enhanced their capacity in entrepreneurship, and improved their infrastructure for processing seed and grain in drying, cleaning, classification, packaging and storage.

PCI innovations: overcoming obstacles created by varietal regulations

The process of releasing crop varieties that are produced through PPB and PGE by farmers' organizations has exposed the limits of the formal system concerning the incorporation of these innovations at a more institutional level. For example, the network of bean breeders in Nicaragua has contributed to national level discussions on the legal and procedural mechanisms that remain a solid obstacle for the future release of more diverse materials produced through PCI. On the other hand, many farmers that belong to our partner CBOs are motivated in their work with PCI because they can contribute to the development of good varieties and produce good-quality seed of those varieties (Aguilar-Espinoza, 2007). In this way, PPBMA has increased the self-esteem of farmers, who are now beginning to see themselves as innovators and breeders, just like their scientific colleagues in PCI.

Access to seed and varieties: motivating farmers to participate in PCI

Farmers have other, rather opportunistic although quite valid, arguments for joining our programme. As seed security at farm household and community levels remains an issue that should not be overlooked in Central America, access to seed of appropriate varieties remains the primary objective for joining in participatory crop improvement. The increased adaptation of their varieties helps farmers achieve this objective. Moreover, our activities result in the increased autonomy of farmers and their CBOs in the seed system. The mechanisms for participatory crop improvement, seed production and marketing that we put in place ensure farmers' access to varieties and seed. In this way, PPBMA improves the food security of both its partners, and the farmers who depend on purchasing seed of the varieties we produce and disseminate through the market, fostering collective action to make the informal seed system work.

Note

The authors would like to thank coordinators of the Participatory Plant Breeding Programme for the Mesoamerican Region – Juan Carlos Rosas (Honduras), Javier Pasquier (Nicaragua), Juan Carlos Hernandez and Rodolfo Araya (Costa Rica), and Carlos Reyes (El Salvador) – for their contributions to this chapter.

5.9 Participatory varietal selection for enhancing farmers' access to quality seed in Ethiopia

Hussein Mohammed, Tadesse Desalegn, Fetien Abay Abera and Marja Thijssen

About 85% of Ethiopia's 80 million inhabitants depend on agriculture for their livelihoods. The agricultural sector is largely characterized by small-scale subsistence farming and low productivity. Farmer access to quality seed of better adapted varieties is of utmost importance for increasing productivity (Bishaw *et al.*, 2008). Public research institutes and universities are primarily responsible for plant breeding. Their varieties generally target high-potential agricultural environments and as such are not specifically adapted to the marginal growing conditions with which small-scale farmers most often have to contend. Participatory varietal selection (PVS) is becoming a common tool for identifying which materials should be incorporated into breeding programmes, and which traits should be addressed, as well as for testing materials in farming conditions before release (Ceccarelli *et al.*, 2009).

The formal seed system takes responsibility for the dissemination of just a few released, high-demand varieties, resulting in the limited accessibility of improved varieties that are adapted to diverse conditions. For example, over the past six years, thirty bread wheat varieties have been released, but the variety Kubsa remains dominant in 60–70% of the area in the country that is planted with seed of improved wheat varieties. The major reason for its dominance is its widespread dissemination through the public seed and extension system. In 2010, the Food and Agriculture Organization of the United Nations (FAO) reported that 410 000 hectares of Kubsa were highly affected by rust, which had a detrimental impact on the livelihoods of over 65 000 farmers. This situation shows the vulnerability caused by the development-oriented structure of the public seed value chain, which focuses on a limited number of released varieties (Louwaars and De Boef, 2012). One of the options for overcoming these bottlenecks is the production of quality seed of a wide range of adapted varieties by groups of small-scale farmers who market this quality seed locally (Neate and Guei, 2011).

PVS is a common strategy for exposing farmers to released varieties, and it can help farmer seed producer groups identify which varieties to select for producing seed. PVS can even be used to identify local varieties that perform well, in case improved released varieties do not do well in the target environment (Abay and Bjørnstad, 2008; De Boef *et al.*, 2010).

The Integrated Seed Sector Development (ISSD) programme in Ethiopia works with farmers' groups that are involved in seed production. The programme supports farmers' groups to become better equipped technically and more commercial in their approach, thereby increasing the autonomy of the groups in their operations, and fostering their development into local businesses. The ISSD programme aims to enhance seed security through the marketing of seed at community and district levels. Thirty-four farmers' groups, mostly organized as seed production cooperatives (SPCs), in four regions of Ethiopia, are being supported by teams made up of three experts from the fields of agribusiness development, farmers' organizational development, and from the technical side of seed production. The SPCs work from units at Bahir Dar University, Haramaya University, Hawassa University, Mekelle University and Oromia Seed Enterprise. Figure 5.9.1 illustrates the locations of the Marwoled, Amard, Kayyo and Habes SPCs. The current chapter shares some examples of how PVS can be used to increase the variety portfolio in local seed business development.

Common set-up of PVS by farmer seed producer groups

We carry out a variation of PVS mother trials with the SPCs based upon the mother and baby design (Snapp *et al.*, 2002; Bänziger and De Meyer, 2002). As researchers associated with various partner organizations in the ISSD programme in Ethiopia, we have been working with the SPCs since 2009, to include released varieties and pipeline materials in their variety portfolio. In the PVS process, the SPC plants the

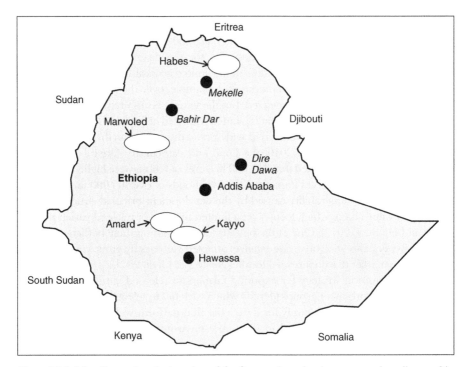

Figure 5.9.1 Map illustrating the location of the four seed production cooperatives discussed in this chapter.

new varieties in three to four fields belonging to fellow SPC members, or in collective fields, either within walking distance from each other in the same community, or in separate communities. We consider each PVS trial to be one replication. Together with members of the SPC, we collect qualitative data, and quantitative data that we analyse statistically. During informal discussions with farmers, we use matrix ranking to identify the selection criteria for vegetative, flowering and maturity crop stages; farmers evaluate the varieties for these traits. Eventually, we work with the SPC to compare the selection made by researchers and farmers, and identify those varieties to be included in their commercial variety portfolio.

Marwoled: bread wheat

Wombera District in the Amhara region is an agricultural area of high potential. Marwoled SPC was officially established in 2010 by a farmers' group that had more than ten years of experience in seed production, through a contractual arrangement with a public seed enterprise. The SPC has 127 members and it specializes in the quality seed production of hybrid maize and wheat. In 2010, a student of Bahir Dar University (BDU), the ISSD partner in Amhara, conducted a PVS trial with Marwoled SPC to evaluate 12 wheat varieties, consisting of five released varieties (Kubsa included) and seven pipeline materials. Farmers ranked the varieties according to qualitative criteria, while the BDU MSc student measured yield-related criteria. The farmers, together with the student, selected one released variety and two pipeline varieties that had outperformed Kubsa for both yield and disease resistance. Some varieties scored well for productivity but were not selected by the farmers since they scored low for other selection criteria. Marwoled SPC will plant the seed of the experimental plot of the released variety as pre-basic seed. The multiplication will also help for demonstration purposes and for promotion. BDU will help the SPC produce seed of high quality and purity. The pipeline varieties that were identified in PVS are from Kulumsa Agricultural Research Centre (KARC). One of the materials selected in Marwoled is a candidate for national verification. Therefore, Marwoled SPC will have to wait for its verification and release. However, BDU has asked KARC to produce pre-basic seed in advance of its release, so that when it is released Marwoled SPC can plant 100 m^2 for the first internal multiplication and for further production, thereby becoming one of the first seed producers in the country to have this variety in its portfolio.

Amard: bread wheat

Amard SPC was established in 2007, in the Lanfro District of the Southern Nations, Nationalities and Peoples' Region (SNNPR). The SPC has 80 members who produce Kubsa on a contractual basis for the Edget Farmers' Union. In 2010, the early onset of rain and subsequent rust infection resulted in low yields. The SPC decided to evaluate five released varieties in order to identify alternatives to Kubsa. In the process of PVS, Hawassa University, the regional ISSD partner, together with the members of Amard SPC, selected three varieties that were 40–85% superior in grain yield compared to Kubsa. Additional traits that were decisive for the selection of these varieties were early maturity, disease and lodging resistance, large seed size and white seed colour. Amard SPC requested basic seed of the three varieties for seed production. It indicated that it was ready to cover its entire area with these superior varieties and market them to replace Kubsa. However, the public entities responsible for basic seed production could only provide seed sufficient for a few hectares. Kubsa basic seed was, on the other hand, abundant. Consequently, Amard SPC must first multiply the three varieties over a number of seasons for it to be able to cater to local demands.

To overcome the time required for the multiplication of seed within the SPC before being able to begin to market it, Edget Farmers' Union has been supplying basic seed of another variety found to be less susceptible to rust. Amard SPC intends to sell this variety through a contractual agreement with the union. It is expected that in subsequent seasons, Amard SPC will start to market the varieties that they identified in the PVS and are now multiplying for basic seed. This will contribute to their autonomy in the seed value chain, while increasing the number of bread wheat varieties available in the area. This example shows that the increasing autonomy of SPCs is directly linked to increasing diversity. Other operators in the seed chain, owing to economies of scale, prefer just a few, rather than a diversity of varieties. As shown by the example of rust in Kubsa, the latter will make the seed system vulnerable to diseases and to stresses related to climate change.

Kayyo: red haricot bean

Kayyo SPC is located in the Boricha District of the SNNPR. The SPC was established in 2007 and its 147 members originate from four neighbouring communities. Since its release in 1974, the red bean variety Wolayita Red has been popular because of its red colour; however, its yields are very low. The identification of more productive red haricot bean varieties follows the strategy of the SPC to produce and package small quantities of quality seed, and market these in neighbouring districts. The SPC decided to test seven released red varieties in PVS, in collaboration with Hawassa University, as part of the ISSD programme. Through matrix-ranking exercises, farmers evaluated traits for early maturity, basal pod height above soil, number of seeds per pod, and upright growth habit. Taking into consideration both yield assessments by researchers and visual evaluation of qualitative traits by farmers, two promising varieties that outperformed Wolayita Red in the PVS trial by more than 175% were identified. During the 2011 season, Kayyo SPC was able to access basic seed of the identified varieties, and now includes them in their variety portfolio for quality seed production.

Habes: local barley varieties

Barley is an important crop for food security in the marginal and drought-prone environments of Tigray, in northern Ethiopia. The recommended barley varieties perform poorly because they are not well adapted. Mekelle University (MU) has been working with PVS in five districts of Tigray since 2007 through its Seed Safety for Diversity project. In this project, MU, together with farmers' groups, evaluated 17 barley varieties, including four improved and 13 local varieties. Some of the local varieties were reintroduced through the national plant genetic resource programme.

In the five districts, farmers selected the variety Himblil, which is a product of participatory plant breeding. Himblil is tolerant to drought and water-logging stress and is more stable than improved varieties under low-input and low-rainfall conditions.

In 2010, as part of the ISSD programme, the farmer research group in Habes was transformed into a legally registered SPC, with 50 members. In Habes, farmers preferred a rare local variety, which was included in the portfolio for seed production and local marketing, along with Himblil and other varieties. We have observed a significant increase in the number of barley varieties cultivated in the communities since the initial PVS trials were carried out in Habes, in 2007. Before PVS, these communities cultivated a maximum of five varieties; now, they grow 9–12 varieties, and all of them use Himblil (Kiros, 2011). Through the analysis of data from an inventory of seed flows, we were able to see that those varieties that were tested and disseminated through PVS, reached farmers within a radius of 100 km from the place where they were initially introduced (Abay *et al.*, 2011).

PVS and SPCs: enhancing community access to genetic resources

Both Marwoled and Amard SPCs initiated bread wheat PVS to identify an alternative to the widely grown but rust-susceptible variety, Kubsa. We identified the availability of released and pipeline varieties, which combine disease resistance and yield potential. It was interesting to see that farmers in SPCs selected for a wide range of locally important traits, and as such farmers of Marwoled SPC and farmers of Amard SPC preferred dissimilar varieties for seed production and marketing. We learned that in the case of wheat there is a market for a wide range of varieties, but that not enough early generation seed is available through the public seed system to meet this demand. The reason for using PVS in Kayyo was to identify red bean varieties that are in demand in the local market. The identification of better-performing released varieties provided the SPC with a new business opportunity. However, we had to link the SPC with the research institute in order to organize the structural flow of the required basic seed of identified varieties. Farmers and seed producers showed an interest in released varieties that had previously remained on the shelf and had not been selected for seed production in the public seed value chain. As a result of the weak link in the seed value chain, the released varieties – the products of plant breeding – did not reach and therefore benefit end users. Consequently, the impact of breeding and research remained limited.

We identified a key bottleneck in the structure and functioning of the (public) seed value chain as being the absence of a well-functioning system for the production of pre-basic and basic seed in demand, resulting in obsolete varieties, such as Kubsa, remaining much longer in production than they deserved. This confirms the limitations of the public sector in many developing countries, as identified by Neate and Guei (2011), who further indicate that it is crucial we support smallholder seed enterprises that operate in public–private partnership with a service- and more market-oriented public sector. The ISSD programme therefore brings together key sector stakeholders in the seed value chain, facilitating the SPCs in the development of a wider varietal portfolio. Furthermore, the ISSD programme supports the SPCs in their autonomy and in their use of diversity, thereby enhancing the resilience of the local production system.

The PVS that was carried out in Habes demonstrates the importance of specific adaptation. Farmers preferred Himblil and local varieties over released improved varieties. Progress is now being made since the Ethiopian government has recognized the importance of specific adaptation, allowing for the release of varieties for specific agro-ecologies. The link with research, through PVS, allows SPCs to have more direct access to a wide range of varieties. The studies in Tigray show that newly introduced, well-performing varieties spread rapidly, once SPCs take up the variety for seed production and initial marketing. As a risk management strategy, especially in marginal environments, farmers often cultivate a number of different varieties. As such, the SPCs may play a role not only in increasing farmers' access to quality seed, but also by injecting multiple varieties into the informal seed system. As Neate and Guei (2011) indicated, in a global discussion on the role of small-scale seed enterprises in seed security, SPCs in Ethiopia are a crucial instrument for increasing farmers' access to quality seed of superior and adapted varieties. Promoting local seed business, therefore, is not just a goal; it is also a tool for enhancing the overall performance of the seed sector, by ensuring that better performing superior varieties, either improved or local, are made available to farmers. Through its role in the seed chain, the seed of those SPCs can contribute significantly to seed security at local but also national level. In our experience, the formal system alone is not able to supply a wide diversity of varieties. However, by partnering with SPCs in Ethiopia, it may be possible to strengthen the farming communities' access to a broader genetic resource base, which will also help them in meeting the challenges posed by climate change.

5.10 Supporting farmers' access to the global gene pool and participatory selection in taro in the Pacific

Tolo Iosefa, Mary Taylor, Danny Hunter and Valerie Saena Tuia

> Taro has always been here; it belongs to us and is part of our people. Taro is our life.
>
> Melanesian taro farmers (Jansen, 2002)

Taro: an ancient food crop

Taro (*Colocasia esculenta*) is a clonally propagated aroid; it is an ancient food crop that was domesticated at least 9000 years ago, largely for its underground corm. All plant parts can be utilized. Its corms are baked, roasted or boiled; the leaves are frequently eaten as a vegetable and represent an important source of vitamins. The blades and petioles of leaves can be preserved or dried, and thereby become an important food in times of scarcity. Various plant parts are used for medicinal purposes (Rao *et al.*, 2010). In common with other tropical roots and tubers, taro cultivars flower infrequently and are highly heterozygous. Although taro originates from South-East Asia, it has spread throughout the world and is now an important crop in Asia, the Pacific, Africa and the Caribbean. South-East Asia and the Pacific represent the two major and distinct gene pools.

The international scientific community has largely ignored taro, despite its local importance. This may be partly because genetic resource collection of taro is difficult and *ex situ* conservation in field gene banks is notoriously expensive (Lebot *et al.*, 2005). In the Pacific, but also elsewhere, farmers remain the custodians of taro genetic and cultural diversity. They maintain thousands of cultivars that are adapted to a wide range of agro-ecologies, covering wetland and dry-land, marginal, complex and often harsh environments (Rao *et al.*, 2010).

Taro: a people's crop

Farmer management of taro genetic diversity is highly dynamic as it is embedded in the social structure of the community, such as kinship obligations that foster exchange. Taro cultivar portfolios are determined as follows: (a) the taros are introduced from outside the village; (b) they are the products of somatic mutations, which farmers observe in their fields and retain after evaluation; or (c) they are found as seedlings in rural and wild environments, the result of sexual recombination, and are retained by farmers after evaluation (Jansen, 2002; Caillon *et al.*, 2006; Camus and Lebot, 2010).

The dynamism results in taro cultivars being exchanged within and among communities, and also within and between countries.

During participatory appraisals of taro diversity in the Solomon Islands, farmers were asked if they ever found taro seedlings (Jansen, 2002). Many farmers described finding taro genotypes that became new cultivars in the bush, primary forest, and along the banks of rivers. Some farmers also reported finding new taro cultivars that had been growing in farm sites under fallow for at least six years (some up to 25 years). Caillon and colleagues (2006) describe similar findings in Vanuatu, where observant farmers cultivated these taros, evaluating and selecting those that performed well. They included the taros that were useful to their agronomical, culinary and social needs. In some Pacific islands, community discussions revealed the presence of taro chiefs in certain communities who conserve a bulk of cultivars as a hobby. Camus and Lebot (2010) refer to them as collector farmers.

Despite the cultural significance of taro and its wide exchange networks in the Pacific, its genetic base in the Pacific is in fact quite narrow compared to that of South-East Asia. As a result, taro cultivars from the Pacific are vulnerable to new emerging pests and diseases, which are having an alarming effect in terms of genetic erosion. Because taro is so important to food and cultural diversity in the Pacific, we need to support farmers' conservation efforts in battling this erosion of genetic diversity and associated traditional knowledge.

Taro improvement project

The narrow genetic base of taro was revealed in 1993 when taro leaf blight (*Phytophthora colocasiae*) arrived in Samoa, with dramatic consequences. Since all traditional cultivars were susceptible, almost overnight, the country's most important domestic and export commodity, major source of income, essential food source and cultural icon was wiped out. The situation had to worsen before researchers began to shift their strategy towards facilitating farmers' access to, and benefit-sharing from, taro genetic diversity originating from distant geographic areas. The problem of taro leaf blight and associated genetic erosion stimulated the establishment of the Taro Genetic Resources Conservation and Utilisation Network (TaroGen) in 1998.

Since 1999, farmers in Samoa have had access to exotic cultivars and breeding expertise through the establishment of a Taro Improvement Programme (TIP) at the School of Agriculture, in the University of the South Pacific (USP). Discussions between researchers and farmers revealed that previously introduced exotic cultivars had shortcomings, including susceptibility to disease in wetter parts of the country, plus low yields, and poor taste and storability. Farmers raised concerns about the delay in research for releasing and providing them with evaluated material. Responding to those complaints, TIP brought together regional and national scientists, extensionists and farmers from Samoa's two islands, Upolu and Savai'i. TIP began to introduce and evaluate taro genetic resources and incorporate participatory varietal selection (PVS) in its taro improvement programme, as visualized in Figure 5.10.1. Participatory appraisals focusing on taro provided vital information on production problems and varietal perceptions. Through various scoring and ranking exercises, criteria for PVS trials were indentified (Singh *et al.*, 2010). Since 2000, we have organized a

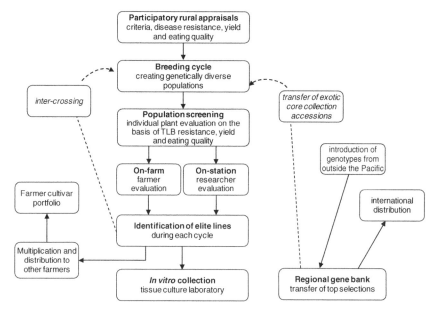

Figure 5.10.1 The Taro Improvement Programme's strategy for participatory taro improvement, illustrating links to the formal plant genetic resources system and informal farmer networks.

number of taro diversity fairs, as part of the TIP project to showcase TIP's breeding and evaluation work. At the USP campus in Samoa, we set up a taro breeders' club to integrate TIP into teaching, encouraging students to participate in germplasm evaluation and interact with farmers and scientists.

During their participation in TIP activities, farmers evaluate and select our clones, which have been produced through crosses between cultivars from Samoa, Palau and the Federated States of Micronesia (Hunter *et al.*, 2001). Although selected clones enabled farmers to start growing and eating taro again, there was concern that breeding from such cultivars with only a Pacific origin would not significantly broaden the required genetic base (Camus and Lebot, 2010). As a next step, farmers obtained access to virus-indexed germplasm from Asia; and TIP brought together two distinct taro gene pools. Taro research is largely neglected by the international community, and before TaroGen, limited data on the genetic diversity of taro was available. Moving taro germplasm internationally was considered unsafe because of the limited availability of information on the presence of viruses and the inadequate methods used for their detection.

TaroGen contributed significantly towards improving this situation. Characterization and diagnostic tools were developed in order to identify taro viruses. The Centre for Pacific Crops and Trees, an international standard gene bank that conserves Pacific staples, and which is located at the Secretariat of the Pacific Community in Fiji, established a regional taro collection and a transit centre for the virus-indexing and safe movement of taro germplasm. The TaroGen gene bank has been facilitating substantial taro collection, characterization and conservation since 1998, including the development of a regional taro core collection (Mace *et al.*, 2010; Taylor *et al.*, 2010). The gene bank has continued to play an active role within TIP since the completion of the TaroGen project, improving farmers' access to exotic cultivars and creating opportunities to distribute any clones developed by TIP to other countries (Figure 5.10.1).

Farmers in the Pacific are now able to access the diversity of the Asian gene pool. In 2005, several accessions from the core collection of a sister regional network in South-East Asia were included in the breeding cycle (Figure 5.10.1). To date, seven breeding cycles have been completed, including researcher and farmer selection (Table 5.10.1). In 2009, five new cultivars were formally released and recommended in Samoa. Top selections from each breeding cycle in Samoa are tissue cultured and subsequently transferred to the regional gene bank, where their virus status is assessed and included in the collection. TIP's top selections are available to farmers and breeders in countries that are Contracting Parties to the International Treaty on Plant Genetic Resources for Food and Agriculture, as a result of the fact that in 2009 the Pacific region agreed that the regional gene bank collections be placed in the multilateral system.

The specific role of farming communities in the conservation and breeding activities of TIP

Forty-five farmers from the two main islands of Samoa, Upolu and Savai'i, became members of TIP in 1999. Over the past ten years, the number of members has increased to over 100 farmers. Since 2010, a core group of 40 members have been actively engaged in the programme through meetings, evaluations and the distribution of clones. In joining TIP, farmer members sign a farmer/researcher agreement that sets out, in considerable detail, expectations and reciprocal roles and responsibilities. Researchers are largely responsible for the crossing, and for some preliminary on-station evaluations to identify potentially useful clones. Farmers are responsible for the planting, labelling, maintenance and selection of those clones, in particular for

Table 5.10.1 Summary of top selections from the Taro Improvement Programme breeding programme over the past 15 years

	Year	No. of parental combinations	No. of seedlings evaluated	Top selections
Cycle-1	1996	4	2 000	10
Cycle-2	1998	5	2 000	26
Cycle-3	2000	26	2 000	30
Cycle-4	2002	45	5 000	30
Cycle-5	2005	30	5 000	42
Cycle-6	2007	$33 + 9\ BCF_1$[a]	11 000	40
Cycle-7	2009	$36\ (17\ BC_1$[b]$)$	12 000	25[c]

a BCF_1, first filial generation of the backcross to taro Niue (most preferred local cultivar, pre-taro leaf blight).
b BC_1, second generation of the taro Niue backcross.
c As of May 2011, 70% of these selections are being conserved at the regional gene bank, 18% of which have been virus-indexed and distributed.

the evaluation of culinary traits. Farmers select and multiply the clones they prefer without any influence from the researchers. They then supply the researchers with the planting materials of their best selections for future breeding and inclusion in tissue culture (for regional distribution); and provide feedback to other TIP members. Future breeding cycles are therefore based on materials identified and selected by farmers, as well as researchers.

Each farmer is usually supplied with up to 100 clones (seedlings) from each cycle. Since 2000, when farmers began to be involved in evaluation on-farm, 6000 clones have been distributed. Some farmers maintain just a few clones, while others maintain many more, provided they produce a good yield and that their eating qualities are good. Farmers have the right to exchange or sell any planting material or corms from their selections. The experiences of TIP illustrate that Samoan farmers are skilful at handling, evaluating and selecting large numbers of clones, lines and cultivars.

The future of taro

In Samoa, prior to the arrival of taro leaf blight in 1993, the main problem was that all taro cultivars were highly vulnerable. This vulnerability still exists in those taro-growing countries in the region that have not yet been affected by the disease, and where the genetic base of the crop remains narrow. The devastating outbreak of taro leaf blight in Samoa resulted in the development of TIP, which led to a better understanding of the disease and of the importance of taro diversity among researchers and farmers. Diversity is now also used to improve taste, nutritional value, yield qualities and to identify clones suitable for value-adding. Farmers and researchers can access diversity through the regional gene bank, which offers them an opportunity to increase the diversity on-farm in order to meet their individual and community needs. Farmers and researchers alike are aware of what can happen when a disease strikes, and they are keener than ever to maintain diversity, a dynamic that in itself keeps interest and awareness alive and ensures greater diversity on-farm. TIP members are aware that they are part of a dynamic process that feeds not only the people of Samoa but, through linking with the regional gene bank at SPC, supports taro cultivation and production in the region.

TIP cultivars have made their way to countries in both the North and South Pacific, and in a sense this provides protection against the future emergence of taro leaf blight in countries that are currently free from the disease. Through the introduction of exotic materials and participatory taro improvement, researchers and farmers in the Pacific have made advancements unforeseen ten years ago. The development of resistant cultivars coupled with advanced virus-testing technology has now made it possible to share and disseminate farmer-selected materials rapidly, under the auspices of the multilateral system, should outbreaks of new diseases occur. Improved lines of taro were transferred to Nigeria because of a recent outbreak of taro leaf blight in the region. This means that unlike the situation in Samoa prior to the establishment of the TaroGen network, countries with new outbreaks of the disease no longer have to wait years to get access to lines with improved resistance. Furthermore, the transfer of these improved taro lines across the world, from the Pacific to Nigeria, reflects the global interdependence in plant genetic resource conservation and use.

Part VI

Community biodiversity management, genetic resource policies and rights

6.1 Genetic resource policies and rights

Opportunities and limitations for community biodiversity management

Niels Louwaars, S.P. Bala Ravi, Pratap Shrestha, Juliana Santilli, Regassa Feyissa and Walter Simon de Boef

Synopsis

Community management of plant genetic resources (PGR) operates through the use of seed and planting material by the farming communities. Apart from being carriers of genetic diversity, which makes seed (in its generic term, also covering other forms of reproductive material) the subject of biodiversity policies, seed has functions that relate it to other policy arenas. Seed is a primary input in any crop production system and a primary determinant of inherent values, such as product quality, yield potential and yield stability, and as such is the target of agricultural policies relating to food security and agricultural development. Seed is also a commodity that is consequently addressed by trade policies. The values and information embedded in seed are subject to innovations that are determined by prevalent innovation policies.

Many policies are translated into rights, such as intellectual property rights (IPRs), national sovereign rights over biological resources, and community and farmers' rights over genetic resources and associated traditional knowledge. These policies and rights establish the rules concerning incentives that target farming communities and other stakeholders in PGR management. Moreover, they define the room for manoeuvre for farming communities and stakeholders to engage themselves in practices for community biodiversity management (CBM); such practices have an impact on the future of conservation and the sustainable use of PGR, at local and global levels. Figure 6.1.1 illustrates how the diversity of areas of policies and rights as defined by the various functions of seed determine the rules of the game for the conservation and use of plant genetic resources for food and agriculture. The figure guides the structure of this section and the current introductory chapter.

In this chapter, we introduce the areas of policies and rights defined by the various functions of seed, provide some examples for each and emphasize their interactions. Each area is introduced, followed by a description of the relevant policy considerations and their legal implementation. We thereby aim to describe the landscape of policies and rights, creating both opportunities and limitations for the implementation of CBM, with its aim to contribute to PGR conservation and livelihood development.

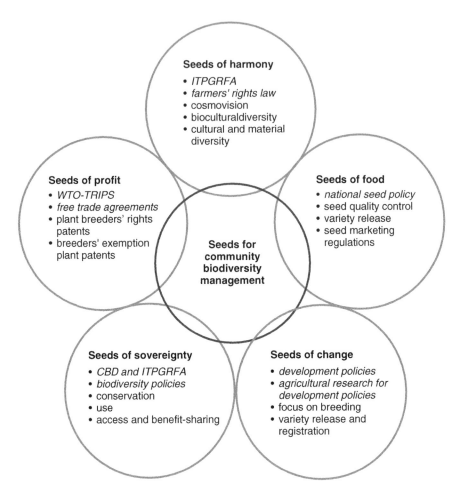

Figure 6.1.1 Functions of seed: key characteristics of genetic resource policies and rights that create limitations and opportunities for community biodiversity management.

The section includes chapters on farmers' rights, access and benefit-sharing (ABS), and seed and variety laws. Andersen (Chapter 6.2) describes how farmers' rights are structured into four elements within the International Treaty on Plant Genetic Resources for Food and Agriculture (ITPGRFA): the protection of traditional knowledge; the fair and equitable sharing of benefits; the right to participate in decision-making; and any rights that farmers have to save, use, exchange and sell farm-saved seed. For each right, the author shares experiences on how farming communities are increasingly being supported in expressing their rights. Bala Ravi (Chapter 6.3) takes a national perspective, explaining how India's Protection of Plant Varieties and Farmers' Rights Act creates a balance in the interests of farmers and breeders through nine farmers' rights. Vernooy and Ruiz (Chapter 6.4) share how multiple forms of benefit-sharing, described through case studies from Honduras, Peru, Cuba,

China, Nepal, Jordan and Syria, provide the grassroots' basis for establishing ABS mechanisms that increasingly meet the actual needs of communities. Albuquerque *et al.* (Chapter 6.5) address the implications of the Brazilian Provisional ABS Act on the interaction between traditional people and scientists, and discuss the fact that even though the Act was developed with the aim of protecting biodiversity and associated traditional knowledge, in practice the mechanisms that have been put in place result in the further appropriation of what was previously common heritage.

This section also addresses seed and variety laws and their potential impact on the farmers' management of local varieties, and in particular on farm-saved seed and informal seed systems. Kastler and Moÿ (Chapter 6.6) describe the evolution of specific regulations within the European Union that aimed to promote on-farm management of plant genetic resources, arguing that the procedures put in place are not practical in terms of farmers' dynamic management, and that they obstruct potential achievements. Santilli (Chapter 6.7) and Kastler (Chapter 6.8) look at the implications of seed and variety laws in Brazil and France, respectively, on the informal seed system, a system that facilitates the implementation of on-farm management of agrobiodiversity. In both countries, seed and variety laws favour the agro-industrial development model. The authors outline the need to secure a legal space for both farm-saved seed and informal seed systems, for the sake of on-farm management of agrobiodiversity and for guaranteeing the livelihoods of small-scale farmers. This bring us back to the definition of farmers' rights as expressed in ITPGRFA; the question remains as to how to find the right balance between recognizing the role of farmers in conservation and in maintaining autonomy in their livelihoods, and supporting the application of intellectual property rights required to promote innovation by plant breeders.

Seeds of food: the right to quality seed

Seed is a crucial input in any form of crop production and is one of the most precious resources in farming. The genetic make-up of the seed determines crop and varietal identity, and potential yield and yield stability, as well as product qualities that may be linked to socio-cultural preferences. The germination percentage and seedling vigour determine the primary plant population in the field, while the seed health status can be a key element in determining the development and severity of a disease epidemic. The choice and handling of the seed thus determine, to a significant extent, the chances of success for the crop. In this context, the two key issues concerning seed for every farmer are accessibility and quality.

Seed has to be accessible for every crop production cycle; it has to be available at the right time, in the right quantities, with the right qualities and at the right price, to allow farmers to access the seed that they need (Louwaars and De Boef, 2012). Seed is both a reproductive unit and product for consumption, in the case of most cereals, pulses and oil crops, and, in principle, it should be readily available. Seed can also be a by-product with little value other than its use as seed or planting material. This is the case in crops like cassava, sweet potato, jute, cotton and tomato. This by-product is either readily available (e.g. cassava, cotton) or seed production needs to be a separate operation (as for many vegetables).

Even where the food grain can be used as seed, availability can be a problem. Poverty may result in individual seed security challenges (Lipper *et al.*, 2010), and ecological or man-made disasters regularly create large-scale seed shortages for such crops (Sperling *et al.*, 2008). Even when seed is available, the price may reduce poor farmers' access to quality seed. Communities that regularly face such conditions commonly develop strategies to cope with them, such as the long-term storage of carry-over seed, and the use of complex, informal bartering and marketing systems (Abay *et al.*, 2008; Sperling and McGuire, 2010).

Seed policies

Seed policies have been framed in many countries to regulate the production and distribution of quality seed; they focus on supporting and controlling the formal seed systems and, more specifically, on promoting private sector seed production and marketing. In order to protect farmers from buying sub-standard seed and to create a level playing field for seed producers, countries want to control seed quality. More recently, the approach of integrated seed sector development (Louwaars and De Boef, 2012) has taken root as a framework for guiding seed policy development (African Union, 2011; Louwaars *et al.*, 2013). This is in contrast with the linear approach to seed sector development, which assumed that within the dominant agricultural development paradigm, all seed used by farmers would be produced and marketed through commercial (formal) systems (Douglas, 1980). The integrated approach fosters a pluralistic model, which matches and responds to a diversity of situations that vary for crops and which target farmers. Such an approach includes supporting international and national seed companies, but also local seed businesses or organized groups of small-scale seed producers. At the same time, it supports the availability of quality seed in several seed systems, as well as the wider use of genetic diversity. Moreover, integrated seed sector development acknowledges the vital functions played in agriculture by informal seed systems, including farm-saved seed. In order to take a pluralistic approach for supporting such diversified seed systems and promoting the use of diversity, it is essential that the different roles and functions of the government are carefully framed (Louwaars *et al.*, 2013).

Seed laws

The seed laws that are developed on the basis of such seed policies commonly establish a seed quality control framework, a variety release procedure, and a regulatory mechanism for seed marketing. Seed quality control is based on seed testing procedures that have been harmonized through the International Seed Testing Association (ISTA). Seed certification (maintenance of variety identity and purity) uses a generation system, from breeders' seed to certified seed, with strict rules concerning field/storage inspections and off-type counts.

Various problems have been identified in the scope and operation of seed laws. The scope of the laws often includes not only the formal seed systems but also, depending on the definitions, any seed that is marketed, transferred from farmer to farmer, or in some cases even farm-saved seed. All seed has to be certified and tested, and thus it has to belong to the officially released varieties. Certification is voluntary in just a few countries, which leaves greater space for informal seed and farm-saved seed systems; a space that is vital for farmers and for on-farm/community management of PGR.

Seed certification procedures commonly do not allow the inclusion of varieties that are not uniform, thus outlawing most local varieties from the market, as for example in the case of local rice varieties in Vietnam, highlighted by Thomas and Anh (Chapter 4.5), and in the case of local varieties and their use by small-scale farmers in Brazil, described by Santilli (Chapter 6.7). This exclusion also applies to varieties produced through various methods of participatory crop improvement (PCI), which may not meet standards of uniformity, as further explained, for example, by Alonzo *et al.* (Chapter 5.8). Furthermore, certification and seed testing procedures may significantly increase the cost of seed, creating a barrier to local or specific adapted varieties, as illustrated by Kastler and Moÿ (Chapter 4.6), for local wheat varieties in France. If seed quality control operations are funded by local or regional government bodies, it may result in poor implementation of the rules or may even invite corruption. Formal and local systems thus follow distinct logics and dynamics, and they cater to the needs of different agricultural and marketing models (Lipper *et al.*, 2010). This diversity of situations and seed systems has so far largely been underestimated by the predominant linear orientation of seed polices and their implementation mechanisms (Louwaars *et al.*, 2012).

Seeds of change: technology transfer

Discussions on agricultural development inevitably lead to the subject of seed (Tripp, 2001). Since seed is the carrier of the genetic make-up of the plant, it is a key tool for technology transfer. Seed and agrobiodiversity are, together with soil and water management, at the centre of the sustainability of traditional farming systems. As recent history has shown, seed of new improved varieties, or even seed of new crops, is capable of changing complete farming systems, or even creating new ones. This can be illustrated by the zero-tillage system that transformed the Brazilian cerrados into an agricultural landscape, facilitated by the introduction of herbicide-tolerant soybean varieties, and by the current rice revolution that is occurring in many African countries, facilitated by the Nerica rice.

The genetic basis of varieties can contribute to increased yield stability, through tolerance to abiotic stress factors or resistance to pests and diseases. It can also result in increased product value, through qualities that are either important for obtaining a good price in the market, such as grain colour of legumes, or that create direct nutritional benefits for home processing or direct consumption. As a result, seed is a key tool for technology transfer and technology-driven agricultural development strategies, and is widely considered a focal point in agricultural progress. It has direct implications on agrobiodiversity, being embedded in diverse agricultural landscapes of different crops and varieties, as well as on the livelihoods of the communities that have created but also depend on them.

The seed policies and laws that facilitate the green revolution mode of agricultural production in many of these countries also constrain the informal (local) seed systems, grossly undermining the role and importance of such systems in local crop production and food security. We will detail this in the following sections.

Crop research and breeding policies

In most countries, the promotion of uniform, so-called 'modern varieties' is of key importance in the modernization of the agricultural development paradigm to increase national food security. A large proportion of national and international agricultural research is devoted to plant breeding. Below, we describe three major weaknesses, which result from agricultural research and development policies that guide public breeding programmes.

First, international research that is carried out by the Consultative Group on International Agricultural Research (CGIAR) concentrates on crops of international importance, such as maize, rice and wheat, and, within those crops, targets just a few important traits. This focus on crops and traits gives the highest return on investment. However, national public breeding programmes often follow, or are destined to copy, this focus. Locally important crops are under-researched and locally important breeding goals, such as yield stability under small-scale farming production conditions, and aspects such as cooking quality of legumes for women farmers in areas where firewood is in short supply are given little priority.

Second, targeting various major areas with one single or a limited group of crops and varieties means that selection for specific adaptation is often neglected. Such selection is particularly important for the development of varieties that are adapted to ecologically diverse environments and that respond to the culinary preferences of culturally divergent communities.

Third, public plant breeders commonly focus solely on CGIAR materials, which are the principal source of improved germplasm. Because of standardization requirements for release, these breeders develop varieties that are genetically uniform. The limited source of improved germplasm, coupled with practices of standardization, are a tremendous force that favour uniformity and monocultures of just a few varieties within the dominant agricultural development paradigm. Within a context of resilience, they disregard the importance of local materials and the potentially adaptive properties that this diversity may have. Alternative breeding methods, including various methods related to participatory crop improvement, may bypass such limitations, as addressed by De Boef *et al.* in Chapter 5.1. Policies which determine that breeders are to be evaluated on their scientific output (articles) and on the number of officially released varieties, instead of on the areas or the number of dissimilar agro-ecologies covered, continue to act as disincentives for breeders to reorient their methods towards creating a diversity of varieties.

Laws for variety registration and release

The policies applied to plant breeding are supported by formal release procedures. These are commonly based on multi-location tests and statistical analysis of the results, and are decided on by a variety release committee. Varieties that are not approved are, in most countries, banned from the market.

Poor representation of stakeholders in the variety release committees, particularly by farmers and the private sector, means little attention is given to traits other than yield and yield components, and commonly leads to the rejection of varieties with specific adaptation or quality aspects. Varieties are tested for release under favourable (high-input) conditions of experimental stations, which creates a bias towards varieties that may benefit commercial farmers and industrial modes of agricultural production (Tripp and Louwaars, 1998). Statistical analysis of the results of those tests focuses on wide adaptation, not recognizing the value of varieties that perform well in ecological niches (Ceccarelli, 1994). Very few countries have explicit policies on the use of genetically diverse varieties. The European Union passed a directive in 2009 on 'conservation varieties', allowing, with strict quantitative restrictions, the marketing of seed of varieties that are naturally adapted to local and regional conditions, and which are threatened by genetic erosion. Brazil's seed law also creates some legal space for local varieties; however, common standardization requirements and the pull to uniformity, driven by the predominant agricultural development paradigm, create obstacles for its practical implementation, as described by Santilli in Chapter 6.7.

Seeds of sovereignty: biodiversity

Seed carries an important component of agricultural biodiversity: plant genetic resources (PGR), that is, the gene pool of plants of present or future value. Since they are also a component of the total biodiversity, they are thus covered by the environmental policies that emerged in the 1980s following concerns over the loss of habitats and species. These were brought together in 1992 by, among other agreements, the Convention on Biological Diversity (CBD). The CBD targets the conservation of biological diversity, its sustainable use, and the fair and equitable sharing of benefits arising out of such use.

But to whom does crop genetic diversity belong? The International Undertaking on Plant Genetic Resources for Food and Agriculture (IUPGRFA) established that such genetic resources are the heritage of humanity and consequently should be available without restriction (FAO, 1983). The CBD, however, takes a much broader outlook, addressing more than just plants and agriculture and concluding that all biological resources fall under the sovereignty of nations. Consequently, countries have the responsibility to conserve and to promote their sustainable use, and also to make them available after prior informed consent and based on mutually agreed terms. The CBD also recognizes the rights of local and indigenous communities over these resources, but places the responsibility for ensuring such recognition with the countries themselves. The Nagoya Protocol on Access to Genetic Resources and the Fair and Equitable Sharing of Benefits Arising from their Utilization is an international agreement associated with the CBD. It aims to ensure that the benefits arising from the utilization of genetic resources are shared in a fair and equitable way, through appropriate access to genetic resources and the appropriate transfer of relevant technologies, taking into account all rights over those resources and technologies; and by appropriate funding, thereby contributing to the conservation of biological diversity and the sustainable use of its components. This protocol raises the bar in terms of what countries are obliged to do vis-à-vis traditional knowledge.

Since bilateral negotiations that are based on CBD principles are difficult to implement in the field of plants and agriculture, the International Treaty on Plant Genetic Resources for Food and Agriculture (ITPGRFA) was negotiated and came into force in 2004. The major argument in favour of establishing the ITPGRFA is that the PGR that are vital to food and agriculture cannot be bound by bilateral agreements. The ITPGRFA includes a multilateral system for facilitated access and benefit-sharing (ABS) for a limited selection of plant genetic resources (i.e. certain crops and resources that are under government control).

Biodiversity policies

Biodiversity policies include biodiversity at landscape, ecosystem, species and genetic levels. Ecosystems are important for PGR conservation when they are repositories of crop wild relatives (Hunter and Heywood, 2011). The conservation of these important gene repositories is essential, as every loss of gene is an opportunity lost for the future. The strategic importance of crop wild relatives for present and future agriculture has increased with recent advances in molecular biology, which now facilitates the mobilization of genes across species and generic reproductive barriers.

Conserving agro-ecosystems or domesticated landscapes may be considered part of strategies for on-farm management, as described by Peroni *et al.* (Chapter 3.1). Conservation of agrobiodiversity targets further development at system, species and genetic levels, through a combination of farmer-led and natural selection procedures. Peroni *et al.* argue that such conservation strategies, which involve vital interactions between people, plants and their environment, can only be sustained and strengthened by building upon CBM. The diversity generated and maintained through such dynamic and complex patterns is vital for achieving resilience and thereby is likely to prove valuable in an era of climate change (De Boef *et al.*, Chapter 7.3).

Gene banks are important for the *ex situ* conservation of genetic resources, and for making them available through their evaluation, documentation and seed management systems (Engels and Visser, 2003). Genetic diversity does not evolve any further in gene banks, but they play a critical role in conserving PGR that would otherwise be lost. Given their different strengths and limitations, the three conservation strategies that focus on the *in situ* conservation of habitats with crop wild relatives; the continued use and evolution of crops and varieties in agricultural eco-systems (through on-farm management, i.e. CBM); and the conservation of PGR in gene banks (*ex situ* conservation), are complementary strategies that strive towards the one goal of maintaining these resources for future use (Hawkes *et al.*, 2000).

Biodiversity law

The CBD stipulates that the Parties to the Convention shall enact national laws in compliance with its principles and articles. Many countries have since established national biodiversity laws that are commonly broad in scope, covering endangered species and protected areas, and which define the national competent authorities that are responsible for handling access and ABS issues. The CBD requires its Parties to respect, preserve and maintain knowledge, innovations and practices of indigenous and local communities that are relevant for the conservation and sustainable use of biological diversity, through their national laws. However, there are many differences between countries in this respect. Even though examples of the implementation of benefit-sharing exist, few cases concerning the use of agricultural genetic resources, where actual benefits have reached communities, have been documented (Visser *et al.*, 2005; Ruiz and Vernooy, 2012).

An increasing number of countries are now implementing the ITPGRFA, and gene banks are making their genetic resources available, by using the standard material transfer agreement, and by including these resources in the multilateral system for facilitated access and benefit-sharing. However, even though countries formally accept the multilateral system, they are slow to implement, or have different perspectives on how to implement, the practice of allowing free access to those resources in their national collections, as outlined in the conditions of the agreement (Frison *et al.*, 2011).

Seeds of profit: intellectual property

The value of seed for crop production, as well as the investments needed to breed new varieties and to manage seed through production, quality maintenance and marketing, add to its commercial value. Notwithstanding the strong, public good characteristics of the seed that arise from its self-replicating nature, seed is a high value commodity. Quality seed may be bartered in local exchange systems for food grain at an equally high rate (Lipper *et al.*, 2010). The actual value of seed for some crops in commercial systems can exceed 50 times the cost of production (Almekinders and Louwaars, 1999) and, in commercial farming, vegetable seed may be more expensive than gold per unit weight. Such prices reflect the actual value of quality seed for farmers and also accommodate the economic return for significant investments by shareholders in research and breeding, and in the production of quality seed.

Intellectual property policies

Supporting innovation is a key element in agricultural policies. Investments in plant breeding and breeding research are considered important, and protecting intellectual property generated from such research is gaining increasing significance in international and national policies. The World Trade Organization (WTO), under its Agreement on Trade-Related Aspects of Intellectual Property Rights (TRIPs), requires member countries to enforce a minimal standard in the protection of intellectual property rights (IPRs) related to new plant varieties (WTO, 1994). However, bilateral trade negotiations and free trade agreements often go beyond this minimal standard.

Plant breeders' rights (PBRs) on varieties, and patent protection on biotechnological inventions, aim to promote the commercialization of newly developed varieties and associated innovated processes. PBRs often go hand in hand with restrictions on farmers concerned with saving, reusing or exchanging and selling their seeds of such protected varieties. Patents are much stronger than PBRs, and offer no room to farmers for saving and reusing seed, and as such the patenting of seed has attracted much discussion and debate (Louwaars *et al.*, 2009).

Plant breeders' rights and patent laws

The concept of plant breeders' rights was developed in Europe in the mid-twentieth century, following recognition that the patent system was not suitable for protecting breeders' work. A major difference between patents and PBRs is that PBRs do not use claims and have different requirements for protection (distinctness, uniformity, stability and novelty instead of inventiveness, industrial use and novelty). The novelty prescribed for PBRs and patents is also defined differently. Unlike patents, PBRs allow some significant exceptions; these are the breeders' exemption and the farmers' privilege. The breeders' exemption permits the protected varieties to be freely used for further breeding and developing new varieties that are also protectable. The farmers' exemption may allow any of the following: the saving and reuse of seed of protected varieties in the farmers' own fields, and its informal exchange and sale. The farmers' exemption, however, varies in countries that are subject to the international conventions to which they are bound, and the national interpretation of such conventions.

Few countries allow the protection of varieties through patents (notably the USA). The emergence of plant biotechnology, and processes and products of molecular research that contribute to plant breeding, has led to an increased demand for patent protection. Several countries allow patents on plants and plant parts, but not varieties (e.g. Europe). Other countries, mainly in Africa, consider the exclusion of plant varieties to mean that protection cannot be granted to plant parts. The exclusive right granted by the patent system is absolute and normally does not allow any scope for farmers' and breeders' exemptions (Louwaars *et al.*, 2009).

Seeds of harmony: farmers' and community rights

Seed is a vital element of farming that is inextricably linked to the culture of the people who developed, selected, maintain and use it. The concept of 'cosmovision', developed in Latin America, approaches agriculture holistically and visualizes it as a continuous interaction between the indigenous culture, the environment and technologies (Reijntjes *et al.*, 1992). It sees the role of people as that of protecting the harmonious coexistence of the spiritual and material worlds, including their agricultural methods. Living organisms, such as seeds, play an important role in bridging the spiritual and material worlds (Ishizawa, 2004). Traditional people, or their representatives, consider themselves the source of biodiversity that has developed in association with their cultural diversity (Pilgrim and Pretty, 2010; Maffi and Woodley, 2010). Such a vision on bio-cultural diversity leads to claims that this linkage between spiritual and cultural diversity, and material diversity, needs to be protected from outside pressures and demands. In the field of PGR, this translates into the fact that varieties and seed must be protected from exploitation by others outside the community to which they belong. In addition, all property claims on seed is rejected, because of the belief that spiritual powers give the seed its special character (Haverkort *et al.*, 2002).

Generally speaking, the recognition that farmers have developed and continue to contribute to the genetic resources that they use is the basis of the concept of farmers' rights, which was first established in 1983 by the IUPGRFA, and later defined in more detail in the ITPGRFA, in 2001.

Farmers' rights policies

The ITPGRFA recognizes farmers' rights as the protection of traditional knowledge relevant to PGRFA; the right to equitably participate in the sharing of benefits arising from the utilization of PGRFA; and the right to participate in decision-making, at national level, on matters relating to the conservation and sustainable use of PGRFA. However, the concluding part of Article 9 of the ITPGRFA, which is devoted to farmers' rights (FAO, 2001), affirms that 'nothing in this article shall be interpreted as to limit any rights that farmers have to save, use, exchange and sell farm-saved seed subject to national law and as appropriate.'

Farmers' rights laws

Several countries have included articles in their national biodiversity laws to safeguard the rights of indigenous and traditional communities, but only a few guarantee to protect their lifestyles and spiritual values. Few countries have explicitly included farmers' rights in their national laws. An exception is India, which has included the right for farmers to save, use, exchange and sell seed (only on a non-commercial basis), and which has established a national gene fund to promote the on-farm conservation of PGR by farmers and indigenous communities. Bala Ravi (Chapter 6.3) describes the farmers' rights law in India in more detail. However, as when farmers' rights are overshadowed by agricultural development and intellectual property, the new seed bill, which is currently pending in India, has provisions which might nullify some of the farmers' rights provisions, as it requires that all traded varieties be registered and that performance evaluation is one of the requirements for such registration. This is likely to negate the right to exchange and sell seed of farmers' varieties.

Seeds for community biodiversity management

The recognition that gene banks are just one component of PGR management, and that PGR have a value not only for breeders but also for farmers in increasing the resilience of their farming system, led to the need to acknowledge the importance of on-farm management of PGR and to design practices to strengthen it. Even though on-farm management strategies were originally designed by scientists, as in the case of the practices in Nepal that are described by Subedi *et al.* (Chapter 1.2), it is now clear that they must be developed as part of the process which is now referred to as community biodiversity management (CBM). In the CBM process, non-governmental organizations and local public development organizations encourage farming communities to assume the conscious responsibility of their agrobiodiversity, as illustrated by Shrestha *et al.* (Chapter 1.3). CBM, as a common approach for contributing to on-farm management of PGR and livelihood development, forms an umbrella over a wide range of practices. It does not have a proper policy arena of its own, but its practices are highly dependent on the range of policies and legal instruments that affect seed systems.

   The wide range of policies that seeds are subject to, and the large number of international and national laws that affect the handling of seed, creates a complex environment that communities, and those parties who assist the communities in managing their biodiversity, should be aware of. On the one hand, communities may have rights over their genetic resources, and a say in negotiating the terms of their use; on the other hand, other entities may also have rights over or may impose restrictions on communities in their use of plants and seed.

A distinction needs to be made between locally developed genetic resources and introduced or commercially improved materials. Local genetic resources, such as local varieties of crops that have been managed by the communities for generations, as well as the crop wild relatives about which communities have knowledge, may fall under the national rules respecting customary rights at community level. Depending on how such rules are framed, these may require communities to establish a way to express their rights over such local genetic resources and to assign or assert their entitlement to negotiate access to these resources. One view is that such rights may also reduce or impede access to local varieties that are held in a particular community, for members of another community in the same country. For example, such restricted access to material may go against public interest, if the material in question is considered important for breeding varieties for the benefit of other farming communities.

Another example can be seen with the legal restrictions that prevent or discourage genetic diversity under cultivation, promoting, instead, genetic uniformity. The restrictions imposed by seed laws on the standard of genetic purity and other aspects of seed quality being commercially transacted, prevent the wider use of local varieties, as well as those commonly grown and disseminated, without circumventing the procedures prescribed by such laws (i.e. non-released varieties and all planting materials that pass through the informal seed system). This situation results in the formal public and/or commercial system, which promotes only the certified seed of a relatively small set of varieties with a narrow genetic base, and which prevents the organized seed production and distribution of many local and diverse varieties by farmers' groups or cooperatives, through its legal and regulatory framework (Louwaars *et al.*, 2012).

An important CBM practice is the enhancement of local varieties through the various methods of participatory crop improvement (PCI), as described by De Boef *et al.* in Chapter 5.1. Methods such as grassroots breeding and participatory genetic enhancement may produce varieties that conflict with patent laws if the parent material used is protected. However, this is not the case in current practices. PBRs with breeders' exemption do not impose such restrictions on PCI methods. However, there are cases where participating farmers or breeders (moreover, the institutions of participating breeders) would like to protect the product of such breeding through PBRs. Several development organizations strongly oppose such protection (Salazar *et al.*, 2007), since a free flow of materials imposes limitations on any form of protection of PCI varieties, as can be seen with the implementation of plant varietal selection that builds strongly on the function of exchange within informal seed systems.

Few systems for PBRs facilitate the protection of non-uniform varieties that are likely to come out of PCI. The formal recognition (registration) of varieties bred though such PCI methods is likely to be faced with a number of problems, including the fact that the variety release system is too expensive for the farmers' groups promoting such varieties, or that such registration systems recommend registration only for those varieties that do well throughout the country, or in all test regions. In other words, a registration system that approves only those varieties that show good average performance across many locations and agro-climatic regions, but which does not approve those that are better performing in specific locations, does not encourage local or diverse PCI varieties; it creates constraints for the use of genetic diversity in the crop production systems. This is antithetical to the CBM approach and to the principles of the ITPGRFA, which both support the sustainable use of plant genetic resources.

Where new and protected varieties form part of the local gene pool, their PBRs may be restricting their use at the local level. This may not become a serious problem in countries where public research generates and provides new varieties without the PBR tag. However, when public institutions start claiming PBRs on their varieties, the space for farmers to apply their rights becomes severely restricted. This becomes serious, particularly if the country adheres to the International Convention for the Protection of New Varieties of Plants, where the farmers' privilege does not include the right to exchange seed among farmers (UPOV, 1991). However, India, through its Plant Variety Protection and Farmers' Rights Act, aims to create the appropriate balance between the multiple interests that define this arena, and to develop a frame-work that is conducive for several elements of farmers' rights and CBM (described in detail by Bala Ravi in Chapter 6.3).

With all these conflicting interests in the PGR arena, the silver lining is that many national laws are not implemented at all levels. For example, many seed laws, while their definitions may include any seed that is sold or bartered, only affect the public and private entities that specialize in seed production and marketing, and do not affect those farmers and their organizations that sell seed among themselves. However, CBM strategies that intend to facilitate the improvement of seed quality at local level may run into problems with the rule of law at some stage of the attempt to professionalize and institutionalize their operations in the informal system.

As illustrated earlier in Figure 6.1.1, a wide range of international and national policies deal with the handling of seed, in relation to the various functions of seed in society. CBM processes and their accompanying practices may be affected by national laws for implementing policies that are developed in response to national needs, or by requirements of international agreements that are strongly embedded in the normal agricultural development paradigm, to which CBM itself is a response and creates new pathways. Since these laws sometimes seem to contradict each other, the rights of the communities and their obligations may not always be clear, thereby transforming these genetic resource policies and rights into seeds of confusion (Louwaars, 2007).

6.2 Farmers' rights in times of change

Illusion or reality?

Regine Andersen

What are farmers' rights all about?

Plant genetic resources (PGR) are probably more important for farming than any other factor, simply because they can adapt to changing environmental conditions. As farmers are custodians and developers of PGR, their rights in this regard are vital for enabling them to maintain this vital role for local and global food security. In order to implement farmers' rights, we must recognize and reward them for their indispensable contribution to the global gene pool and associated traditional knowledge (ATK).

The implementation of farmers' rights is a precondition for PGR conservation, which is the basis of food and agriculture. For this reason, farmers' rights constitute a cornerstone of the International Treaty on Plant Genetic Resources for Food and Agriculture (ITPGRFA); their realization is a precondition for achieving its three objectives of conservation, sustainable use, and fair and equitable benefit-sharing.

Article 9 of ITPGRFA identifies four elements of farmers' rights as being (i) the protection of traditional knowledge; (ii) the fair and equitable sharing of benefits; (iii) the right to participate in decision-making; and (iv) any rights that farmers have to save, use, exchange and sell farm-saved seed. In this chapter, I will describe in more detail each element, sharing, when available, experiences where those elements are put to practice.

The Fridtjof Nansen Institute in Norway has supported the ITPGRFA in its efforts to address farmers' rights, through research-based guidance, by conducting a range of surveys and case studies, developing a collection of success stories, and carrying out dissemination activities and rounds of consultations. With our research work, we aim to establish a common understanding of the contents of these elements, associated rights and what it takes to implement them. A range of organizations in different countries have carried out local, national and regional consultations on farmers' rights and are active in implementing these rights at local and national levels. Information about the project is available at www.farmersrights.org.

Farmers' rights in relation to the protection of traditional knowledge

Traditional knowledge comprises the knowledge of how to select, store, use and manage seed. This knowledge is vital for understanding the properties of plants and varieties, their uses, cultural significance, and how to cultivate them (Brush, 2004). Traditional knowledge can be approached in two ways, first, for protection against extinction, and second, for protection against misappropriation (Andersen, 2006).

Protection of traditional knowledge against extinction is all about ensuring that it is kept alive and can further develop. Traditional knowledge is disappearing along-side genetic erosion at an alarming pace. Measures for its protection are considered crucial by farmers who are engaged in farming practices in which the use of PGR, or agrobiodiversity in general, is promoted through community biodiversity manage-ment (CBM). The best way of protecting traditional knowledge against extinction is to use it and share it. Thus, the motto here is: 'protection by sharing'. Measures for sharing traditional knowledge can include the organization of seminars, conferences and gatherings among farmers, to share knowledge associated with diversity. In Nor-way, a number of activities have been initiated to help disseminate traditional knowl-edge (Andersen, 2011). In Nepal, ATK and PGR are shared through diversity or seed fairs as discussed by Shrestha *et al.* in Chapter 2.2. Another way of sharing knowledge is to document it in seed catalogues and registries, for example through community biodiversity registers in Nepal (described by Subedi *et al.* in Chapter 2.4).

Community seed banks are vital for sharing not only seed and varieties, but also the associated knowledge. A good example of this is shown by Dias (Chapter 2.7), who describes an extensive network of more than 200 community seed banks in the state of Paraíba in north-eastern Brazil. There are many examples of documentation of knowledge in catalogues, books, magazines, videos and websites, including the potato catalogue from Huancavelica, Peru (CIP and FEDECH, 2006). Finally, gene banks facilitate sharing when they make accessions and related knowledge available to farmers and communities, as can be seen in the case of the national gene bank in Ethiopia, described by Feyissa *et al.* (Chapter 1.4); and in the work of the national gene bank in Ecuador, highlighted by Tapia and Carrera (Chapter 2.3).

A different approach is required for the protection of traditional knowledge against misappropriation, which is based on the fear that local varieties, together with ATK, could be discovered and further developed by commercial actors. Such actors, in reward for their discovery and exploitation, 'appropriate' the knowledge, and may use intellectual property rights (IPR) without benefit-sharing mechanisms. In order to protect traditional knowledge from such misappropriation, its sharing should only be facilitated following recognition of and agreement on benefit-sharing. Methods to protect against misappropriation first include the regulation of access to PGR and ATK, with measures on prior informed consent and mutually agreed terms. The inclusion of legal clauses in catalogues of PGR and ATK also ensure against misap-propriation, and could be used, for example, with the aforementioned community biodiversity registers (Nepal). Another way to protect knowledge is through the appli-cation of 'user country measures', like conditions for IPR, such as certificates of origin of genetic resources, which follow the legal procedures for access to genetic resources in provider countries when providing access internationally.

On a more local level, the application of geographic indications or regional brand-ing protects products of specific varieties produced in specific locations. Examples of this can be seen with the case of a special flour and bread that is produced with a particular variety of wheat in one specific region in France, as detailed by Kastler and

Moÿ (Chapter 4.6), and with the case of Kalajeera rice produced by tribal farmers in Jeypore, India, described by Chaudhury and Swain (Chapter 4.4).

The regulation of access, where some communities limit access to and use of a particular set of plants for a particular group as part of a cultural practice, is a more endogenous way of protecting knowledge and biological resources. Examples of this can be seen with the cultivation of coffee in the Afromontane forests of Ethiopia, as outlined by Feyissa *et al.* (Chapter 3.6), and with the collective management of the natural resources from sacred groves and forests found across Africa, Asia and the Pacific, as described in more detail by Borrini-Feyerabend *et al.* (2007).

In some countries, measures taken against the misappropriation of traditional knowledge make the sharing of such knowledge difficult, and contribute to distrust among farmers. This has made *ex situ* conservation of crop varieties difficult, for example, in Peru (Andersen, 2005b). In addition to this, such measures mean that even biological scientists who take a fair and ethical approach, within a context of CBM, must first prove that they are not 'biopirates', an unfavourable issue that communities of traditional people are now being faced with. An important question to address is how great the risk of misappropriation of farmers' PGR and ATK is, and whether this risk is worth the measures taken to avoid misappropriation. So far, there are very few known examples of misappropriation of farmers' knowledge. The challenge is to balance these concerns in such a way that traditional knowledge can still be shared to the greatest extent possible.

An impressive number of examples concerning documentation and sharing are provided above. Together they constitute good practices for the implementation of farmers' rights related to the ATK. Much more is required to keep such knowledge alive and to promote its further development. In many countries, there is a need to raise awareness of the importance of traditional knowledge related to PGR, and to develop strategies on how to maintain and disseminate traditional knowledge in a systematic way, in order to halt the loss of such knowledge. In our global consultation on farmers' rights, we found that many stakeholders stressed the vital importance of documenting remaining traditional knowledge, and spreading awareness about its importance (Andersen and Winge, 2011).

Farmers' right to participate equitably in the sharing of benefits

In the South, policies on benefit-sharing, if any, are provided in laws and regulations on access to biological resources, sometimes in legislation on the protection of biodiversity, as described in more detail by Ruiz and Vernooy (Chapter 6.4). In India, benefit-sharing is addressed by its Protection of Plant Varieties and Farmers' Rights Act of 2001, which is further detailed by Bala Ravi (Chapter 6.3). Countries with legislation on indigenous or traditional peoples' rights often include provisions on benefit-sharing in these laws, which then also cover indigenous farmers, or, as in the case of Brazil, traditional farmers, as discussed by Assis *et al.* (Chapter 3.2).

Most of these regulations foresee forms of direct benefit-sharing between the 'owners' and 'buyers' of genetic resources, often upon prior informed consent on mutually agreed terms, as set out in the Convention on Biological Diversity (CBD, 1992a).

Despite all these efforts, today, 20 years since the CBD was agreed upon, no examples of direct benefit-sharing between providers and recipients of plant genetic resources for food and agriculture (PGRFA) have resulted from such legislation (Andersen, 2008).

Other ways of sharing benefits, often referred to as indirect ways, are feasible and have been elaborated over the past two decades, recalling the early days of the negotiations on farmers' rights in the Commission on Plant Genetic Resources for Food and Agriculture. Benefits should be shared between the custodians of PGR and society at large. This is based on the idea that it is the legitimate right of those custodians to be rewarded for their contributions to the global gene pool, from which we all benefit. Furthermore, it is seen as an obligation of the international community, and individual nations, to ensure such recognition and reward. The most important forms of sharing include the following:

- conservation activities, including local gene banks and community seed banks in all their variations, as elaborated on in Parts I and II of this volume;
- access to seed and propagating material, and related information, and the strengthening of farmers' or informal seed systems, as further detailed by De Boef *et al.* (2010) and Lipper *et al.* (2010);
- enhanced utilization of farmers' varieties, including value adding and market access, as described in Part IV;
- participatory crop improvement, and its variations in methods that facilitate the collaboration between farmers and scientists, as discussed in Part V;
- CBM practices that aim to contribute to the empowerment of communities for assuming responsibilities in the conservation and use of PGR, which are in turn transformed into actions that guide the communities towards asserting their farmers' rights.

The benefit-sharing fund of the ITPGRFA has disbursed around US$10 million for the biennium 2010–2011, after an initial disbursement of US$543 000 in 2009. The fund is relevant for the provisions on farmers' rights, as its financial resources are to be distributed directly and indirectly to farmers who are conserving and sustainably using PGR. Many of the organizations that share experiences in this volume are engaged, in one way or another, in projects that will be implemented under the second disbursement of the fund. Despite this progress concerning a relatively small international funding mechanism, there is still a long way to go before the fund reaches a size that will enable the third objective of the ITPGRFA to be truly achieved.

Bilateral and non-governmental development cooperation represents another source of benefit-sharing, and it supports many promising projects at local level in developing countries. Many of the success stories as shared in this book have been financed through such mechanisms. Also in the North, as highlighted in several chapters concerning the seed network of small-scale farmers in France, a number of projects have been made possible with support from regional governments.

The major challenge today is to find ways and means to scale up, or rather institutionalize such practices in such a way that they become independent from specific project or development funding, and are embedded in government or other more sustainable programme structures. Nevertheless, it is important to remain aware of incentive structures that are in fact counterproductive to benefit-sharing, such as agricultural policies which hamper the production and marketing of farmer-produced seed and the products of local crops and varieties. Louwaars *et al.* (Chapter 6.1) address these constraints and opportunities for the whole of genetic resources and agricultural development policies.

Farmers' right to participate in decision-making

The participation of farmers in the development of laws, regulations, policies and programmes related to PGR management is important for the simple reason that farmers are key actors. Ideally, policies and programmes that target farmers should take their situation and perspectives as points of departure. Legislation for regulating mainstream agriculture is relevant, as such legislation tends to produce incentive structures that may be detrimental to farmers' rights or to CBM in general, and that may lack any measures of compensation, as discussed in more detail by Louwaars *et al.* (Chapter 6.1).

Mechanisms for participation include extensive use of public hearings at various stages in the process of policy development. Farmers' participation is also relevant in the implementation of laws and regulations, or in what can be referred to as PGRFA governance. In normal situations, boards and institutions are established through such acts and regulations, to oversee and/or administer implementation and appoint line ministries. In assuming farmers' rights, farmers' representation and participation in such bodies becomes central. The process by which farmer members are selected is of crucial importance for ensuring that farmers take proper responsibility for assuming their rights.

There are five important preconditions for achieving the increased participation of farmers in decision-making and governance. First, decision-makers need to be aware of the important role played by farmers in conserving and developing PGRFA, in order to understand why their participation is required. Second, many farmers are not in a position to participate effectively in complicated decision-making processes without prior capacity-building. Third, central measures to be taken are awareness-raising among decision-makers on the role of farmers in PGR management and governance, and capacity-building in farmers' organizations. Fourth, the capability and skills of policy- and decision-makers need to be strengthened, in order to create more transparent and democratic bodies for PGR management, where farmers can express their rights. Fifth, for new governance in plant genetic resources management, De Boef *et al.* (Chapter 7.1) argue that a process of institutional learning is required to facilitate such a change in professionalism, structures and processes in PGR management, for accommodating CBM and farmers' rights. While there are several examples of the awareness raising and capacity-building of farmers and their representative groups, for participating in the decision-making processes (first and second precondition), there are only a few examples of the latter preconditions, which focus on decision-makers and governance (third, fourth and fifth).

In general, we find few examples of legislation on farmers' participation. Even so, the actual participation of farmers in decision-making processes seems marginal and is often limited to large-scale farmers who are normally not engaged in PGR conservation and use. In the North, the participation of farmers in decision-making processes is more common, even if those farmers that are engaged in PGR conservation and use are rarely represented in PGR governance and management structures. It should be noted that where there is farmer participation, it usually does not promote the development of specific policies or laws relevant to PGR. Farmers in the North claim that their influence is decreasing, due to their countries' commitments to regional and international organizations and agreements, such as the World Trade Organization (WTO) and the European Union (EU) (Andersen, 2005a). While the process of implementing participation has been slow, there have been a few success stories, such as various capacity-building measures to prepare farmers for participating in decision-making in Malawi, Zimbabwe, the Philippines and Peru, and several successful advocacy campaigns regarding the implementation of elements of farmers' rights, where farmers have been directly involved, for example, in India, Norway and Nepal (Andersen and Winge, 2008).

Farmers' rights to save, use, exchange and sell farm-saved seed

Article 9.3 of the ITPGRFA states that nothing in the relevant article 'shall be interpreted to limit any rights that farmers have to save, use, exchange and sell farm-saved seed, subject to national law and as appropriate', which is vague, except for labelling these farming practices as 'rights'. The preamble notes that 'the rights . . . to save, use, exchange and sell farm-saved seed and other propagating material . . . are fundamental to the realization of farmers' rights'. This indicates the importance of those practices, but does not give much guidance. Despite the lack of precision, the general line of thought is clear. Farmers are granted rights in this direction, although the individual countries are free to define the legal space they deem sufficient for farmers regarding their rights to save, use, exchange and sell farm-saved seed.

The question of farmers' rights to save, use, exchange and sell farm-saved seed is the most contentious in the whole ITPGRFA, as it directly enters the domain of interest of the seed industry and their prospects of remuneration for their investments. This element of farmers' rights also has important ramifications for farmers and their ability to conserve and sustainably use PGR, moreover to produce food, and sustain their family and livelihood. Typical issues concern how farmers' rights can be protected and promoted in IPRs, such as through patents and plant breeders' rights, and regulations concerned with variety release, registration and the marketing of seed. Louwaars *et al.* (Chapter 6.1) describe in more detail this confusing interface between PGR, variety and seed policies, with CBM and farmers' rights. A basic question is how to strike the best balance between farmers' and breeders' rights. What is needed is the application of measures to ensure that farmers can continue their vital contribution to the conservation and sustainable use of PGR to the greatest possible extent, and also that the seed industry has the income required to continue its pivotal work in providing agriculture with the best possible varieties. Both are crucial to food security.

In general, legislation on IPR, variety release and seed certification is most restrictive in the North and least restrictive in Africa, while countries in Asia and Latin

America can be placed somewhere in the middle. In the EU, for example, farmers are not allowed to use farm-saved seed from protected varieties on their own holdings, or they must pay a licence fee to do so. With respect to non-protected varieties, they are not allowed to exchange seed or even give it away, due to variety release and seed marketing regulations. Kendall and Gras (Chapter 1.7) provide more practical insights into how a seed network in France manages to avoid restrictions that limit the farmers' dynamic management of local varieties; and Kastler (Chapter 6.8) addresses more policy and legal aspects related to the farmers' dynamic use of PGR. The EU is currently in the process of reviewing this legislation with a view to its compatibility with the ITPGRFA. Several regulations have been issued to remove some of the barriers to the conservation and sustainable use of PGR, but much remains to be done before the EU policy, in this regard, can be said to be conducive to CBM. Nevertheless, farmers in Europe continue to exchange seed and propagating material informally, or through experimental mechanisms, as can be seen in the case of the aforementioned seed network in France. Furthermore, several EU member countries, like Italy and Austria, still tolerate the farmers' exchange of local varieties. In other parts of the world, variety release and seed distribution policies are less restrictive, but trends are developing in the direction of stricter regulations.

On a positive note, informal seed exchange networks, seed fairs and a broad range of other practices are found in many countries, in the South as well as in the North. Many of these only cover local areas or communities, but some seed exchange networks and practices have a broader coverage, such as the nationally operating network in France, and also Seeds of Diversity in Canada. These activities seem to be carried out mainly by non-governmental or civil society organizations (Shrestha *et al.*, Chapter 2.8).

A major concern is the lack of awareness among both farmers and decision-makers with regard to this incompatibility between farmers' rights to save, use, exchange and sell farm-saved seed, and other PGR policies and laws. The more restrictive the legislation, the more it limits the ability of farmers to further contribute to the on-farm management of PGR, and to commit to and sustain CBM. This situation calls for the need to support, promote and, above all, recognize informal seed systems.

Farmers' rights: illusion or reality?

Farmers' rights are being implemented in various forms, and awareness regarding the need to put those rights into practice is increasing among many stakeholders. There are many examples at national and local levels that can be regarded as models for the further up-scaling and implementation of farmers' rights. Much has also been achieved internationally, with regard to developing a joint understanding of farmers' rights, their importance, and the steps required for recognizing them and putting them into practice.

Nevertheless, major incentive structures and regulations, for example in seed and variety laws, are often detrimental to the conservation and sustainable use through CBM, and thereby represent serious hurdles to the full implementation of farmers' rights. Consensus-finding is also developing slowly under the ITPGRFA. Does this mean that that the full implementation of farmers' rights is an illusion?

Based on the knowledge of what it takes to ensure on-farm management of PGR or community biodiversity management, it is not possible to see the implementation of farmers' rights as an illusion: it is a necessity. Without the implementation of these rights, we will not be in a position to maintain and further develop our plant genetic heritage and ensure that future generations will enjoy the benefits of this treasure, which is so vital for food security and survival.

Note

The author is grateful to Wilhelmina Pelegrina, formerly of the South-East Asia Regional Initiatives for Community Empowerment (SEARICE), for her valuable comments on this text.

6.3 Farmers' rights, their scope and legal protection in India

S.P. Bala Ravi

Farmers' contributions to the development of plant genetic resources

Today, we understand and appreciate the importance of plant genetic resources (PGR) as forming the building blocks for continuous crop improvement. Farmers may have little or no understanding of the scientific basis of genetic diversity, but they certainly understand its paramount importance to agriculture, and the need for promoting variability in agricultural practices. The autonomy that every farmer exercises in selecting, saving and maintaining seed for re-sowing has been fundamental for the agronomic transformation of plant species into crops, and their further selection. Any realistic valuation of PGR generated by farmers could well run into trillions of dollars, which is many times higher than the value that modern plant breeding has contributed.

Social construction of farmers' rights

Despite the intellectual efforts needed to create improved variability in a wide array of local varieties, the concept of intellectual property rights (IPR) extended to new plant varieties has virtually ignored the contributions of farmers. For example, the patent or plant breeders' rights systems are heavily influenced by commercial considerations, where PGR are dealt with as common property. This ignores the aforementioned monumental contributions that have been made by farmers, leading to the social and political construction of the concept of farmers' rights.

Farmers' rights and intellectual property rights

The basic principle underlying IPR on plant varieties is the recognition of human innovation in developing a new plant variety through selection, with or without recombination, which is novel and distinct from the pre-existing varieties. Unlike the innovations that are made in many non-biological domains, life forms such as crop varieties are not completely invented, but are always created from pre-existing life forms and propagated by natural processes. Thus, the creation of a new variety has two components: the use of pre-existing varieties and the knowledge required to select a new variety by recombining the pre-existing ones or by other processes. Equity demands that the recognition of innovations made on the newly bred varieties should also include the similarly innovative component invested in the source varieties (i.e. plant genetic resources). The latter essentially represent the far greater cumulative intellectual inputs contributed by generations of farming communities over a long period. The fact that those communities lack identity and institutional backing, unlike the present commercial plant breeders, should not mean that they are given less importance or recognition for their intellectual inputs.

While IPR on plant varieties are upheld, the demand for free access to varieties developed by farmers, without the payment of royalties applicable to varieties protected by intellectual property (IP), can be seen as a double standard concerning rights. Moreover, the granting of exclusive rights over the seed or propagating material of an IP-protected variety marks a turning point from the traditional unrestricted right farmers had enjoyed over seed. This restriction on the seed of a patent-protected variety is rigorous, allowing no flexibility for farmers and minimal flexibility for breeders, depending on the jurisdiction.

Construction of farmers' rights in the international policy arena

The context of plant variety protection

IPR that are granted to breeders of plant varieties are referred to as plant breeders' rights (PBRs). The International Convention for the Protection of New Varieties of Plants (UPOV Convention) is the earliest system for plant variety protection and is currently adhered to by 69 countries. PBRs allow a plant breeder to exclude others from the production, processing, stocking, distribution, marketing, sale, export and import of propagating material of a protected variety for a specified number of years. It also allows the breeder to license such rights to others, and to receive royalties generated from the authorized use of the propagating material. These rights may in some countries also include harvested material, such as cut flowers, fruits or foliage of the protected variety, in cases where the breeders do not have reasonable opportunities to exercise their rights over the planting materials. The legal space available to farmers concerning the seed of a protected variety under such a system for plant varietal protection takes the form of farmers' rights, together with PBRs; or that of the farmers' privilege within PBRs.

The UPOV Convention of 1978 had crafted the exception to PBRs as a private and non-commercial action on propagating material of the protected variety, which allowed – according to certain interpretations – farmers to use the propagating material accessed as the product of the harvest obtained by planting (UPOV, 1978). The UPOV version of 1991, however, mentions the farmers' privilege which limits the exemption to certain crops (by choice of the member state) and only to seed produced by farmers for planting back on their own holdings, thereby dramatically reducing the traditional farmers' rights for use.

The context of the International Undertaking and the International Treaty

While recognizing PBRs on plant varieties, the inter-governmental Commission on Genetic Resources for Food and Agriculture (CGRFA) also approved farmers' rights in a resolution on the interpretation of the International Undertaking on Plant Genetic Resources for Food and Agriculture (IUPGRFA) in 1989 (FAO, 1989). Its primary objective for recognizing farmers' rights was to ensure that farmers would continue to contribute to the conservation and sustainable use of PGR for strengthening the global food and nutritional security.

In 1993, the CGRFA initiated negotiations that concluded almost ten years later in the International Treaty on Plant Genetic Resources for Food and Agriculture (ITP-GRFA; FAO, 2001). The ITPGRFA provides legitimacy to farmers' rights by recognizing the enormous contribution that local and indigenous communities and farmers, particularly those in the centres of origin and crop diversity, have made and will continue to make for the conservation and development of PGRFA, which constitute the basis of food and agriculture. It has, however, left the responsibility of implementing these farmers' rights, as they relate to PGRFA, with national governments, in accordance with their needs and priorities, as appropriate, and subject to national legislation. Andersen (Chapter 6.2) describes the relationship between the ITPGRFA and farmers' rights as being made up of the following four elements: the right to protect traditional knowledge; the right to equitably participate in benefit-sharing; the right to participate in decision-making; and the right to save, use, exchange and sell farm-saved seed. The current chapter focuses on how we have been able to incorporate farmers' rights into the Protection of Plant Varieties and Farmers' Rights Act (PPVFR Act, 2001) in India.

Farmers' rights in the Indian PPVFR Act, 2001

Recognition of farmers as users, custodians and breeders

As is evident from the title, the Protection of Plant Varieties and Farmers' Rights Act (PPVFR Act) seeks to address the rights of plant breeders and farmers on an equal footing. It affirms the necessity of recognizing and protecting the rights of farmers with respect to the contribution they make in conserving, improving and making PGR available for the development of new plant varieties. The PPVFR Act also deems it equally necessary to protect PBRs to stimulate investment for research and development, both in the public and private sector, for the development of new plant varieties. Under the Act, PBRs allow breeders to hold exclusive rights to produce, sell, market, distribute, import or export the propagating material of a registered variety.

The PPVFR Act recognizes the multiple roles played by farmers in cultivating, conserving, developing and selecting varieties. With regard to developing or selecting varieties, the Act refers to the value added by farmers to wild species or traditional varieties through selection and identification of their useful traits. Accordingly, farmers' rights encompass the roles of farmers as users, conservers and breeders. Farmers are granted nine specific rights, which are briefly described below.

Right 1: Access to seed

Farmers shall be entitled to save, use, sow, re-sow, exchange, share or sell their farm produce, including seed of protected varieties, in the same manner as they were entitled to before the coming into force of the PPVFR Act. However, farmers shall not be entitled to sell branded seed of a variety protected under this Act. The Act does not specify the quantity of seed that farmers can save from a crop cultivated in their own farms from a protected variety.

Right 2: Benefit-sharing

All Indian legal entities who provide PGR to breeders for developing new varieties, including farmers, shall receive a fair share of the benefits from the commercial gains of the registered varieties. Out of all the national plant variety protection laws enacted since 2001, the PPVFR Act is the first that integrates a provision for access and benefit-sharing (ABS) along with PBRs. Legal accession of the genetic resource used in breeding is not addressed in the Act; this falls instead under the Biological Diversity Act, 2002. However, the PPVFR Act requires a breeder to make a sworn declaration on the geographical origin of the genetic resources used in the pedigree of the new variety, and how they were accessed.

Right 3: Compensation

Registered seed must be sold with full disclosure of their agronomic performance under recommended management conditions. When such seed is sold to farmers but fails to provide the expected performance under recommended management conditions, the farmer is eligible to claim compensation from the breeder through the office of the PPVFR Authority.

Right 4: Reasonable seed price

Farmers have the right to access seed of registered varieties at a reasonable price. When this condition is not met, the breeder's exclusive right over the variety is suspended under the provision concerning compulsory licensing, and the breeder is obligated to license the seed production, distribution and sales of the variety to a competent legal entity. Most of the laws for plant variety protection have provisions on compulsory licensing of protected varieties to ensure adequate seed supply to farmers, and several of them also use unfair pricing as grounds for compulsory licensing.

Right 5: Farmers' recognition and reward for contributing to conservation

Farmers who have been engaged in PGR conservation and crop improvement, and who have made substantial contributions in providing genetic resources for crop improvement, receive recognition and rewards from the national gene fund. The gene fund receives resources from the implementation of the Act, which in turn are complemented by contributions from national and international organizations. The expenditures of the fund are earmarked to support the conservation and sustainable use of PGR, and in this way it can be considered to be a national equivalent to the global benefit-sharing fund operating within the ITPGRFA, as described by Andersen (Chapter 6.2). Since 2007, the plant genome saviour award, associated with the national gene fund, has been rewarding farming communities and individual farmers for their contribution to *in situ* conservation on-farm and to the selection of PGR (Bala Ravi and Parida, 2007).

Right 6: Registration of farmers' varieties

The Indian PPVFR Act allows for the registration of existing farmers' varieties that fulfil requirements for distinctness, uniformity, stability and denomination, but does not include that of novelty. This right provides farmers with a one-off opportunity for a limited period of time, from the moment when a crop species is included in the crop portfolio under the PPVFR Act for registration. Once registered, these varieties are entitled to all PBRs.

Right 7: Prior authorization for the commercialization of essentially derived varieties

When farmers' varieties, whether extant or new, are used by a third party as source material for the development of an essentially derived variety, the farmers need to provide prior authorization for its commercialization. Such a process can allow farmers to negotiate the terms of authorization with the breeder, which may include royalties, one-off payments, benefit-sharing, etc.

Right 8: Exemption from registration fees for farmers

Under the PPVFR Act, farmers have the privilege of being completely exempt from paying any kind of fees or other payments that are normally payable for variety registration; tests for distinctness, uniformity and stability (DUS), and other services rendered by the PPVFR Authority; as well as for legal proceedings related to infringement or other causes.

Right 9: Farmer protection from accidental infringement

If a farmer can somehow prove before court that he or she was not aware of the existence of any rights at the time of an infringement on any such rights, as detailed in the PPVFR Act, he or she will not be charged. This provision is made in consideration of the centuries-old unrestrained rights that the farmers had over the seed of all varieties, the novel nature of the PPVFR Act and the low legal literacy of farmers.

Current status of the implementation of PPVFR Act

The PPVFR Act has been in operation since 11 November 2005. Since then, the PPVFR Authority has brought 43 crop species into its fold. By the end of December 2010, the PPVFR Authority had received applications for the registration of 841 new varieties belonging to 27 crops, 1222 applications for the registration of extant varieties belonging to 24 crops, and 55 applications for the registration of farmers' varieties of seven crops. Table 6.3.1 illustrates the species distribution of applications to register new varieties.

Farmers' rights in the context of rural poverty in India

The relevance of farmers' rights, as provided for under the PPVFR Act, becomes more evident when considering the fact that about 660 million people in India are subsisting

Table 6.3.1 Species-grouping of new varieties registered by the PPVFR authority, India, up until December 2010

Cereal crops	No.	Legumes	No.	Oil crops	No.	Fibre and other crops	No.
Rice	99	Pigeon pea	19	Sunflower	38	Jute	5
Wheat	11	Black gram	10	Ground nut	1	Cotton	241
Maize	118	Green gram	6	Mustard	1	Sugarcane	2
Sorghum	74	Common bean	1	Castor bean	5		
Pearl millet	77	Lentil	1	Soybean	1		
		Green pea	1				
Subtotal cereals	379	Subtotal pulses	38	Subtotal oil crops	46	Subtotal fibre and other crops	248

Source: Protection of Plant Varieties and Farmers' Rights Authority, India.

largely on agriculture for their livelihoods. It is notable that during the early days of IPRs, food security was still one of the top priorities in Europe and the USA, leading to the exemption of food crops from IPRs, and the application of farmers' exemption. It has been argued that the Indian PPVFR Act, with its extensive farmers' rights, is not conducive to promoting private investment in plant breeding. However, since private investment in Indian plant breeding, including food, horticultural and commercial crops, is rapidly increasing, and the use of private-bred seed in Indian agriculture (Freeman and Barwale, 2010) is spreading, such arguments lack evidence.

While the private sector is growing, farmers are also being promoted to maintain their rights to conserve, use and further develop their PGR. In this way, India is providing the world, emerging economies and developing countries in particular, with a viable alternative that balances the private sector's and farmers' rights.

New balance between farmers and breeders

One often notices a paradox when considering the high level of poverty in India that exists side-by-side with a wealth in PGR. The commercial entities that freely access PGR seldom realize that conservation is often associated with poverty. The Indian PPVFR Act includes important provisions for promoting the on-farm management of PGR. The national gene fund supports this conservation strategy as implemented by poor rural communities. The provision for fair and equitable benefit-sharing from plant breeders using the PGR conserved by farmers is a major incentive for linking conservation with livelihood development, in the process of community biodiversity management (CBM). In order to ensure the continued contributions of farmers through CBM, farmers across the world must be granted liberal rights over the seed of every variety that they develop. These rights should be accompanied by the provision of monetary, moral and technical support, through the application of CBM practices within seed systems including those using modern technologies. Such rights and support must be provided as a priority, and must be equal to the exclusive rights that have been granted to plant breeders, who forage on the PGR that once originated and continue to be conserved and developed by those farmers.

6.4 Access and benefit-sharing

Putting a global legal concept into practice through local initiatives

Ronnie Vernooy and Manuel Ruiz Muller

Benefit-sharing: a global concept that requires local implementation

Since the adoption of the Convention on Biological Diversity (CBD, 1992a) the notion of fair and equitable sharing from access to, and use of, genetic resources has received increasing attention. Social and political scientists, and representatives of development organizations, have begun to support policy-makers in exploring how to put this relatively new policy and legal framework into practice. One challenge is to link the global concept on access and benefit-sharing (ABS) to farmers' activities in the conservation and sustainable use of biodiversity at local level (i.e. community biodiversity management, CBM). Together, farmers, researchers, companies and conservation professionals can look at how ABS, in relation to plant genetic resources (PGR), can be taken from the negotiation table and brought to a level where it can achieve its objective (i.e. to share in a fair and equitable manner any benefits derived from the use of genetic resources).

The growing importance of benefit-sharing as a global legal concept

Initial efforts to implement the ABS concept at national level involved a very narrow interpretation (Glowka *et al.*, 1996), focused on genetic resources (i.e. seed, genes, genetic material) in high-tech research and development, with particular attention to biotechnological processes and their resulting products (i.e. pharmaceuticals, cosmetics and industrial products). Bio-prospecting for new compounds was at the heart of initial debates and controversies regarding rights and control over genetic resources and associated traditional knowledge (Pistorius, 1997; McMannis, 2007).

High-profile cases involving questionable patents granted over inventions based on biodiversity and associated traditional knowledge (ATK) that originated in the South (e.g. neem in India, quinoa in Bolivia, maca in Peru) have sparked an intensive reaction against the intellectual property (IP) system and its growing role in granting rights over biodiversity components. These cases have highlighted the very complex interrelations (and tensions) between the IP system, biodiversity and ATK.

Inspired by the CBD, but addressing a more specific set of biological resources, the International Treaty on Plant Genetic Resources for Food and Agriculture

(ITPGRFA) focuses on benefit-sharing based on access to, and use of, plant genetic resources for food and agriculture (FAO, 2001). The heart of the ITPGRFA is a multilateral system on ABS that is applicable to a closed set of species: 35 crop and 29 forage species. The implementation of farmers' rights in the ITPGRFA has opened the way to reflect upon and design new activities, measures and projects to secure farmers' interests in PGR. The ITPGRFA has incorporated benefit-sharing as the second element of farmers' rights. Andersen (Chapter 6.2) further elaborates on these and other elements related to farmers' rights, providing some examples for each of them.

Over time, as a result of national efforts to implement the CBD, provisions on ABS have become more accommodating of a broader understanding of benefit-sharing. A number of countries have, for example, extended the notion to include access to, and use of, derived products (e.g. the Andean community) and, in some cases, to the use of biodiversity and its components in general (e.g. Cuba, Venezuela, Costa Rica).

The recent Nagoya Protocol on ABS (Secretariat of the Convention on Biological Diversity, 2011), also applies to derivatives of genetic resources, thus broadening the scope of the concept to include biochemical and similar compounds resulting from cell metabolism. Within the ongoing debates on environmental and ecosystem services, sharing the costs and benefits of these services is, to some extent, based on the original CBD debates, and on the classical economic cost-benefit theory.

Finally, the concept of benefit-sharing also permeates discussions regarding the legal protection of traditional knowledge. At present, the World Intellectual Property Organization (WIPO), with its Intergovernmental Committee on Intellectual property and Genetic Resources, Traditional Knowledge and Folklore (IGC), is negotiating an international regime for the protection of traditional knowledge, which seeks to ensure the equitable participation of indigenous peoples in the benefits derived from access to, and use of, their traditional knowledge, expressed in innovations and practices.

What is clear today is that countries and communities want to participate in how benefit-sharing is defined. It is fundamental that they are involved in the design of practical mechanisms for sharing the actual benefits that are generated by providing access to, and facilitating the use of, a wide range of biodiversity-related components, including: genetic resources, derivatives, environmental services, and even ATK.

ABS: putting the concept into practice

Few examples have been documented of how benefit-sharing permeates local level activities. This limited number is based on the fact that studies initially focused on policy and legal frameworks. We have identified a number of examples that are part of normal, regular activities undertaken by small-scale farmers and farming communities, with development and PGR organizations, throughout the world, avoiding those activities that were specifically initiated to contribute to ABS. The activities were analysed, focusing on characteristics that may contribute to putting ABS to practice. The detailed case studies, and an in-depth comparative analysis, can be found in Ruiz and Vernooy (2012). Some examples are provided below.

Targeting the recognition of farmers in PGR conservation and use in south-west China

Informal seed systems continue to play a major role in meeting the seed demands of farmers in marginalized areas of the Guangxi province in south-western China, which in turn contributes to the maintenance of PGR. The Centre for Chinese Agricultural Policy at the Chinese Academy of Sciences has been carrying out community-based action research since 2000, in collaboration with the Crop Research Institute (CRI) of the Chinese Academy of Agricultural Sciences, and the Guangxi Maize Research Institute (GMRI). They aim to contribute to PGR conservation and participatory crop improvement (PCI) by linking local reality and demands to national level policy- and law-making processes (Song and Vernooy, 2010). Key decision-makers in the rural development policy arena are now directly engaged in multi-scalar analysis and debates concerning ABS. These collaborative efforts led to the signing in 2010 of two novel ABS agreements between farmers from 12 communities, and scientists from CRI and GMRI, concerning the conservation and improvement of local varieties of maize and rice, which could serve as concrete examples of the implementation of the Nagoya Protocol (Song *et al.*, 2012).

Recognizing the value of innovations within informal seed systems in Cuba

Currently, agriculture is struggling to survive under difficult conditions in Cuba. Farmers, together with a number of young agricultural researchers associated with the National Institute of Agricultural Sciences (INCA), the Agrarian University of Havana, and the Sociological and Psychological Research Centre of Havana, have been (re)discovering that necessity is the mother of innovation and creativity. Since 1999, they have been revitalizing agriculture by embracing informal seed systems as a basis for production and reproduction. New forms of participation and cooperation have emerged in which benefit-sharing arrangements are evolving. No pre-designed plan guided this process, much less the legal obligations under the CBD, or any other international or national instruments. A more flexible, open and dynamic view of how social change can be brought about is guiding the development of ABS in Cuba, which involves strengthening the seed systems, increasing farmers' access to PGR, and allowing agriculture to regain some of its dynamic nature.

Supporting the participation of farmers in crop improvement in Jordan and Syria

Syria and Jordan are pioneering countries of PCI in the Middle East. Farmers benefit from their contributions by obtaining access to the global gene pool and scientific knowledge. Thus, PCI can be seen as an instrument for benefit-sharing. Moreover, it fits especially neatly under farmers' rights as defined by the ITPGRFA. In Jordan, agricultural policies and laws, and international agreements, are being implemented at national level, reflecting the country's efforts to introduce and institutionalize PCI.

Designing informal ABS mechanisms in an unfavourable setting in Honduras

The free trade agreement between Central American countries, the Dominican Republic and the United States (DR-CAFTA) of 2006 has had a profound impact on the implementation of ABS. One of the member countries, Honduras, aims to meet its obligations to the ITPGRFA relating to farmers' rights, for example, by supporting the protection of traditional knowledge. However, this conflicts with the DR-CAFTA, which considers local varieties and ATK patentable commodities. Prior informed consent, disclosure of origin and benefit-sharing are not required for patent applications. Since 1999, locally organized small-scale farmers, with support from researchers from the Foundation for Participatory Research with Farmers of Honduras (FIPAH) and the University of Guelph, Canada have been trying to develop their own, parallel and informal practice to achieve ABS, for protecting PGR and ATK. They have had some degree of success, through the spread of PCI and the formal recognition of farmers' contributions to new varieties, as further described by Alonzo *et al.* (Chapter 5.8), but with almost no policy or legal support. The experience in Honduras highlights the complex relationship between biodiversity and intellectual property, in light of growing pressures to adopt regional and bilateral trade agreements with provisions that approach plant variety protection in ways that bypass both the CBD and ITPGRFA (Musungu and Dutfield, 2003).

Applying CBM and PCI practices in Nepal

Innovative practices within a larger context of CBM, as implemented by Local Initiatives for Biodiversity, Research and Development (LI-BIRD) and its partners, are giving concrete meaning to the concept of ABS. Through their work, LI-BIRD is bringing farming communities and policy-makers together in the creation of a policy and legal environment in Nepal, in favour of both biological and socio-cultural diversity. CBM practices that contribute directly to the implementation of ABS include the community biodiversity registers, as illustrated by Subedi *et al.* (Chapter 2.4), and community seed banks, as described by Shrestha *et al.* (Chapter 2.8). These practices are complemented by various PCI methods that contribute to ABS, including grassroots breeding for neglected and underutilized species, and for rare rice varieties, as detailed by Sthapit *et al.* (Chapter 5.2); and the participatory genetic enhancement of local varieties, for example Jethobudho rice, as outlined by Silwal *et al.* (Chapter 5.5). The CBM fund detailed by Shrestha *et al.* (Chapter 2.9) was designed by LI-BIRD and, like other CBM practices, is currently being mainstreamed in Nepal (Paudel *et al.*, 2010). This community fund can be seen as a way to institutionalize CBM processes within community structures, and as a mechanism for putting ABS into practice at local levels and beyond.

Protecting ATK against misappropriation at local and national levels in Peru

Peru has been very active in the development of legal frameworks regarding ABS and the protection of biodiversity and ATK. These efforts were made to address strong concerns regarding the misuse, illegal use and misappropriation (biopiracy) of PGR and ATK of indigenous peoples. In this way, Peru has developed a framework for ABS that is closely associated with the first element of farmers' rights in the ITP-GRFA, as described by Andersen (Chapter 6.2). Governmental, non-governmental and civil society organizations in Peru, such as the Ministry of the Environment, the National Commission for the Prevention of Biopiracy, the National Institute for Natural Products, the Peruvian Society for Environmental Law (SPDA), and the Association for Nature and Sustainable Development (ANDES), have been engaged in the design and implementation of mechanisms to protect biodiversity and ATK, regulate access, and facilitate the sharing of benefits derived from their use. Such mechanisms include, as in Nepal, the development of biodiversity and traditional knowledge registers at community level, but also at national level. In follow-up to Peru's pioneering role in the ABS arena, its legal environment has become a useful enabler and trigger for local activities and projects designed to institutionalize and strengthen benefit-sharing mechanisms.

The current status of ABS and its local implementation

For almost a decade, decision-makers, advisors and advocates ardently debated the precise wording of ABS policies and laws at international and national fora; while at the same time, farmers, indigenous peoples, researchers, staff of non-governmental and civil society organizations, and local government agents around the world worked hard to implement, test and assess mechanisms for implementing fair and equitable ABS at local level. The results and lessons learned from the experiences described in this chapter share the local advances that have been made, independent from national and international policy- and law-making processes. It should, however, be noted that a wide gap exists between innovative practices and local processes for ABS, and the definition and institutionalization of ABS in the policy domain.

The development of ABS practices outside the legal context

A number of benefit-sharing practices have developed as a result of interactions between farmers and other actors, regardless of existing ABS policy and legal contexts, including the CBD and the ITPGRFA. Benefits arising from the access and use of PGR are shared in different ways and are part of specific circumstances that many would subscribe to be practices within a CBM framework, as described by Shrestha *et al.* (Chapter 1.3), and included in most of the other chapters in this volume. These benefits are shared between farmers and researchers, farmers and breeders, among farmers and communities, or between national institutions and farmers.

Monetary versus non-monetary benefits

Whether benefits that are shared are equitable or not depends on a wide range of factors, and on how the process is measured, valued and quantified. Vogel (2005) stated, for example, that monetary benefit-sharing in the context of bio-prospecting agreements under the CBD and national ABS regulations are scandalous and unacceptable. Others parties engaged in the debate on ABS have noted that the ITPGRFA proposes a set of non-monetary benefits and defines a specific percentage on seed sales in terms of monetary benefits, which have been well accepted by farmers, the private sector and, ultimately, governments (Halewood and Nnadozie, 2008; Frison *et al.*, 2011). Practices in CBM and PCI can provide us with new, effective and diverse mechanisms to achieve ABS. Actors engaged in PCI generate benefits that are distributed according to a wide range of criteria, which often go beyond the narrowly defined ABS policy and legal criteria.

ABS practices within the context of policy and legal frameworks

Constructing ABS and related protective policy and legal frameworks is a relatively new process in comparison with the development and consolidation of the IP system, which has taken more than a century to evolve. Many countries, including China and Peru, have rapidly developed their own particular ABS laws, regulations and approaches (Ruiz and Vernooy, 2012). More interactions with farmers and communities are required to ensure that ABS policies and norms appropriately reflect reality and do not rapidly become obsolete.

Experiences in several countries (e.g. Brazil, Costa Rica, Peru and the Philippines) have demonstrated that speeding up the development of these frameworks, although positive in some respect, tends to result in laws and regulations with limited possibilities of implementation. Another reason stems from differentiating between sectors and activities, and from making ABS laws all-embracing, covering cosmetics, nutraceuticals, agriculture, pharmaceutical and natural ingredients, all together, at the same time.

Although ABS and its new principles, such as prior informed consent, are becoming better known as formal concepts, they are often very difficult to understand and hard to give a practical form at local levels, where traditional and customary practices are based on collective identities, knowledge and customs. This then points to the need to continue searching for alternative ABS policy and legal options beyond the narrow approach that prevails at the moment, such as is reflected in the Nagoya Protocol.

Grassroots design for better reflection of reality

At the very minimum, local benefit-sharing options and practices within agriculture in particular should be respected (i.e. through customary practices) and used as the grassroots basis for the development of national or international policy and legal frameworks, instead of squeezing them into narrow, top-down and legally defined options. Understanding the local context is also crucial for facilitating efforts to implement ABS, and for developing or identifying ways to put it into practice.

Building dynamic and supportive institutional and formal legal structures bottom-up, especially in the case of access to resources and traditional knowledge, and the resulting benefit-sharing, is necessary to ensure practical results. Reflecting on local initiatives, and developing bio-cultural protocols or even codes of conduct, with direct input from farmers and communities, are means and tools that could facilitate multi-scalar synergies and help the implementation processes. As illustrated in the examples, many farmers are consolidating their conservation and sustainable use efforts through strengthening local practices within the wider CBM and/or PCI frameworks but, at the same time, they realize that they require an enabling and complementary national policy and legal framework to continue and sustain this work. This is where the CBM and PCI practices and ABS meet. Improvement of the communication between local-level ABS practitioners, and national and international ABS designers and decision-makers, may lead to demands from both sides.

To this day, many of these practices are being implemented with international donor support. National governments should set up funds, such as the national gene fund in India (Bala Ravi, Chapter 6.3), or seek mechanisms for community trust funds, such as the CBM fund in Nepal (Shrestha *et al.*, Chapter 2.9). Such efforts to institutionalize ABS would allow for its wider adoption and adaptation, involving more farming communities who could reap their benefits. The organizations that have acquired a certain expertise in this field could be invited to play an advisory role to set-up appropriate mechanisms, thereby ensuring that such future ABS mechanisms better match reality.

Note

The authors would like to acknowledge with thanks contributions from Kamalesh Adhikari, Adnan Al-Yassin, Salvatore Cecarrelli, Teresa Cruz Sardiñas, Alessandra Galié, Omar Gallardo, Marvin Gomez, Stefania Grando, Sally Humphries, José Jiménez, Li Jingsong, Yasmin Mustafa, Bikash Paudel, Humberto Ríos Labrada, Pitambar Shrestha, Fredy Sierra, Song Yiching and B.B. Tamang.

6.5 Access and benefit-sharing in Brazil

Towards the appropriation of the commons

Ulysses Paulino de Albuquerque, Natalia Hanazaki and Juliana Santilli

Access to biodiversity and to associated traditional knowledge

The Convention on Biological Diversity (CBD, 1992a) resulted in the formalization of the relationship between scientists and traditional people, concerning access to biodiversity and associated traditional knowledge (ATK). For scientists, access to biodiversity includes two basic components:

1   access to natural or agricultural biodiversity, which includes collecting samples of animals, plants or other organisms;
2   access to traditional knowledge associated with that biodiversity at all levels (i.e. from genetic diversity up to landscape level, using sampling tools such as interviews, for example).

In the pre-CBD era, such interactions were characterized by their lack of specific concerns in relation to access to biodiversity and ATK, and in relation to securing benefits from their use. Biologists used to operate within their own fields of research, with little concern for the social, economic and/or political implications of their work. In the post-CBD era, the concept of access and benefit-sharing (ABS) has become an integral element of their professional world, since they must now follow clear guidelines in their work with indigenous and traditional communities. Consequently, they cannot hide from discussions of a more anthropological or sociological nature, or from those concerning the political economy of biodiversity and ATK, and ABS issues, and as such an awareness of the social science dimension of biodiversity has become almost imperative today.

ABS: regulating interactions between scientists and traditional people

More than two decades ago, the International Society of Ethnobiology (ISE), in its Declaration of Belém (1988), highlighted the need for mechanisms to recognize indigenous specialists as competent authorities, and compensate native populations for the use of their knowledge and their biological resources. Furthermore, it altered the view of the global community concerning the value of ethnobiological knowledge for the well-being of humankind (ISE, 2011). Consequently, professionals working in the fields of ethnobiology, ethnoecology, ethnobotany and other related areas, in recognition of the Declaration of Belém, as a way of self-regulation, began to follow the sound ethics outlined in most policy guidelines that were later imposed on them by ABS laws. It should, however, be realized that most mainstream biologists (zoologists, botanists, geneticists, and so on) are completely unaware of the ethical code of the Declaration of Belém, and as such the affirmation is only true for ethnobiologists.

Brazilian ABS Act: formalizing scientists' access to biodiversity and ATK

The provisional Act no. 2186-16 (MMA, 2001) in Brazil, which we refer to as the ABS Act, regulates how biologists, scientists and other professionals should access biodiversity and ATK. A special agency, with its own set of procedures, has been set up within the Ministry of Environment, thereby further formalizing and bureaucratizing the voluntary code of conduct of the ISE for such research activities. With this formalization, we face many challenges from the regulatory side, addressing the professionalism of researchers, and their relationship with communities engaged. One challenge is to contribute to the design and implementation of policies that do not consider all interactions between scientists and traditional communities that relate to biological resources as criminative (i.e. to avoid the development of regulations that force biologists and other professionals to prove that they are not biopirates). Ethnobiologists, who previously followed the professional ethical code outlined in the 1988 Declaration of Belém, now need to prove their ethical integrity and legitimate work aims before carrying out any activity. The diversity of situations that can be found in a country of so many contrasts, like Brazil, poses another challenge for putting ABS into practice. In addition, there must also be compatibility with other public policies, such as those that guide scientific production or even the dominant agricultural development model.

Concern that information was being used to generate benefits for third parties led to the development of the ABS law. However, many are now worried about the impact of the ABS regulation on the access to information produced by public universities and research institutions. The question arises as to whether such knowledge can remain public. Debates on this issue reiterate that the resources sustaining universities remain public, and therefore academic production continues to serve the society in general. Academic biological research therefore cannot serve individuals or corporate interests. However, many believe that this originally public orientation can be circumvented by the increasing appropriation of biological resources and ATK, which may be further accelerated by the formalization and implementation of legal ABS procedures. A key question remains as to whether that is the original intention of the CBD and most ABS laws, in relation to the protection of traditional communities and fair benefit-sharing.

ABS and its relation to traditional people

As a result of the great diversity of agricultural landscapes and the existence of different kinds of farmers with distinct livelihood and production strategies, Brazilian rural and agricultural issues are extremely complex. It is common to refer to the duality of agricultural models in Brazil – agribusiness and small-scale farmers. Yet Brazil's agricultural diversity is reflected not in duality, but in a multitude of agricultural systems that are associated with multitudes of livelihoods of indigenous, Quilombola and traditional people, and their relations with their natural environment (i.e. biocultural diversity), as addressed in more detail by Assis *et al.* (Chapter 3.2).

Decree No. 6040 (Brazil, 2007) defines traditional communities as 'culturally differentiated groups that identify themselves as such, have their own forms of social organization, occupy and use territories and natural resources as a condition for cultural, social, religious, ancestral and economic reproduction, and use knowledge, innovations and practices generated and transmitted by tradition'. In addition, the decree formally recognizes the rights of traditional people over their territories. It establishes a national policy framework for the sustainable development of traditional communities, which is being implemented by a commission composed of representatives of public agencies, and of traditional communities that include rubber-tappers, forest gatherers, fisher folk, and small-scale and traditional farmers.

The aforementioned provisional Act on ABS states that prior informed consent is required when accessing genetic resources and ATK in areas that are recognized as traditional, indigenous or Quilombola territories, since these communities have clearly established rights over their natural resources and cultural heritage. However, no consensus in the interpretation of the provisional Act exists for ensuring access to genetic resources and ATK for communities that are not (yet) formally recognized as traditional, or for those that are generally considered to be local communities.

ABS: common resources and knowledge

As mentioned by Alexiades (2003), in the present neoliberal world, with the politicization of genetic resources and the rediscovery and appreciation of traditional knowledge, we realize that the flow of information, although fast, is unequal. The greatest risk is that of treating science or rather knowledge as a commodity, where the market establishes research priorities. In this sense, it is also important not to lose sight of which mechanisms influence the establishment of public policies. Many of these mechanisms are grounded in this neoliberal model: international treaties on commerce, patent laws, and the internationalization of intellectual property rights. It is not surprising, therefore, that we often find ourselves in a paradoxical situation, seeking mechanisms for benefit-sharing motivated by the CBD and the ABS Act in Brazil, but cutting away important roots of traditional knowledge that is grounded as science in free circulation (Cunha, 1999).

The ABS Act in Brazil states that the resource providers and users should negotiate the conditions for access and benefit-sharing. Consequently, identifying traditional or local communities that are providers of genetic resources and ATK has created difficulties. Genetic resources and ATK become commodities or goods to be negotiated at market prices, which subverts the logic of how these resources and traditional knowledge are generated, and how, over history and today, they continue to be shared within and among local communities. Since farmers are accustomed to sharing and promoting the exchange of genetic material, knowledge and agricultural experiences through social networks that are regulated through local norms, how will such farmers define the ownership of these resources? Also, by requiring these communities to establish who owns these resources, it seems to us that the provisional Act on access to genetic resources might actually stimulate rivalries, which could restrict the free movement and exchange of genetic materials that is fundamental to the conservation of agrobiodiversity and community resilience. The ABS Act treats genetic resources like economic, utilitarian and fragmented goods, out of context of the biological and socio-cultural processes of constructing agrobiodiversity and ATK (Santilli, 2012), in that it:

- ignores local perceptions and values associated with agrobiodiversity resources and ATK;
- overlooks the local norms on ownership rights over these resources;
- disconnects PGR and ATK from the environment and culture;
- takes no notice of the circulation and exchange of PGR and knowledge within and among communities;
- does not consider the complexity of the social and cultural processes that generate agrobiodiversity;
- limits the free circulation of PGR and stimulates monopolies;
- restricts the public domain;
- impacts negatively on sustaining local agricultural systems and associated livelihoods.

If the legal treatment given to the genetic resources of wild species by the ABS Act has raised several questions, the difficulties created by the legal regime for species used for food and agriculture are even more serious and difficult to surmount. Brazil has ratified the International Treaty on Plant Genetic Resources for Food and Agriculture (ITPGRFA), but it is still in the process of putting it into practice and defining how to address farmers' rights, which is becoming one of the most controversial issues, in particular as it conflicts with many aspects of this process of appropriation of genetic resources and ATK. Santilli (Chapter 6.7) addresses this debate in more detail.

Scientists and the paradox created by ABS

Considering the scenario that has been sketched out here, the relationship between biologists and professionals from other disciplines and traditional people, and also the biological matter that they use, is changing, with major implications on the professionalism of the scientists. In other areas also, scientists are faced with discussions on ABS and need to re-adjust their professionalism to interact with communities. Yet, curiously, we live in a paradoxical situation, since now more than ever scientists need to deal with those legal aspects related to access despite the fact that such aspects were not, and are still not, systematically covered during their professional training. More important still, is the fact that professionals not only understand how to apply such mechanisms, but also that they comprehend why they should apply such mechanisms of ABS at all. To address this paradox, the academic training of scientists should include, first, discussions on ethical and philosophical principles concerning ABS to show why it is required, and second, critical thinking that questions whether the policy and legal instruments that aim to respect and transform those principles are appropriate, and result in the desired situation or not.

In the first case, scientists are prepared to deal with issues related to biodiversity and ATK, but inexorably find themselves unprepared to manage the ethical issues that arise from working in the interface between biological and cultural diversity. Undoubtedly, we need professionals who take into account a set of ethical principles in their research, which include informed consent, confidentiality, and sensitivity and respect when dealing with issues related to benefit-sharing (Engels *et al.*, 2011), one of the CBD's greatest goals. Although this is an important aim, it needs widespread discussion as it is not yet clear how to attain it. The concept of 'fair and equitable' still seems somewhat vague, despite its recurrent use in international documents and treaties. Accordingly, biologists and other professionals can no longer avoid their responsibility to understand these legal principles when working with, and being guided in their interactions with, local communities (De Jonge, 2011).

The second case can be exemplified by free informed consent. In principle, local communities who are engaged in research by providing access to PGR and ATK should have the autonomy, freedom and security to make appropriate, conscious and collective decisions. However, Albuquerque *et al.* (2010) have demonstrated that many communities sign consent forms without necessarily fully understanding their rights, even when they have been informed of them. The issue remains as to who the representatives of communities actually are and what structures ensure collective consensual decision-making. This raises an important point to consider: following a legal rule that requires signed consent may not have the expected result, especially if scientists are unprepared to face the fact that they need to share the benefits of the results derived from the research. Furthermore, where access to, and the level of, education is poor, which is often the case in rural areas where traditional people live, reading a (legal) document does not guarantee that it has been understood.

Thus, it becomes clear that despite advances in the ABS discussions, several polemical topics continue to surround the issue, including farmers' rights, the biological and socio-cultural processes of constructing agrobiodiversity, the appropriation of the commons, and the need to train professionals to operate in accordance within this new ABS framework.

6.6 The European Union directive on conservation varieties and its incompatibility with on-farm management of plant genetic resources

Guy Kastler and Anne-Charlotte Moÿ

Variety registration and the loss of agrobiodiversity in Europe

The European community implemented a regulation in 1966 that made it obligatory to register varieties in an official variety list before their seed could be commercialized. Before being accepted in the variety list, varieties had to meet criteria of distinctness, uniformity and stability (DUS). Cereal species also had to meet the specific criteria set for value for cultivation and use (VCU). These requirements resulted in the spread of just a few varieties that performed well in uniform growing conditions. Practices that had commonly been used in more traditional farming systems were replaced with those of adapting the production environment to varieties, and of encouraging the intensive use of chemical fertilizers and pesticides, as well as an increasingly disproportionate amount of mechanization and, in some cases, irrigation. This process further encouraged the use of an ever-decreasing range of crops and varieties (i.e. farmers were encouraged or required to cultivate only those few varieties or crops as required or demanded by the food industry and major retailers). In this way, the requirement for registration in the variety list can be considered part of a larger process to create uniformity and promote the industrialization of agriculture, which as we now know caused the massive erosion of plant genetic resources (PGR). In each member state of the European Union (EU), a dramatically lower number of uniform and stable, and commonly used modern varieties has replaced a diversity of local varieties.

Variety list: obligations and exceptions

Each country handles the registration of varieties and the compilation of related information in lists in its own way. The common European catalogue is based on these national variety lists. The guidelines of the European regulation specify that only seed of those varieties that are included in the variety list are allowed to be commercialized, and as such farmers can only purchase seed of registered varieties. The guidelines do, however, provide space for the exchange of varieties for the purpose of research, breeding and PGR conservation. Furthermore, they allow for the purchase and cultivation of seed of non-registered varieties by amateur gardeners, placing restrictions on the quantities of seed marketed, and on the use of those varieties for commercial production.

Vegetable variety list for amateur gardeners in France

An explosion in the sales of seed of traditional varieties that were not included in the list led to the development of another exclusion related to amateur gardeners in France. In 1997, with the aim of structuring the popular informal seed market, the French government introduced an annex, which is not obligatory, to its official vegetable variety list, allowing for the inclusion of old varieties used by amateur gardeners. The costs of registration were lower, and the registration criteria were much less rigid, than those of the official variety list. The registration of seed in this annex placed several conditions on the seed producers and traders targeting this market. The guidelines only permitted the sale of varieties in small packages. In addition, traders needed to maintain a strict administration of the sales of each variety. These conditions resulted in higher costs, which in the end had to be covered by the clients (i.e. amateur gardeners). The motivation to establish this rigid system was that of preventing seed producers and traders from circumventing the common vegetable variety list. Despite these restrictions, the amateur gardeners' vegetable variety list quickly attained success, notwithstanding the creation of many bureaucratic obstacles by official seed agencies, including their common refusal to register old varieties.

The Kokopelli Association, which sells seed of its heirloom or traditional (non-registered) varieties without any specific labelling, was condemned for its lack of consistent labelling of vegetable seed. Its president was deemed guilty of having marketed the seed of unauthorized varieties; unauthorized because the Kokopelli Association had on principle decided not to register its seed in the annex. Widespread media coverage of this case throughout Europe reinforced the misconception that registration in this amateur gardeners' vegetable variety list had become mandatory, regardless of the purposes for which the seed was sold. The amateur gardener' vegetable variety list was discontinued in France, as from 1 January 2012. Those varieties that had been included in the list were transferred either to the list of conservation varieties or to the list of those varieties have been developed to meet specific growing conditions.

EU directives for conservation varieties

First non-binding directive

In line with the Convention on Biological Diversity (CBD, 1992a), and later with the Global Plan for Action for the Conservation and Sustainable Utilization of Plant Genetic Resources for Food and Agriculture (FAO, 1996b), the European Union (EU) introduced the issue of *in situ* conservation of PGR into its seed legislation in 1998. EU Council Directive 98/95/EC thus states that 'Specific conditions may be established

... to take account of developments in the areas of . . . conditions under which seed may be marketed in relation to the conservation *in situ* and the sustainable use of plant genetic resources, including seed mixtures of species, which . . . are associated with specific natural and semi-natural habitats and are threatened by genetic erosion'. The specific stipulations of this directive require a 'known provenance approved by the appropriate Authority in each Member State for marketing the seed in defined areas [and] appropriate quantitative restrictions'.

It should be noted that the directive is not binding, and leaves the member states free to decide whether they want to impose these specific conditions on marketing or not. Given that PGR conservation was not considered to involve commercial seed marketing, and was therefore not covered by the obligations set by such directives that relate to the variety lists, the majority of member states continued to tolerate the trading or marketing of seed for the purposes of conservation or for amateur gardening, as well as for research and breeding. In this way, the directive followed a similar strategy to that of France, prior to its inclusion of the annex for vegetable varieties used by amateur gardeners.

Introduction of mandatory directives and administrative acts

Fearing that the Administrative Act of 1997 did not help the cause of maintaining its vegetable variety list for amateur gardeners, and protecting the seed industry, the French government approached the European Commission at the end of 2003 to ask it to clarify the mechanisms by which the EU directive should be implemented. After lengthy negotiations, the EU published a directive on agricultural crops and potatoes, in 2008 (EC, 2008), and on vegetables, in 2009 (EC, 2009).

These two EU directives, which are both mandatory rather than optional, define two new variety categories in the European variety catalogue: 'conservation varieties' of agricultural crops, potatoes and vegetables (EC, 2008), and 'varieties created to meet specific growing conditions' for vegetables (EC, 2009). Neither of the two categories of varieties can be protected by plant breeders' rights (PBRs), and as such they remain in the public domain. This means that the cost of registration cannot be compensated by a marketing monopoly of the seed company or licence holder as is the case when PBRs are applicable. Some member states have consequently decided to take on all or part of the costs of registration.

The directives define conservation varieties as 'landraces and varieties, which are naturally adapted to the local and regional conditions and threatened by genetic erosion'. Compared with the normal procedure for variety registration, the directives have broadened the criteria for uniformity, allowing up to 10% of off-type plants. They state that no official testing is required if the variety can be shown to be a conservation variety, by providing the following information: (a) the description of the conservation variety and its denomination; (b) the results of unofficial tests; (c) knowledge gained from practical experience during cultivation, reproduction and use, as notified by the applicant to the member state concerned; (d) other information, in particular from the plant genetic resource authorities or from organizations that are recognized by the member states (EC, 2008, 2009).

Constraints of the directives for conservation varieties

Despite the aforementioned relaxed criteria, the directives have numerous constraints. The standards of distinctness and stability are the same as for the normal EU catalogue and the national variety lists. As regards uniformity, the diversity of a population, by definition, covers all its components and not just the 10% that are off-types, as stated in the directives. The proportions of each of the components of a population may vary from one year to the next. No population, let alone a group of populations, can meet the proposed uniformity criteria. In order to register a variety as a conservation variety, the applicant must show that the variety is relevant to PGR conservation, and must prove that the variety is under threat of genetic erosion. Furthermore, proof must be provided that the variety has been cultivated historically in a region or regions, to which it is naturally adapted. Such regions are referred to as the 'region of origin'. Seed of conservation varieties cannot be marketed outside that region. These restrictions exclude any recently created local varieties and confuse the justified protection of a defined region of origin with an unjustified restriction on their distribution. The production of seed of a conservation variety is limited annually by a quantum, in order to limit the total quantity of seed of those varieties put on the market. The varieties created to meet specific growing conditions, the second category, are not subject to these constraints, but their seed can only be marketed in very small packages corresponding to the need of amateur gardeners, with the stated aim of restricting their spread by increasing their unit price.

Limited implementation by member states

A few European countries have now published regulations on the implementation of these directives. Italy registered a number of conservation varieties in its national and regional variety lists, which had already been established by the authorities in several of its regions. These regional lists operate without altering the conditions for registration and distribution already in place. Switzerland, which is not an EU member, began a conservation list in 2010, but without any quantitative or geographical restrictions on the distribution of seed of the conservation varieties. The fact that shortly after the variety list was initiated a number of varieties had already been listed shows that those restrictive components of the EU directives were limiting their implementation.

Implementation in France

The French government began a conservation list in 2009, retaining the constraints of the European directive but without including any exemptions for small-scale farmers, amateur gardeners or local sales, as permitted under the European regulations. The potential applicants for registration of conservation varieties were put off by all these constraints, and by the associated bureaucracy, and none of them have as yet registered varieties at all in this variety list, despite the promise made by the French state to cover all registration costs.

Despite many efforts, the French government failed to establish its vegetable variety list for amateur gardeners at a European level, and was equally unsuccessful in making it mandatory. It therefore cancelled the specific annex of the vegetable variety list and intends to register all varieties in an annex to the list of 'varieties created to meet specific growing conditions', thus following the second category of the EU directive. But it is not very likely that organizations or seed producers and 'informal' traders of such varieties will be prepared to register new varieties in this register. Why should they bother with a variety list, if PGR that are exchanged for the purposes of conservation, and seed that is marketed for the purposes of amateur gardening, are not obligated to register varieties? When farmers produce for the agricultural market, how can they accept a quantum that limits them to produce seed that is just enough for a few acres? It would force them to eliminate off-type plants in their diverse populations of local varieties, and insert onerous approvals and checks on the location of seed production. Farmers interested in cultivating conservation varieties do not want to keep them mummified in a small corner of their farm or garden. Conservation is only achieved if farmers cultivate and use the varieties and, above all, contribute to the continuous dynamic renewal of agrobiodiversity by re-sowing a proportion of their crop each year and by exchanging their seed with others. Each year, this dynamism contributes to the continued generation of new diversity. Only in that manner can farmers guarantee adaptation to the soil and progress in agricultural techniques, and meet the requirements of consumers in potentially changing climatic conditions. Why should seed dealers impose additional costs that result from excessively strict geographical, quantitative or packaging limitations, if registration in the common variety list is still simpler, despite the cost?

It can be concluded that the EU directives and their regulations have been designed within the normal paradigm of conservationists, rooted in *ex situ* conservation. They approach conservation and diversity disconnected from production and consequently, they consider those farmers who are interested in conservation solely as amateur gardeners or farmers. The design and formulation of the directives hinder the dynamic management of local and traditional varieties by professional and mostly small-scale farmers that operate within distinct regional or specialized markets, which we now know is vital for achieving *in situ* conservation (Jarvis *et al.*, 2011; De Boef and Thijssen, Chapter 1.8). This mismatch is confirmed by the fact that the new European directives on conservation varieties fail to meet the stated objective of favouring *in situ* conservation through the implementation of new rules for marketing seed of local varieties. The EU directives can in fact be considered failures; and the variety list for conservation varieties has only been a success in one non-member state, Switzerland, where it has been implemented with significant adaptation.

The need for a legal system that promotes dynamic on-farm management

As we can see from those efforts that were made to create a legal space for local varieties in the highly organized and regulated formal seed system of Europe, the informal seed system, with its dynamic use and exchange of local varieties, cannot be satisfied with just a small niche within the mainstream regulatory set-up intended for the industrial seed system. The EU directives and the French regulations created a niche that is too narrow for the conservation and renewal of PGR on farms. More importantly, they created a system that undermines the dynamic processes required for *in situ* on-farm conservation, or for those practices that contribute to community biodiversity management, as described by various authors in this volume.

The use of standardization procedures that impose uniformity and stability will gradually halt the continuous evolution of varieties, which is part of the goal to achieve *in situ* conservation of PGR. What is required is a system that promotes the use of diversity and high intra-varietal variability that allow for dynamics and continuous local adaptation. Based on our experiences, such as with the Maison de la Semence Paysanne, as described by Kendall and Gras (Chapter 1.7), and on the multitude of other experiences shared in this book, we suggest that the following limited information should be required for the registration of a conservation variety:

- the variety name, which of is often associated with the place of origin;
- the name of the producer who has propagated the variety;
- indications on the methods of cultivation required;
- the breeding methods used;
- the provision of guarantees concerning the reproductive properties (such as seed quality and viability);
- the guarantee that there has been no contamination by genetically modified organisms or patented genes.

To assist farmers in contributing to the conservation of PGR in a dynamic manner on their farms, the EU should shelve its directives for conservation varieties and transform it to allow for more flexible variety lists. It should accommodate local varieties that are characterized by being populations and therefore are less stable and less uniform. In this way, they can be better adapted to regional and organic cultivation methods. The EU and its member states should move away from restrictive regulations and top-heavy bureaucracy towards creating an enabling environment for farmers who contribute to conservation. The informal seed system and farmers' practices for conserving, utilizing, exchanging and selling the seed of local varieties from their farms should be recognized, thereby facilitating their role as custodians of PGR. Such recognition would add another dimension to the EU commitment to the International Treaty on Plant Genetic Resources for Food and Agriculture, in terms of contributing to the implementation of farmers' rights in Europe.

6.7 Local varieties, informal seed systems and the Seed Law

Reflections from Brazil

Juliana Santilli

Agrobiodiversity and law: a relatively new encounter

The concept of agrobiodiversity has emerged in many disciplines over the last 10–15 years, but it has still not been fully recognized by, or found its place in, the juridical world. The loss of agricultural diversity is associated with changes in agriculture, particularly those that followed the green revolution or industrialization of agriculture. The green revolution took place between the 1950s and 1970s, and it was characterized by the introduction of chemical, mechanical and biological (quality seed of improved varieties) inputs. Industrial inputs and mechanization have spread throughout different parts of the world, but particularly so across Brazil's agricultural frontier. This model has generated various negative socio-environmental consequences that are well known. The increasingly artificial nature of agricultural ecosystems has also caused a reduction in the diversity of crops and varieties, as well as in the variation in agricultural ecosystems, in many cases leading to their complete disappearance. Evidently, the loss of agrobiodiversity cannot be attributed to legal frameworks only, but their effects need to be considered. Seed laws, for instance, were adopted as a legal support for (or component of) the industrialization of agriculture. In the domain of seed and varieties, the agricultural development paradigm was translated into a linear approach to seed sector development, in which formal and commercial systems would eventually replace the sometimes considered 'inferior' informal or farmers' seed systems (Louwaars, 2010). Furthermore, it was considered only a matter of time until these 'inferior' local varieties would be replaced by continuous flows of improved ones. Many of the seed and varietal laws that were developed in the 1970s and 1980s reflect this agricultural development paradigm (Louwaars *et al.*, 2013), which does not take into account the capacities of farmers to deal with seed, and which considers farmers as mere 'recipients' of new technologies, who can be persuaded to adopt new seed that is produced by formal systems.

In this chapter I will analyse the Brazilian Seed Law and the ways in which it has negatively affected the conservation and sustainable use of agrobiodiversity and agricultural development in the domain of small-scale farmers in Brazil. As such, this chapter complements the chapter on the Brazilian provisional ABS Act 2001, authored by Albuquerque *et al.* (Chapter 6.5).

Brazil: multiple agricultural systems and one single legal framework

As highlighted in several other chapters in this volume, the agricultural sector in Brazil is often characterized by two sectors that operate side by side: agro-industry, with large-scale and modernized production, and small-scale farming (*agricultura familiar*). Small-scale farming, as a sector, also includes communities that were settled through agrarian reform, and farming communities belonging to indigenous peoples and traditional communities, which are described in more detail by Assis *et al.* (Chapter 3.2). It is logical that both sectors use different seed systems that are appropriate to each, even though in some areas, or for some crops, there is some overlap. The agro-industrial sector is served by a public and private complex of technology development, which uses limited diversity and is oriented towards the market (Louwaars and De Boef, 2012). The second sector, which comprises small-scale farmers, settlers, indigenous peoples and traditional communities, practices more diversified forms of agriculture, in which farmers use seed primarily, but not exclusively, from the informal system. I refer to this sector as small-scale farmers; they primarily use local varieties, but to some extent they also use seed of improved varieties that are distributed or marketed through the public–private formal system using a development orientation (Louwaars and De Boef, 2012). In Chapter 2.7, Dias discusses in more detail the formal distribution systems that link the agro-industrial sector to the small-scale farming sector in Brazil, through a politically motivated system of seed distribution, and their impact on farmers' autonomy and food security. The second sector also includes the agricultural systems in which agrobiodiversity is used as a livelihood asset by farmers and their communities, thereby contributing to the on-farm management of plant genetic resources (PGR).

The policy arena, as defined by the dualistic approach to agriculture, is shaped by three ministries: the Ministry of Agriculture, Livestock and Food Supply (MAPA), which relates to the predominant agricultural model and focuses on agro-industry; the Ministry of Agrarian Development (MDA), which is associated with small-scale farming and rural reform; and the Ministry of the Environment (MMA), which is mainly responsible for genetic resources and biodiversity, but also addresses several aspects concerning traditional people, owing to their strong association with nature conservation.

A number of laws in Brazil have had a direct impact on agrobiodiversity, including the Seed Law (Brazil, 2003), the Law on Plant Breeders' Rights (Brazil, 1997), and the provisional Act concerning the access to, and benefit-sharing of, genetic resources and associated traditional knowledge (MMA, 2001). However, it should be noted that their effect on small-scale farmers' livelihoods and seed systems has been largely underestimated. With the diverse landscape of laws, responsible ministries and their associated dissimilar interests, and the multiple realities in agriculture, the legal framework established by the Brazilian Seed Law (2003) is a challenge for the country. In the current paper, I aim to assess how the current Seed Law relates to informal seed systems and the farmers' management of local varieties, and explore the barriers and opportunities it poses for community biodiversity management.

Importance of informal seed systems in Brazil

In Brazil, informal seed systems are responsible for the supply of a large part of the seed used by small-scale farmers. Informal seed systems in many circumstances cover the development, production, adaptation and distribution of seed of local varieties, as well as the use of improved and commercial varieties, and their adaptation to local conditions. Farmers maintain seed for use in the following year (Almekinders and Louwaars, 1999). In these informal systems, the extensive and complex social networks, which promote the exchange of seed, varieties and agricultural knowledge, play a fundamental role in ensuring food security and farmers' sovereignty over their genetic resources, while at the same time contributing to on-farm management of PGR.

Comparing data collected by the Brazilian Seed Association, which represents the largest producers of seed in Brazil (Abrasem), for the 2007–2008 agricultural year, with historical data estimates for the period 1991–2003, we can see that the use of seed originating from informal systems increased (Table 6.7.1). In this case, the situation in Brazil is similar to that of many other countries, whether they have developing, emerging or industrialized economies (Byerlee *et al.*, 2007). In other words, informal systems are responsible for the supply of seed for most crops, and the use of seed produced by the formal/commercial system has decreased. Abrasem (2008) noted that the reasons farmers save seed for use in the following year include: family or regional traditions, cost reduction, lack of availability of, or access to, seed of improved varieties, over-priced commercial seed, and low or unreliable quality of commercial seed.

The Brazilian Seed Law and its relationship with informal systems

In spite of the dominance of informal seed systems, the current Seed Law (Brazil, 2003) essentially focuses on the formal system. However, it does create some legal space for local varieties and informal seed systems regarding the following points (Santilli, 2012):

*   It establishes that small-scale farmers (referred to in Brazilian law as family farmers, agrarian reform settlers and indigenous peoples), 'who multiply seed . . . for distribution, exchange or marketing among themselves, are exempted from registration in the national registry of seed and seedlings'; that is, as long

Table 6.7.1 Estimate of the percentage of seed originating from informal seed systems as used in agriculture in Brazil

Crop	Average 1991–2003[a]	Estimate 2007–2008[b]
Common beans	81%	86%
Rice	52%	60%
Soya	28%	46%
Maize	25%	17%
Cotton	23%	56%
Wheat	11%	34%

Sources: a Carraro (2005); b Abrasem (2008).

as the distribution, exchange and even the sale of seed is carried out among the farmers themselves, there is no need for registration. However, the Seed Law imposes some restrictions when seed distribution is carried out through farmers' organizations (even for non-commercial purposes). In addition, farmers' organizations cannot sell seed without being registered.

- It acknowledges local varieties and exempts them from rules that require variety registration, so that their seed can be produced, improved and commercialized. However, as it is still not clear, from a legal point of view, which varieties can be considered 'local', there are doubts concerning which varieties do not need to be registered.
- It allows (Article 48) the inclusion of seed of local varieties in publicly funded programmes, and in policies aimed at supporting seed exchange and distribution by small-scale farmers.
- It defines farm-saved seed, and safeguards the right of farmers to save part of their production, each year, for sowing in the future; this practice is traditionally used by farmers and forms the root of the informal seed systems. However, the Brazilian Seed Law imposes some limits on the quantity of farm-saved seed that can be used each year.

The exceptions made for local varieties by the Seed Law, as described above, represent important achievements of social movements and of organizations within civil society, and they deserve to be highlighted in spite of impasses, described below, that prevent their full enforcement.

Socio-technical and legal definition of local varieties

Many definitions have been proposed for local varieties. Local varieties are also referred to as traditional or creole. Correa and Von der Weid (2007) refer to 'creole seed' as the seed of varieties that have been improved and adapted by farmers by applying their own methods and management systems in a process that started with agriculture itself, over 10 000 years ago. They emphasize the fact that there are hundreds of varieties of each crop, and each one of them evolved under specific environmental conditions, systems of cultivation and cultural preferences. Hardon and De Boef (1993) define local varieties as 'varieties that undergo continuous management by farmers, according to dynamic cycles of cultivation and selection, (not necessarily) within specific agro-ecological and socio-economic environments'.

According to the Seed Law, a local (traditional and/or creole) variety is a 'variety developed, adapted or produced by small-scale farmers (in Brazilian terms family farmers, agrarian reform settlers or indigenous peoples), with clear phenotypic characteristics acknowledged by their own communities'. The Seed Law also requires that local varieties be described for their socio-cultural and environmental value, and as such they must be clearly distinguishable from commercial varieties. In spite of the advances that have been made in the recognition of local varieties, the law leaves it up to MAPA to define whether local varieties are characterized as 'substantially similar to commercial cultivars' or not. However, this does not make sense, since the law itself defines local varieties as those with 'phenotypic characteristics acknowledged by their own communities'. Thus, it should be left to the local communities, with the support and involvement of government representatives and agricultural technicians if necessary, to define the criteria for the identification and characterization of local varieties, and to differentiate them from commercial varieties. Consequently, the definition of local varieties and the distinction between commercial and local varieties remains rather complex and open for debate.

The Seed Law does not only require the agronomic and botanical characterization of local varieties, it also requires their socio-cultural and environmental description (i.e. the contexts in which these varieties were developed). MAPA has still not issued any regulation to define local varieties and the criteria to differentiate them from commercial varieties, as originally determined by the Seed Law back in 2003, which shows that this topic is not being prioritized.

However, the ministry that is most closely associated with small-scale farming, the Ministry of Agrarian Development (MDA), issued a ruling in 2007, which determined that local varieties are to be understood as including:

- varieties that have been developed, adapted or produced by small-scale farmers (e.g. family farmers, agrarian reform settlers or indigenous peoples);
- varieties with clear phenotypic characteristics that have been acknowledged by their own communities;
- varieties that have been used by farmers in these communities for more than three years;
- varieties that are not the product of genetic engineering, or of any other process of industrial development or manipulation in laboratories; and that are not either genetically modified or the product of hybrid seed, which would hinder their management under the control of local communities of small-scale farmers.

An important point to note is that this description of local varieties is only used within the context of the MDA, and not yet within the larger context of the Seed Law, or by MAPA. This paradox is symptomatic of how Brazilian policies that are related to the diversity of agricultural production systems often lack proper integration. Different ministries frequently promote contradictory and conflicting policies, as they represent the political and economic interests of different agricultural systems, as explained above (Santilli, 2012).

Legal definition of farm-saved seed

Another aspect of importance in the Seed Law is the way in which farm-saved seed is defined and regulated. Farm-saved seed is the seed that is saved by farmers after every harvest, exclusively for sowing or planting over the following year, in their own properties, or in other properties of which they have possession, for as long as the parameters registered in the national register of cultivars are observed in the calculation of the amount stored. The storing of seed for sowing in the following planting season is a tradition and a necessity for most small-scale farmers. The legal safeguarding of this practice is essential for informal seed systems and for the conservation of agrobiodiversity, for varieties, crops and agro-ecosystems.

Legal space for farm-saved seed in policies and publicly funded programmes

The legal provision in the new Seed Law of 2003, which allows the inclusion of seed of local varieties in funding programmes, as well as in public programmes for the distribution or exchange of seed, is a major improvement on the prior Seed Law (1977). Previously, the Seed Law treated the seed of local varieties as mere grains, making it harder for policies to support the initiatives of civil society organizations (CSOs) that aimed, in partnership with farmers, at the recovery, improvement and reintroduction of informally produced seed of local varieties. Dias (Chapter 2.7) describes how the Seeds of Passion network in the state of Paraíba, in Brazil, has been able to produce seed and grain for the institutional market, because of such a policy change. Consequently, the legal acknowledgement of local varieties and informal seed systems allows policies to support a number of such initiatives, thereby creating many opportunities for income generation for small-scale farmers in harmony with the production and seed systems, without having to convert to 'formal' industrial agriculture.

Seed laws, informal seed systems and on-farm management

Although seed laws cannot be held wholly responsible for the loss of genetic and socio-cultural diversity, they have certainly contributed to it. The exceptions made by the Seed Law in Brazil to informal seed systems and local varieties aim to reduce any possible negative effects on agrobiodiversity and farm-level seed security. However, they do not change the principles and general concepts on which the law is based (i.e. the promotion of industrialization and consequent standardization of agriculture, as well as the reduction, or rather elimination, of the farmers' role in selection and innovation) (Santilli, 2012).

Seed laws, and the relatively small amount of legal space that is devoted to informal systems, make it difficult to adopt or promote models of agricultural development that are based on small-scale farming. In order that farmers can select, store and exchange their seed freely, which is essential for community biodiversity management. Policies ought to promote a broad diversification of legal frameworks dealing with the innovations and development of the informal seed systems, and the development and use of local varieties. Policies should devote more legal and institutional space to informal systems. Instead of arbitrarily imposing a single system (the formal system), policies ought to support a structure that could facilitate the coexistence of, or promote pluralism in, the various seed systems (Louwaars *et al.*, 2013).

Integrated seed sector development fosters a pluralistic rather than linear approach, within one enabling and evolving policy framework that responds to a reality with multiple situations (i.e. seed systems). When such an approach is taken, the development of the seed sector can be complemented by promoting the use of agrobiodiversity by small-scale farmers, traditional communities and indigenous peoples. In this way, informal seed systems will become recognized as seed systems that serve small-scale farming, in parallel to the formal public–private system that primarily serves large-scale, agro-industrial production. With this recognition, the provision of services by the formal public sector to the informal sector becomes more visible, facilitating the provision of seed and varieties to small-scale farmers also, and thereby making it clear that the formal sector (both public and private) may have to follow dissimilar pathways to provide services to the two agricultural sectors. Seed policies and laws should more explicitly accept informal systems, in which locally adapted varieties are used, distributed and sold. Additional measures that support on-farm or rather community biodiversity management should be included, by issuing laws specifically concerned with conservation and sustainable use. In this way, seed legislation for protecting agrobiodiversity would contribute not only to the conservation and use of agricultural diversity, but also to the food security of human populations, to social inclusion and to sustainable rural development.

The Brazilian Constitution, seed systems and agrobiodiversity

Agricultural laws have not taken into account the fact that biodiversity, and the cultural diversity associated with it, is protected by Brazil's constitution, and that laws and policies dedicated to this sector, including agricultural policies, ought to promote its conservation and sustainable use. In this way, seed policies and laws should acknowledge rather than disregard the principles established in the constitution concerning the conservation and sustainable use of biodiversity, and the safeguarding of national cultural heritage, which includes local varieties (which are the product of cultural intervention), farmers' knowledge and innovations developed by them. Seed laws, at the very least, should not harm efforts to conserve and use agrobiodiversity. Furthermore, they should be coherent with the constitutional principle, which establishes that public authorities must preserve the diversity and integrity of Brazil's genetic patrimony.

6.8 Seed and variety regulations

Obstructing informal seed systems and the use of local varieties by farmers in Europe

Guy Kastler

The importance of informal seed systems

The seed of varieties that is selected, propagated and stored on-farm still represents the bulk of crop diversity that is available and utilized worldwide today. However, this is less and less the case in western Europe. Fifty years of direct or indirect public subsidies that promoted quantity over quality in agricultural production, and encouraged the use of modern industrial inputs to meet those quantities, have resulted in the use of only a handful of elite varieties of just a few crops, by a dominant group of large-scale farmers. As such, these farmers have abandoned the broad diversity of local varieties that are adapted to local and low-input agriculture and that require labour-intensive rather than capital-intensive production systems.

The decision to rely so heavily on these few varieties that are associated with industrial production systems is questioned today by many, both farmers and consumers alike, who are concerned with the following issues:

- environmental damage;
- climatic and health related aspects of the inputs;
- exhaustion of natural resources, such as water, soil fertility and biodiversity;
- the explosive growth of unemployment in rural areas and the resulting rural exodus;
- the rising costs of inputs and fossil energy;
- the frequency and magnitude of climatic stresses and economic shocks, such as those associated with fluctuations in prices of agricultural products and food.

A growing number of farmers are reintroducing greater flexibility and diversification into their agricultural strategies. They aim to reduce energy consumption in agricultural production, and lessen the environmental impact of farming. At the same time, these farmers are trying to adapt to unpredictable variations in the climate, exploring ways to mitigate their emissions, for example, through the use of practices that increase carbon accumulation in soil humus, and aiming to contribute to the conservation of agrobiodiversity. Farmers engaged in such efforts are abandoning the strategy of adapting their environments to fit the elite varieties. Instead, they are reversing this strategy, seeking to adapt the varieties to the diversity and variability of the natural environment. Once again, they appreciate the robustness of their varieties (i.e. specific adaptation to their location, the cultivation of mixed crops, the use of intra-varietal diversity, and the cultivation of a diversity of varieties). However, while making progress in such efforts, these farmers will also meet several obstacles in the seed and variety legal and regulatory frameworks, which will be further described in this chapter.

Creation of a legal space for informal seed systems

It is essential to explore ways in which European regulations can include and promote informal seed systems. Such rules should be distinct from those used for variety lists, which focus on regulating the market for commercial seed. Seed companies would like to see a watertight fit between the informal system and the commercial system, so that neither one offers possibilities for circumventing the rules of the other. One option for balancing the informal system with the commercial system is to restrict the quantities involved in each informal exchange. This option builds upon the assumption that informal exchange concerns a large number of local varieties that are not widely disseminated. However, such a limitation can only be acceptable if they do not impose quantities that are only appropriate for flowerpots or garden allotments. The system must still remain suitable for farmers who cultivate the varieties through dynamic management in their fields for commercial purposes also.

Regulations that obstruct informal seed systems and farmers' dynamic management

Obstacle 1: The variety list

Today European farmers cannot find varieties on the market that suit their changing and diversifying production systems because of the way the official variety lists operate. In effect, the variety lists drive genetic progress towards the use of varieties with a wide adaptation. Such varieties are consequently dependent on the use of external inputs to reach their potential. This situation is in contrast with farmers' practices in breeding, maintenance and reproduction of their seed stock, which encourage the adaptability and adaptation of local varieties. The organization of regular collective seed exchanges between farmers would contribute to increasing the adaptive capacity of the varieties and the overall seed system. Under current regulations, before exchanging seed, the varieties need to be included in the variety list, implying cumbersome, time-consuming procedures. However, local and traditional varieties are often populations that will not meet required criteria for distinctness, uniformity and stability (DUS). In addition, the duration and costs of testing hinder the inclusion of local varieties, and also obstruct the exchange of varieties among farmers within the informal seed system. In another chapter in this volume, co-authored together with Anne-Charlotte Moÿ (Chapter 6.6), I look at the efforts made by the European Union (EU) to create a legal space for local and traditional varieties within a specific European variety catalogue and subsequent national lists for local and traditional varieties. As with the case of the EU, most national seed regulations limit the use of local varieties and the informal seed system. A different framework, built upon principles for seeking diversity and adaptation is required.

Obstacle 2: Plant breeders' rights

Europe favours plant breeders' rights (PBRs) over patents for plant variety protection. However, the EU does recognize patents on genes, for which the protection extends to the whole biological organism (plant and product) containing the patented gene that expresses the function described in the patent. Patents on production technologies may also provide the same protection when associated with the presence of genes or other markers.

Since 1994, the farm-saved seed of many species has been banned, including maize, soybean, a majority of vegetables, and so forth. For 21 species, including most cereal grains, the seed is subject to farmers' privilege, on condition that royalties are paid to the breeders annually. However, the majority of farmers do not pay these royalties because they do not consider them legitimate. Farmers take advantage of the difficulty breeders have in distinguishing between breeders' varieties and varieties that have developed in fields. An interesting perspective on the farmers' privilege is the fact that when breeders first collected the varieties that became sources for their breeding programmes, they paid nothing to the farmers who had bred and conserved varieties for many centuries. The breeders feel that they are meeting their obligations to keep sharing the benefits by leaving their PBR-protected varieties free to use by other breeders. But why should farmers themselves not be entitled to such breeders' exemption for those varieties that their ancestors so graciously granted free to the breeders? When farmers re-sow a particular crop for several years after open-pollination, aren't they selecting for new, locally adapted varieties? Nevertheless, the European Commission, and some of its member states, are examining ways to penalize farmers who use farm-saved seed. They are also exploring ways to support breeders in collecting their royalties (e.g. by providing conditional subsidies on the use of certified seed).

Obstacle 3: The protection of patented genes

The patenting of genes in commercial varieties is gradually increasing. Their presence in a seed or a crop can be easily identified by molecular markers. These patents pose a threat to breeders in that they remove the breeders' exemption, which allows the exchange and use of protected materials among breeders in breeding programmes. This threat is less obvious in cases involving patents on techniques or combinations of non-transgenic genes. Such patents can generate a risk of blocking innovation, threatening both breeders and also farmers, for the benefit of just a few multinational companies that hold the largest portfolios of patents.

Obstacle 4: The biosafety laws

Biosafety laws constitute a new threat to informal seed systems and the dynamic management of local varieties by farmers. Farmers are unable to analyse all farm seed batches that might have been contaminated by neighbouring genetically modified (GM) crops. Consequently, farmers can be forced to abandon the farm-saved seed of local varieties in favour of certified seed (that is tested to ensure it is free of genetically modified organisms) in order to avoid liability in the event of their crop being contaminated. The annual accumulation of contamination by genetically modified organisms (GMOs) in the fields of farmers, who regularly re-sow part of their crop from the previous year, can rapidly reach levels well above the thresholds for mandatory GM labelling (currently 0.9%). Such contamination deters farmers from cultivating any traditional or non-GM varieties of the same species as GMOs that are grown in their region. This situation has already happened with maize in Spain and oilseed rape in Canada, and is currently taking place in Brazil.

Obstacle 5: Biopiracy

Lastly, biopiracy could also come into play and affect informal seed systems and the farmer management of local varieties if the European legal framework for the protection of industrial property is not changed. PBRs can be claimed on a local variety discovered in a small-scale farmer's field, in order for it to be developed (i.e. made stable and uniform, and then described without any obligation to disclose the origin of the resources employed). The best way to oppose this practice would be to claim a PBR on the local variety beforehand. However, the definition of a variety as required for a PBR is based on the same DUS criteria that guides acceptance in the variety list, which is not applicable to local varieties. Another way of protecting against appropriation by third parties would be to prove that the local variety existed before its discovery by others; but there is often no way to prove this. Small-scale farmers do not spend their time making written records or digitizing everything they do; their knowledge is passed down through practices and is shared verbally. As for patents, companies aiming to appropriate local varieties circumvent prior consent and will never allow any sharing of benefits whatsoever, as there is no associated obligation to disclose the origins of the resources employed.

Legal advances made in the EU in light of the ITPGRFA

Breeders, research laboratories, genetic resource centres and conservation networks regularly exchange seed of varieties that are not registered in any national variety list. The registration requirements only apply to marketing 'for the purposes of commercial exploitation'. Selection, research or conservation are not seen as the commercial exploitation of a variety and are therefore not illegal. The International Treaty on Plant Genetic Resources for Food and Agriculture (ITPGRFA) creates the legal space for such exchange of varieties or rather transfer of materials. The EU ratified the ITPGRFA in 2004. Its implementation in Europe remains rather informal because neither the EU nor its member states have so far translated the commitments of the ITPGRFA into domestic legislation and regulations.

The ITPGRFA recognizes the past, present and future contributions of farmers to the conservation of PGR (FAO, 2001). Furthermore, it does not separate the purpose of use of local varieties, be it for conservation, or production for the marketplace.

Their importance to farmers' livelihoods, value in regional or local markets, and specific characteristics, motivate farmers to continue to cultivate local varieties. In this way, local use and markets are associated with the continued use and conservation of local varieties, such as the local bread wheat variety, Meunier d'Apt, in the Lubéron region in France (Kastler and Moÿ, Chapter 4.6). Even so, some European countries are attempting to prohibit the informal seed system on the pretext that the seed that is traded within that system is also used for crop commercialization. This means that they are prohibiting any contribution by farmers to the conservation and renewal of PGR, which is contrary to the commitment made when signing the ITPGRFA.

In 2009, the European Commission began to evaluate its legal framework, with the aim of simplifying procedures for marketing seed. This evaluation was then followed by a similar exercise in 2010, which focused on plant breeding. Several member states, and the majority of commercial seed companies, are calling for a revision of the legal framework regarding patents in order to make them subject to the breeders' exemption, like the PBRs. The question thus emerges as to whether this revision of European law will recognize and acknowledge the contribution made by farmers to the conservation and development of PGR as indicated in the ITPGRFA. It should, however, be realized that the ITPGRFA subjects the farmers' right to save and exchange seed to national law, thus putting it in the hands of the EU and its member states. Thus, the EU has the option within the ITPGRFA to limit the rights of breeders and make them subject to the positive recognition of farmers' rights.

Part VII

Community biodiversity management and resilience

7.1 New professionalism and governance in plant genetic resource management

Walter Simon de Boef, Marja Thijssen and Abishkar Subedi

Community biodiversity management in a wider context of plant genetic resources

Most countries in the world have developed their own national plant genetic resource programmes (NPGRPs). Even though they usually intend to take a complementary approach to conservation, their emphasis is and remains on *ex situ* conservation. In the introduction to this book Thijssen *et al.* (Chapter 1.1) indicate that the Convention on Biological Diversity (CBD, 1992a) and the International Treaty on Plant Genetic Resources for Food and Agriculture (ITPGRFA, 2001) identified *in situ* conservation through on-farm management as one of the strategies. NPGRPs, however, lag behind in implementing this strategy (FAO, 2010). With this book, we have compiled a vast range of experiences on CBM and its practices, as implemented by conservation and development organizations or sometimes non-governmental organizations (NGOs) that are not necessarily associated with their NPGRP. De Boef and Thijssen (Chapter 1.8), as well as a number of other authors, argue that the CBM methodology is instrumental for the implementation of on-farm management. NGOs and their development professionals are key drivers for promoting the CBM methodology. Consequently, on-farm management is implemented by organizations that are not necessarily recognized by, or associated with, the NPGRP. As such, CBM and its practices are not usually part of what constitutes the national framework governing the management of plant genetic resources (PGR), or the normal professionalism associated with those responsible for the conservation in NPGRPs.

In the current chapter, we use two perspectives to meet the challenge of how to introduce CBM as a methodology for contributing to on-farm management, and link it to those organizations and their professionals that are responsible for PGR management within the NPGRP concerned. The first perspective is rooted in discussions initiated by Robert Chambers (1993) concerning new professionalism in rural development. The second and more recent perspective focuses on the new governance that is emerging in environmental management (Holley *et al.*, 2012). Inspired by Nobel Laureate Elinor Ostrom (1990), we have designed a set of principles that may offer guidance for creating ample space for CBM within PGR professionalism and governance. With those design principles we target key decision-makers engaged in PGR management at local, national and global levels. In addition, we target those in charge of programmes for capacity development and higher education, ensuring that future professionals have the necessary knowledge and skills to make CBM work within the larger framework of PGR management.

Normal and new professionalism in rural development

Chambers (1993) introduced the distinction between normal and new professionalism in rural development, referring to 'normal professionalism' as the processes, values, methods and behaviour that dominate a profession or discipline. Furthermore, he links it to structures of power and knowledge, which are reproduced through teaching. In criticizing the dominant development model, he maintains that normal professionalism in rural development is principally concerned with the needs and interests of the rich. Furthermore, it is biased towards an urban, industrial model, which is male-dominated, highly technological and focused on quantity rather than quality. Its actors include international, national and local organizations, particularly specialized departments or programmes of the government (agriculture, community development, forestry, nature conservation, plant genetic resources, and so on). Normal professionalism is closely linked to the research paradigm of positivism, which is dominant in agricultural research and education (Lincoln *et al.*, 2011). Following David Korten (1980; Korten and Klauss, 1984), Chambers contrasts 'normal' and 'new' professionalism associated with blueprint and learning processes in a context of rural development, as shown in Table 7.1.1.

Although it has been two decades since Chambers' book on professionalism in rural development was first published, 'normal' professionalism remains dominant, particularly in agricultural development. This is despite the enhanced understanding of the

Table 7.1.1 'Normal' and 'new' professionalism contrasted in a context of rural development

Character	Normal professionalism	New professionalism
Dominant research paradigm	Positivism	Constructivism, participation
Analytical preposition	Reductionism	Systemic, holistic
First steps	Data collection and planning	Awareness and action
Design	Static, by experts	Evolving, involving
Main resources	Central funds and technicians	Local people and their assets
Human resource training and development	Classroom, didactic, one way flow of knowledge and information	Field-based learning, reciprocal flows of knowledge and information
Management focus	Spending budgets, completing projects in time	Sustained improvement and performance
Content of action	Standardized	Diverse
Communication	Vertical: orders down, reports up	Lateral: mutual learning and sharing of experience
Evaluation	External, intermittent	Internal, continuous
Errors	Buried as expressions of failure	Embraced as options for learning
Effects	Creation of dependencies	Enhanced autonomy and empowerment
Rural development approach	Blueprints	Learning processes

Source: Adapted by the authors from Korten (1980); Korten and Klauss (1984) and Chambers (1993).

continuously changing complex processes that shape and influence rural development, and the increased use of new paradigms that have been developing in human sciences, such as critical theory, constructivism and participation (Lincoln *et al.*, 2011), which provide more informed insights into social change and thereby contribute, through knowledge and information, to new pathways to rural development (Lincoln *et al.*, 2011).

Normal and new professionalism in plant genetic resource conservation

Before addressing professionalism in PGR, we approach the field of nature conservation and natural resource management, in which Pimbert and Pretty (1997) applied the perspective of normal and new professionalism. Their insight is relevant in further refining the professionalism perspective on PGR management. Pimbert and Pretty indicate that 'normal' conservation scientists perceive ecosystems through a single window of their discipline (i.e. they have been formatted to do so by their scientific training as ecologist, botanist or zoologist, or simply as biologist). Consequently, their criteria for decision-making on setting priorities or measuring the performance of conservation actions are often based on a single species (i.e. the number of migrating birds, or density of medicinal plant species), which are then translated to the number of local varieties of a specific crop species used by farming households or communities. Such reductionism, when applied in the field of conservation, results in more traditional models of ecosystem management where conservation and development are considered contrary forces. Pimbert and Pretty use their understanding of normal and new professionalism in rural development (Table 7.1.1) to address the changing relationships between professionals and communities in biodiversity conservation and natural resource management. De Boef (2000) further elaborated on the insights of these authors, examining normal and new professionalism in PGR management. Our involvement in the development of the CBM methodology, and in professional training and post-graduate education in the field of PGR over the past decade has enabled us to further refine the differences between normal and new professionalism in plant genetic resources (Table 7.1.2).

New PGR professionalism as a condition for enabling CBM

In contrasting normal and new professionalism in PGR, we can see that when we aim to promote CBM we are not simply addressing ways to achieve *in situ* conservation; we are approaching PGR in a different manner, or even through another paradigm. As such, the guiding paradigm has implications on what we perceive to be the functions of PGR conservation, and even of farmers and agriculture in society. Consequently, it is only logical that CBM developed as a methodology, particularly within NGOs such as Local Initiatives for Biodiversity, Research and Development (LI-BIRD) in Nepal, and M.S. Swaminathan Research Foundation (MSSRF) in India. These organizations, while developing the methodology, were primarily concerned with livelihoods of their community partners. Other NGOs, such as South-East Asia Regional Initiatives for Community Empowerment (SEARICE), and civil society network organizations like the Seeds of Passion network of community seed banks in Brazil, and the Réseau Semence Paysanne in France, strengthen the autonomy of farmers and farming communities in their dynamic management of local crops and

Table 7.1.2 'Normal' and 'new' professionalism contrasted in a context of plant genetic resources

Cluster name	Character	Normal professionalism	New professionalism
Research orientation	Research paradigm	Positivism	Constructivism and participation
	Disciplines	Genetics, crops science, conservation biology	Idem, complemented by human sciences
Characterization of interactions	Between professionals and people	Controlling, policing, inducing, motivating, dependency-creating	Enabling, supporting, empowering
	Approach to farmers	Clients or beneficiaries	Actors and partners
	Role of professionals in participatory appraisals	Extractor of information	Facilitator and co-learner
	Types of participation	Contractual, consultative	Interactive, self-mobilization
Approach to conservation and diversity	Perspective between humans and biodiversity	Masters and guardians of biodiversity	Members and participants in biodiversity
	Management approach	Control; conserve diversity for research and use	Management and facilitation; conserve and support management for use
	Approach to diversity	Diversity in conservation and uniformity in production	Diversity as a principle of production and expression of culture
	Predominant conservation strategies	Ex situ conservation, national parks	In situ conservation, on-farm management, sustainable development reserves
	Direct beneficiaries of conservation efforts	Researchers and breeders	Farming communities
Relation to varieties and seed	Approach to plant breeding	Productivity, wide adaptation, market demands	Productivity, location-specific adaptation, user group demands
	Approach to seed systems	Formal seed systems will replace the 'inferior' informal seed systems	Informal and formal seed systems operate in a complementary manner
	Approach to varieties	Local varieties are replaced by improved varieties	Local and improved varieties are options for farmers
Market orientation	Food market system	Globalized agro-industrial	Local livelihood-based

Source: Authors, adapted from Pimbert and Pretty (1997); De Boef (2000).

varieties, and, where appropriate, seek ways to link up with formal research and PGR structures. We learn from these examples, that non-governmental and civil society organizations put farmers' dynamic management *first and last*, using the terminology of Chambers *et al.* (1989). As such, new professionalism in PGR defines the relationship between its professionals and rural communities.

This perspective offers insights into why it has been difficult for normal PGR professionals, who are generally occupied with gene bank management, to assume the responsibility and acquire the necessary capabilities to make CBM work. Gene bank curators are responsible for their collections of accessions; within their professionalism, losing an accession is considered the ultimate defeat. Consequently, they control in a rigid manner the conditions for the multiplication, seed processing and storage of their accessions. One can only imagine these same professionals working with those farming communities who manage their local varieties in a dynamic manner, where gaining and losing diversity is part of the game. We should not necessarily assume that professionals responsible for gene bank management have the skills required to facilitate the CBM processes, nor is it likely that they would be the ones expected to facilitate them.

Furthermore, with this perspective on new PGR professionalism we can learn lessons about the role of education and professional development. Where most PGR professionals have a background in agronomy, genetics or biology, they lack the communication and facilitation skills required for working with communities in a context where conservation and development meet. Walter Simon de Boef, as Visiting Professor, faced this challenge in designing course work for the Post Graduate Programme on Plant Genetic Resources of the Federal University of Santa Catarina in Brazil. His challenge was to create a space for new professionalism in an academic environment that is still grounded in normal professionalism. It is, however, encouraging that within the current generation of students joining the programme, an increasing number have the appropriate mindset and interest to be guided towards new professionalism.

In the PGR training programme, which is being implemented by the Centre for Development Innovation of Wageningen University and Research centre, in partnership with MSSRF, we have responded to this challenge. We work mostly with conventional PGR professionals and approach new PGR professionalism as means to working with CBM, which results in a rather different dynamic. From our experiences in education, we have learned that addressing new professionalism in capacity development is vital. We have been engaged in such programmes for more than ten years; however, we now face the challenge to create a space within PGR institutions and governance structures for such new professionalism. We address this challenge to complement the perspective on new professionalism that was developed in the 1990s, with the more present-day perspective of 'new governance'.

New governance in environmental management

New environmental governance (NEG) is increasingly used to address complex socio-environmental issues (Karkkainen, 2004; Lockwood *et al.*, 2010; Holley *et al.*, 2012). We refer to 'normal governance' as being characterized by representative democracy, singular authority, centralized and hierarchal commands, rigidity, and uniform regulatory rules. New governance is characterized as being a series of innovative experiments intended to develop forms of public governance that are 'flexible, less prescriptive, committed to diverse outcomes, and less hierarchal in nature' (Karkkainen, 2004; Walker, 2006; Holley *et al.*, 2012). It is inspired by placing policy, management and research in a setting of social learning instead of control, and is guided by the principles of adaptive management (Gunderson *et al.*, 1995; Berkes *et al.*, 1998). Holley *et al.* (2012) define the following characteristics of 'new governance': collaboration, participation, deliberation, accountability and learning (Table 7.1.3).

New governance in institutional PGR frameworks

We use the five key characteristics of new governance to explore ways in which CBM as a methodology can be incorporated into the institutional framework of NPGRPs for contributing to the implementation of on-farm management. We address this issue guided by the features of new governance (i.e. collaboration, participation, deliberation and, finally, accountability and learning).

Collaboration

In an NEG process, the first objective is to form a multi-stakeholder collaborative body that develops a plan to guide policy implementation. A wealth of knowledge and experiences exists concerning the collaborative management of ecosystems, watersheds and natural resources (Borrini-Feyerabend *et al.*, 2007). In NPGRPs, conservationists in particular collaborate with researchers and plant breeders in the public and private sector, with policy-makers from ministries of agriculture and often, in some manner, with representative bodies of farmers' organizations. In most NPGRPs, collaboration is approached more as an instrument to implement on-farm management than as a way to ensure the involvement of various stakeholders, including NGOs, community-based organizations (CBOs), or even farming communities in decision-making on PGR policies and management practices. However, for NPGRPs to work with the CBM methodology, they need to create the conditions for such collaboration, by strengthening and recognizing farming communities as partners in PGR management.

Reflecting upon the body of experiences with CBM, we can see that most experiences shared in the book have been, or are being, implemented by NGOs or university groups operating outside mainstream NPGRPs. In some case studies, leading actors have an informal relationship with public researchers or conservationists working within the

Table 7.1.3 Key characteristics and associated guiding objectives of new environmental governance

Characteristic	General guiding objective in new governance
Collaboration	To form a multi-stakeholder collaborative body that develops a plan to guide policy implementation
Participation	To develop processes that will deepen the ways in which non-governmental actors are able to contribute to governance
Deliberation	To design decision-making processes that allow inclusiveness and representation, and flexibility in implementation at different levels
Accountability	To ensure that public resources are used appropriately and to secure performance expectations
Learning	To value learning in facilitating adaptation to continuing processes of change

Source: Based on Holley *et al.* (2012).

NPGRPs. For example, the CBM work of both the large NGO MSSRF in India (see King *et al.*, Chapter 4.3 and Chaudhary and Swain, Chapters 4.4 and 5.6) and the small NGO AgroBio Périgord in France (see Kendall and Gras, Chapter 1.7 and Lassaigne and Kendall, Chapter 5.4) is carried out autonomously, though with some informal linkage to their NPGRPs. In Nepal, since the evolution of its CBM work, LI-BIRD has made sure that it works closely with the NPGRP; however, it is LI-BIRD that drives the collaboration and innovation (see Subedi *et al.*, Chapter 2.4 and several other chapters in the book). SEARICE operates in a similar manner in Bhutan, Thailand, and Vietnam; it promotes practices modelled on CBM within public entities associated with NPGRPs (Doctor, Chapter 1.6). In Ethiopia, the NGO Ethio-Organic Seed Action (EOSA) sustained the *in situ* work that was initiated by the NPGRP (Feyissa *et al.*, Chapter 1.4). The experiences involving the participation of universities from Brazil and India are illustrations of CBM activities that are carried out in collaboration with NGOs and CBOs rather than with member institutions of respective NPGRPs (see Zank *et al.*, Chapter 3.8; Ogliari, Chapter 5.7; Reis *et al.*, Chapter 3.5, for examples from Brazil; and Ramesh, Chapter 3.7 for an example from India). There are very few examples of where the NPGRPs are responsible for initiating CBM work; in the current book, they are restricted to efforts in the use of CBM practices in several locations in Brazil (Dias *et al.*, Chapter 2.5), and a more consolidated CBM process leading to the establishment of a conservation-oriented CBO in Ecuador (Tapia and Carrera, Chapter 2.3).

Participation

With NEG, the aim is to develop processes that enforce ways in which non-governmental (or non-state) actors contribute to the governance of human impacts on their environment. Holley *et al.* (2012) indicate that inclusion and representation are vital criteria for achieving participation. Inclusion relates to eligibility to participate in decision-making and management. In PGR frameworks, participation is relevant for identifying those who are eligible to participate in the implementation of on-farm management; those who are affected (farmers and farming communities); those who contribute to conservation (NGOs or CBOs); or those who use PGR (researchers, other communities and farmers). Representation relates to mitigating imbalances in power relationships and addresses the capacities that limit the participation of actors in decision-making at the level of policy formulation, in management and in the practical implementation of strategies for PGR conservation. In PGR frameworks, representation is about balancing interests and involvement of relevant actors. Within CBM processes, time and financial costs for actors, including individual farming households, define participation. The same applies to ways in which farming communities, CBOs, NGOs and other actors assume their responsibility to join as representatives in NPGRPs.

As indicated when addressing collaboration, we realize that most organizations that shared CBM experiences in the book have an informal relationship with NPGRPs. For example, the Seeds of Passion network of community seed banks, in the state of Paraíba in Brazil, operates within civil society and is supported by a number of NGOs, and informally by some individual researchers (Dias, Chapter 2.7). The Maison de la Semence Paysanne in France operates at the periphery of what is formally considered PGR conservation, even though the network and its farmers contribute to the implementation of on-farm management at a national level (Kendall and Gras, Chapter 1.7). The network of organizations engaged in participatory crop improvement in Central America is driven by farmers' organizations that are informally associated with plant breeders in public research organizations (Alonzo *et al.*, 5.8). The community seed banks associated with the network operate in a rather autonomous manner (Fuentes and Alonzo, Chapter 2.6). The experience of increasing the access of farmers in the Pacific to the taro gene pool is a good example of where community-based activities are embedded within a NPGRP in a more structured manner (Iosefa *et al.*, Chapter 5.10). The question remains, however, as to who took the initiative to form the group for introducing germplasm, and who or what drives the process. In this case, scientists in the PGR project took the initiative; the demand for new germplasm, rather than community management of PGR, drives the process.

We conclude that most national PGR programmes are hardly representative or inclusive in terms of non-governmental or civil society actors that are engaged in CBM. Government representatives in these NPGRPs, or even the NGOs or CBOs, would question whether such representation and inclusion is necessary. It should be taken into account that one of the components of farmers' rights as described in the ITPGRFA is participation in decision-making. Moreover, in order to create an enabling framework for CBM this requirement should be enforced by encouraging both NPGRP members and CBM practitioners to collaborate not only at practical level but also in decision-making. With reference to farmers' rights and the demand for embedding CBM within NPGRPs, ensuring such participation can certainly be achieved; however, there is still a long way to go.

Deliberation

With deliberation, NEG aims to design decision-making processes that promote inclusion and representation. Major challenges to equal and fair forms of deliberation are imbalances in power and capacity, which is referred to as 'agency' in the subsequent chapter on empowerment (De Boef *et al.*, Chapter 7.2). When using the CBM methodology, conservation and development organizations design processes that target a situation in which communities make decisions and assume responsibilities for managing their agrobiodiversity in a collective and purposeful manner. This situation, which we refer to as CBM, is reached following complex multi-step processes of decision-making involving several actors and, most importantly, the establishment of CBOs; and through enhancing the capacity of the communities in decision-making (i.e. agency). As we have learned from experiences in Brazil (Seeds of Passion network), France (Maison de la Semence Paysanne), Ethiopia (EOSA's network of community seed banks), India (CBM work of MSSRF) and Nepal (CBM work of LI-BIRD), such deliberative processes take years, and the variation between and within countries illustrates that these processes require adaptive implementation. For NPGRPs, incorporating such processes in a more structural manner, and creating the enabling environment necessary for groups of NGOs and CBOs engaged in motivating farming communities and farming households to assume such responsibilities, requires careful and adaptive procedures that address inclusion and representation. It would involve the re-configuration of the NPGRPs, accompanied by the inclusion and representation of non-governmental and community-based (or civil society) actors. Moreover, creating a space for CBM will affect the existing power base that governmental actors (research, conservation), but also the corporate sector (seed business) currently have in the NPGRPs.

The challenge therefore emerges as to the manner in which NGOs and CBOs are embedded within NPGRPs, and also in the larger setting of PGR policies and rights, as described by Louwaars *et al.* (Chapter 6.1). As can be learned from addressing collaboration and participation as features of new governance in PGR, within the current formal structures of NPGRPs the space for NGO and CBO actors is still rather limited. As such, deliberation at this point in time should focus on the creation of this space, while also motivating NGO and CBO actors to move towards contributing to and joining the NPGRPs. When this space has been created, NPGRPs can build upon the CBM experiences of organizations, such as those shared in many chapters in this book. Therefore, at the practical level, experience with CBM exists; the challenge is of an institutional nature.

Accountability and learning

When addressing adaptation within NEG, accountability and learning as features are closely linked. The fundamental importance of associating accountability with learning, in a context of NEG, is their value in facilitating adaptation to continuing processes of change. As such, accountability and learning are inevitable features of policy engagement in environmental management, and in this way they link PGR governance and CBM processes to resilience.

Traditionally, accountability in governance is considered important to prevent abuse by government authorities, and to ensure that resources are used appropriately and that performance expectations are fulfilled. Accountability in such a traditional setting is a formal, largely hierarchical relationship within which a legislature (the principal) confers on an administrative or executive agency (the agent) responsible the obligation to explain, justify and be held accountable for its actions (Holley *et al.*, 2012). In NEG, such a formal and hierarchical form of accountability may be supplemented and/or altered by 'new' forms and mechanisms of accountability, when an increasing number of non-governmental actors begin to assume important roles. Holley *et al.* (2012) argue that the above-described traditional hierarchical relationship between principal and agent in NEG is no longer applicable. It is too simplistic to capture in such a relationship the dynamic of multiple interests in a multi-stakeholder collaboration. When starting to promote the use of the CBM methodology within NPGRPs, new forms of accountability within PGR management are required. These relate to the division of responsibilities between governmental and non-governmental actors, and mechanisms used for the implementation and sustenance of CBM processes. In such a manner, mutual forms of accountability emerge where collaborators or participating actors monitor each other's behaviour.

Traditional mechanisms for accountability tend to constrain the prudence of agencies and other actors simply because they are embedded in complex administrative and legal procedures. Consequently, accountability is often placed outside management processes, and as such may lack the necessary flexibility associated with the step-by-step processes of decision-making that makes NEG work. In a process of redesigning accountability in NPGRPs, measures that primarily focus on *ex situ* conservation strategies, which are implemented by governmental agencies, should be complemented with procedures that address, in a flexible manner, non-governmental actors and the dynamic management of PGR by farmers and farming communities. New mechanisms for accountability result in a greater deference to decision-making by both governmental and non-governmental actors. From the case studies shared in this book, we learn, however, that so far there has been little interaction between actors engaged in CBM and those within formal conservation programmes.

New mechanisms for accountability in new governance tend to address process and performance. If accountability guarantees that processes are reflective in terms of the impact of management and policies, it connects with learning and adaptation. We can conclude that in a setting of new governance of PGR accountability guides governmental and non-governmental actors to focus on outcomes of PGR management practices, and motivates them as to how they can best achieve the expected results. Since we face the challenge of embedding CBM within NPGRPs, PGR professionals and organizations are now learning to work within such a framework of accountability and, where necessary, are seeking ways to adjust management practices and policies. In new governance, accountability is embedded within the process of adaptive management, which includes mechanisms for setting performance goals and effective monitoring, fosters effective and mutual accountability, encourages actors to implement accountability processes, and designs and promotes adaptive and systemic learning.

Learning in order to achieve adaptation is increasingly recognized as a vital feature of NEG. It strengthens community resilience in times of change, and emerges from the cumulative process of other new governance features, such as collaboration, deliberation and accountability. Designing and implementing a deliberative process, as with new professionalism, facilitates learning. What is clear is that owing to their dependence on monitoring systems, accountability and learning are entangled. When accountability and learning meet in new forms of governance, this impacts the way in which accountability is achieved, and motivates collaborators to be open and continuously share information and monitor impacts of management.

A key question that echoes throughout this book is what conditions are needed to foster CBM as a methodology within NPGRPs? The two final features of new governance have the strength of facilitating a process that integrates CBM work within NPGRPs. With accountability, new governance has the potential to enforce governmental and non-governmental actors to gain legitimacy in actions that contribute to the conservation and use of PGR. This accountability, however, has direct implications on building multi-actor and multi-level arrangements for the implementation of complementary conservation strategies. New professionalism and new governance can thereby support NPGRPs in such a process of change.

Design principles: towards new PGR governance

New governance should be approached as a continuous social experiment. Within the context of this book, the focus of that experiment with new governance is to find mechanisms for promoting CBM and contributing to PGR conservation; and to engage non-governmental and community-based actors within the institutional frameworks in PGR management. Following Elinor Ostrom (1990), and using the five features of new governance as set out by Holley *et al.* (2012), we formulated a set of nine design principles to guide both decision-makers in NPGRPs, and CBM practitioners, given the complex multi-actor and multi-level dynamics they face.

The first design principle addresses the location where the CBM work will be carried out. The boundaries of CBM at national and community levels, as well as its affiliation to the NPGRP, must be well defined. Consequently, when we design CBM within NPGRPs it is important that we are mindful of size and scale. In relation to location, it is vital to promote CBM where local crops and varieties, or agrobiodiversity management in general, are embedded in the livelihood strategies of farming communities.

The second design principle recommends congruence between formal appropriation and provisory rules, and local or customary mechanisms in PGR management. New PGR governance needs to address issues in the context of the appropriation of PGR, such as access and benefit-sharing; the ways in which informal seed systems operate; or how farmers exchange and share seed of local varieties. As such, it concerns farmers' rights, or more specifically, their right to save, share and produce seed. This design principle addresses the interface between the community management of local crops and varieties, and genetic resources, variety and seed laws, which impact on the PGR management by farmers and CBOs. In this way, local and national mechanisms for regulating PGR management converge.

The third design principle emphasizes the need to develop legal structures that make CBM and PGR management function. NPGRP management is formally embedded in biodiversity laws, addressing seed and variety laws when necessary (Louwaars *et al.*, Chapter 6.1). In order to incorporate CBM, with its strong basis in non-governmental and civil society actors, a legal space must be created within which those actors can contribute to, assume responsibility for, and be recognized in PGR management. As such, this principle also contributes to achieving another element of farmers' rights. The success of CBM depends on farming communities devising their own institutions and organizations, remaining in this way autonomous in their PGR management. The challenge for governmental organizations within NPGRPs is to be supportive of such autonomy and flexible in motivating CBOs to define procedures for local PGR resources that match local demands and realities.

The fourth design principle places an emphasis on flexible structures for monitoring accountability. The promotion of CBM and involvement of its practitioners in an NPGRP requires that mechanisms for accountability be defined for governmental, non-governmental and community-based organizations. When rights are granted to those actors operating beyond the original organizations in charge of PGR management, these rights clearly come with obligations. The NPGRP is required to develop flexible mechanism for monitoring and accountability for actors operating at different levels of PGR management.

The fifth design principle reflects the need to design incentives carefully, and to ensure significant investments and clear funding arrangements for promoting CBM. These incentives need to be designed in deliberative but also participatory processes. In Nepal, LI-BIRD has been successful in the establishment of CBM funds that are operated by CBOs. Mechanisms need to be designed to channel financial resources in PGR management from national (and global) levels to the level of farming communities, while guaranteeing CBOs' and farmers' autonomy in their dynamic management of agrobiodiversity.

The sixth design principle focuses on the development of innovative mechanisms to engage non-governmental actors in new PGR governance. With CBM, the type and number of actors involved in NPGRPs will increase significantly. As such, a learning environment is required that promotes innovation and stimulates new actors to engage themselves in PGR governance. Careful facilitation is needed as such processes will meet many challenges, such as the reconfiguration of responsibilities, including that of the role played by state agencies in PGR management. Non-state actors (NGOs, CBOs) need to take on more responsibility, which is a prerequisite for achieving the situation referred to as CBM. The state is not just one of the actors; it remains active in new PGR governance, but in a more supportive role to promote community empowerment in CBM.

The seventh design principle places critical conditions on the process of changing to new governance; it emphasizes the incorporation of effective horizontal, as well as vertical, information-exchange into structures of new PGR governance. New governance creates conditions for experimenting with new institutional arrangements. To facilitate innovation and learning, NPGRPs need to invest in and facilitate continuous multi-directional information flows.

The eighth design principle recommends that new governance in PGR requires identification, nurturing and maintenance of governance capacities. A challenge is that new governance in PGR requires the appropriation of a new methodology and new actors within existing institutional frameworks, in which governmental organizations and primarily 'normal' professionals have assumed certain responsibilities and gained interest. As with any such change, a strategy is required to facilitate this change. New governance and its features is a powerful way of supporting visionary and strategic 'new' NPGRP professionals who can act as catalysers or facilitators in such change processes.

The ninth and final design principle also addresses the process of the institutional changes that are required to make new PGR governance work. New implementation mechanisms, tailored to foster CBM within NPGRPs, should be set up from the start. This places a clear condition adding CBM components within NPGRPs. Certain obstacles that are inherent in the structure of normal PGR professionalism and governance can be avoided. From their inception, CBM programmes operating with NPGRPs should thus be guided by features of new governance (e.g. collaboration, deliberation, accountability).

The ninth design principle is coherent with the strategy adopted by the CBM and resilience project, which is described briefly by Subedi *et al.* (Chapter 1.2). The aim of this project is to support organizations with a proven record with CBM, in developing strategic actions plans for linking their existing work to national and regional PGR programmes. In addition, the partners in the project incorporate aspects of capacity development, thereby contributing to the development of new professionals with the knowledge and skills to facilitate and support CBM. In this way, the project aims to contribute to the objectives of the benefit-sharing fund of the ITPGRFA for promoting on-farm management; design strategies for enhancing community resilience; and support new professionalism as well as new governance in PGR management.

7.2 Community biodiversity management and empowerment

Walter Simon de Boef, Karèn Simone Verhoosel and Marja Thijssen

Introduction

Empowerment can be defined as a process of transformation that involves enhancing the awareness of poorer people of imbalances in power relations (Chambers, 1993); only then can they make effective choices to transform those imbalances and put these choices into practice. Empowerment is reached when these poorer people conclude such a transformation, are capable of reflecting upon the effectiveness of the choices, and thereby further engage themselves in this process of empowerment.

The complexity of empowerment challenges our basic notions of power, participation, professionalism, achievements and sustainability, and as such coming up with a single and simple definition is not easy. The community biodiversity management (CBM) methodology and its practices are designed and implemented by agronomists and biologists, plant genetic resource (PGR) scientists and rural development workers. Because of their involvement in CBM, empowerment has become part of their day-to-day work and professionalism. Even though for those working with CBM empowerment has become part of their vocabulary, what it means and how it defines our work, remains a challenge. Our goal in this chapter is to gain a better understanding of what empowerment means in the context of CBM.

In order to achieve our goal, we rely on understanding empowerment within development studies. We use as guiding references the works of Andrew Bartlett (2008), and Ruth Alsop and Nina Heinsohn (2005). Their articles guide us through concepts and theory; we learn lessons from them by directly applying them to experiences and/or aspects of CBM. We use their considerations on intrinsic and instrumental empowerment for making a distinction between participation and empowerment. The insights gained by understanding the degrees of empowerment confirm that empowerment within the CBM methodology is a step-by-step process for achieving sustainable community-based structures, which in the end are capable of transforming imbalances in power relationships. The concepts of agency and structure help us to understand that in CBM it is crucial to find the right balance between strengthening community-based capabilities and institutions, and addressing relevant policy and legal frameworks, as well as traditional and informal institutions, norms and values. We explain concepts on empowerment by using examples from the five focal countries addressed in the book (Brazil, Ethiopia, France, India and Nepal). The examples from previous chapters are used again to provide insights into concepts of empowerment.

Case study 1: Community grain, seed and gene banks in Jeypore, Orissa, India

Jeypore, in the state of Orissa in India, is a centre of origin for rice, and is also known for the diversity of its tribal people. It is a place where diversity in people is associated with diversity in farming systems and genetic resources. However, although they maintain valuable rice genetic resources, the tribal people of Jeypore are very poor, partly because of the harsh and insecure conditions in which they practice agriculture. In times of drought, floods or other frequent natural disasters, many households are not able to maintain the seed of their varieties. As a result of the disasters that occurred over the past few decades, the amount of local rice varieties being cultivated by farmers in Jeypore has decreased dramatically. The loss of these local rice varieties meant that the communities had no choice but to cultivate improved varieties.

The M.S. Swaminathan Research Foundation (MSSRF) began to work in Jeypore in the late 1990s, with the aim of halting the genetic erosion associated with poverty and vulnerability. The Foundation organized communities by setting up grain, seed and gene banks. Through their enhanced social organization, communities in Jeypore began to get involved in activities that supported their livelihoods. One example that reflects how the communities make use of their richness in local rice diversity is the marketing and promotion of the local rice variety Kalajeera, which is greatly valued in urban markets (Chaudhury and Swain, Chapter 4.4). Following the definition of empowerment, MSSRF facilitates a process of transformation that increases the capabilities of poor tribal people to make effective choices in their livelihoods, put these choices into practice, and reflect upon the effectiveness of their choices in alleviating their poverty, while sustaining their cultural and agricultural diversity, distinguishing them from other poor rural farming communities in India.

In situ conservation, communities and empowerment

In the example from Jeypore, MSSRF approaches agrobiodiversity in a process aimed at alleviating poverty. The grain, seed and gene bank is a practice through which MSSRF seeks to contribute to the empowerment of rural communities in their management of agrobiodiversity. This practice is successful because it enhances food, seed and varietal security, and strengthens the autonomy of communities in maintaining their livelihoods. Organized communities are better able to respond to those common uncertainties that previously kept them poor. Furthermore, through such collective organization and the implementation of practices like the seed bank, tribal communities are able to maintain their diversity of local varieties.

In CBM projects, conservation and development organizations, such as MSSRF in Jeypore, often refer in their strategies to the empowerment of farmers and farming communities, which contributes to livelihood development and *in situ* conservation of agrobiodiversity. In order to refine this common understanding, we formulated the following four assumptions, which describe this relationship between CBM, empowerment, and *in situ* conservation (see also Figure 7.2.1):

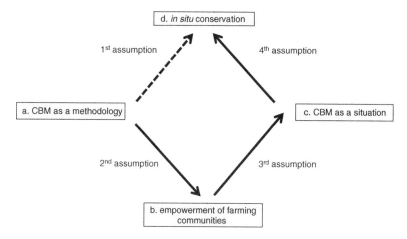

Figure 7.2.1 Visualization of the assumptions underlying the relationship between community biodiversity management (CBM), empowerment and *in situ* conservation.

Note: assumptions refer to the four assumptions underlying the conceptual framework that describes the relationship between CBM, empowerment and *in situ* conservation of agrobiodiversity.

1. The CBM methodology contributes to *in situ* conservation of agrobiodiversity.
2. The CBM methodology contributes to the empowerment of farming communities and their local institutions.
3. Empowerment leads to a situation where communities manage agrobiodiversity in a collective and purposeful manner.
4. The situation in which farming communities collectively and purposefully manage agrobiodiversity in a way that is sustainable contributes to the implementation of *in situ* conservation.

Our focus in this chapter is on the second assumption (that the CBM methodology contributes to empowerment) and the third assumption (that empowerment leads to CBM as a situation). Walter de Boef and Marja Thijssen examined the relationship between CBM and *in situ* conservation of agrobiodiversity in Chapter 1.8. Before addressing the more conceptual aspects of empowerment, the following example from Ethiopia illustrates the relationship between CBM, empowerment and *in situ* conservation.

Case study 2: Participatory breeding and seed entrepreneurship in Habes, Tigray, Ethiopia

For several years, Mekelle University in Tigray, northern Ethiopia, has been working with participatory action research in assessing the diversity of local barley varieties; understanding and documenting informal seed systems; understanding gender aspects in innovation and seed systems; and conducting participatory barley breeding. In Habes, a farmers' research group has been operational for more than five years, participating in seed sector assessment and contributing to participatory varietal selection (PVS) as part of the barley breeding programme.

Since 2008, Mekelle University has been working with the Integrated Seed Sector Development Programme, which aims to strengthen the capacities of farmers' groups so that they are technically equipped, autonomous and market-oriented in their seed entrepreneurship (Hussein *et al.*, Chapter 5.9). One of the first actions of the farmers' research group in Habes was to establish a seed producer cooperative (SPC). In response to the demand for seed of local varieties, which is not covered by the public seed dissemination system, the SPC began to produce and market seed of several local barley varieties. SPC members also participated in PVS, and identified a barley variety that is well adapted to local production conditions, and which has recently been released. Instead of using the common practice of producing seed under a contractual arrangement for public entities or non-governmental organizations (NGOs), the SPC began to sell and market the seed in several markets, and exchange barley seed for wheat and teff grain at traditionally set market ratios. The SPC now sells the food grains in the market, and women members are engaged in value addition through processing the grains into *kollo* and *tihni* (traditional food products) in order to further maximize their profit. Seed prices set by the public system are fixed at 15% of the grain price; however, the farmers in Habes are able to obtain a much higher rate for their grain (Tedla, 2012).

We can conclude that SPC members in Habes set out their own path in seed entrepreneurship, including local varieties and a new released variety in their variety portfolio. In this way, they assumed responsibilities for the conservation and use of local varieties. The fact that the farmers' group in Habes was involved in participatory action research for several years led to an increase in their knowledge and expertise in identifying a variety portfolio and a well-defined and innovative marketing strategy within the informal seed system.

In order to learn lessons from the experiences in Habes about empowerment and *in situ* conservation, we will analyse the following issues:

- To what degree has the SPC been guided in its social organization by participatory action research?
- In what manner did the insights in their own traditional seed system, gained through their involvement in the seed sector assessment, and their experience with PVS, contribute to their decision to include local varieties in seed business and develop a traditional but innovative marketing strategy?
- To what degree has the collaboration with Mekelle University contributed to the empowerment of the farming communities in their management of local varieties (second and third assumptions)?
- To what degree does this empowerment result in purposeful and collective actions that contribute to the conservation and use of agrobiodiversity (fourth assumption)?

Mekelle University is not directly engaged in *in situ* conservation; however, its involvement in participatory action research and in the Integrated Seed Sector Development Programme have contributed to the empowerment of the farmers' group in Habes, by enhancing their operational and organizational capacities, including their management and use of local varieties. We can conclude that through this process of empowerment, the university has contributed to *in situ* conservation of local barley varieties in Habes.

Intrinsic and instrumental empowerment

In order to enhance our understanding of empowerment, we need to be able to distinguish between its intrinsic and instrumental values. Andrew Bartlett (2008) indicates how the empowerment of farmers was and often still is part of a struggle of social movements to free farmers from oppression by landlords, colonial regimes and corrupt governments, and from exploitation by multinational companies. In the field of PGR, scientists, activists, NGOs and social movements, and farmers and their organizations follow the intrinsic discourse on empowerment in the debates on farmers' rights and access and benefit-sharing. This discourse has resulted in the recognition of farmers' rights as a principle in the International Treaty on Plant Genetic Resources for Food and Agriculture (ITPGRFA), as described by Andersen (Chapter 6.2). Access and benefit-sharing is addressed in the third objective of both the Convention on Biological Diversity (CBD) and the ITPGRFA, as described by Vernooy and Ruiz (Chapter 6.4). Both of these international legal instruments are gradually being put into practice through national legislation.

In the 1990s, development agencies began to approach empowerment as a means to reduce poverty (World Bank, 2001), and empowerment soon became associated with participatory approaches in agricultural development. Empowerment was thus no longer merely of intrinsic value, a topic of social struggle and transformation; it became an efficient instrument for alleviating poverty. Although it remained part of the language of activists and, primarily, social and political scientists, empowerment also became part of the discourse of governments and development programmes, with the World Bank as its representative. Farmers and farming communities were supported in their capabilities to take greater control over their livelihoods. Despite the fact that empowerment was mainstreamed through participatory approaches, many actors have maintained their position in seeking 'real empowerment', which for them entails intrinsic empowerment.

When we approach empowerment in CBM to achieve *in situ* conservation, we follow the instrumental rather than intrinsic discourse. However, due to the complex social, economic, and rather politicized context of PGR, farming communities continue to be challenged by imbalances in power in their efforts to manage and utilize agrobiodiversity. We realize that the intrinsic discourse on empowerment remains vital in the motivation of many professionals and their organizations supporting those communities. As can be concluded in many chapters of this book, this is what motivates them to continue enhancing farming communities' awareness and capacities; enforcing their organization; and supporting their access to knowledge and information in order face and transform those imbalances in power that effect them in their management of agrobiodiversity.

One of the distinctions between instrumental and intrinsic empowerment lies in how we relate to participation in CBM. Bartlett (2008) makes it very clear. When farmers are given a greater role in the agenda of professionals, for example in defining criteria and selecting material during PVS (such as the farmers' research group in

Habes), we often refer to 'participation'. When farmers take control over their own agenda, as the farmers' group in Habes did with their seed production and marketing, we often refer to 'empowerment'. It is perhaps simplistic to equate participation to instrumental empowerment, but doing so does help disentangle participation and empowerment in agrobiodiversity management.

Case study 3: Facilitating the exchange of local varieties in Aquitaine, France

AgroBio Périgord is a farmers' group in Aquitaine, in south-western France. Since 2002, it has been engaged in promoting the use of local varieties of maize and several other crops. One of its activities involved establishing and managing a seed bank, as described by Kendall and Gras (Chapter 1.7). The strict variety and seed laws of France, however, prohibit the exchange and sale of local varieties (Kastler, Chapter 6.8). The national farmers' seed network, Réseau Semences Paysannes (RSP), supported AgroBio Périgord in developing a clever way to operate within this rigid seed system. If farmers want to access a local variety from the seed bank they have to become a member. They are then able to access the seed in legal terms for experimentation. In relation to CBM, the farmers are contributing to the dynamic management local varieties. They must return some seed to the seed bank and, since farm-saved seed is allowed, they can continue to use it even when they are no longer members. In 2011, the seed bank distributed seed to more than 250 farmers covering many regions of France.

Degrees of empowerment

Alsop and Heinsohn (2005) define empowerment as the capacity of an individual or group of individuals to make effective choices, and the capacity to transform those choices into desired actions and outcomes. They formulated this definition in a working paper for the World Bank in which they developed degrees of empowerment as part of an exercise to design mechanisms to measure empowerment. They characterize three degrees of empowerment as follows:

1. whether the person or group has the opportunity to make a choice;
2. whether the person or group actually uses the opportunity to choose;
3. once the choice is made, whether it brings the desired outcome.

We will use the example from France to translate these three degrees of empowerment into the context of CBM. For the first degree of empowerment, AgroBio Périgord decided to promote the dynamic management of local varieties by establishing a seed bank that contributes to farmers' autonomy and the conservation of agrobiodiversity. For the second degree of empowerment, AgroBio Périgord, through its linkage with RSP, was able to access resources and information to establish a seed bank and develop a clever mechanism that allows it to disseminate seed of local varieties among farmers, even though such exchange with seed and varieties is prohibited. For the third degree of empowerment, AgroBio Périgord developed a simple monitoring system where farmers could provide feedback about the dynamic management of the varieties they obtained; this feedback can then be discussed, for example, during field days and diversity fairs in order to learn whether the modalities used result in the desired outcomes. This example illustrating the degrees of empowerment shows that farmers associated with AgroBio Périgord are empowered in their dynamic management of local varieties. A CBM case study in Brazil shows where imbalances in power relationships did not result (yet) in desired outcomes, as further detailed here below.

Case study 4: Community struggle for maintaining access to land in Santa Catarina, Brazil

A coastal community of traditional farmers and fisher folk have been maintaining and utilizing their unique landscape, the Areais de Ribanceira, to maintain their livelihood for many years. The community is collectively managing the *restinga*, a salt marshland that is characteristic of this coastal area, in which they cultivate local manioc varieties, harvest fruits from the butiá palm and gather medicinal plants. The restinga is maintained as a result of the relationship between the farming community and its landscape. However, both the farming community's livelihood and the landscape are currently under threat of urbanization, owing to the gradual transformation of the area into an industrial zone. In response, the community formed an association, which, in collaboration with the Federal University of Santa Catarina, NGOs and social movements, is seeking ways to transform the landscape into a sustainable development reserve (SDR), a unit within Brazil's national system of conservation units. Zank *et al.* (Chapter 3.8) provide more insight into this case study, with the SDR as a conservation unit of relevance to institutionalizing CBM at landscape level.

With regards to degrees of empowerment, the community in Areais de Ribanceira has chosen to engage itself in trying to establish an SDR (first degree of empowerment). It aims to get protection for its livelihood, which is associated with the restinga landscape. Various actors, including the government agency responsible for establishing SDRs, the university, NGOs and social movements, provided technical and legal advice to support the community in the process to establish the SDR. As a result of this support and advice, the community was able to access the required information and resources, which strengthened its capacity to make the choice (second degree of empowerment). Because of strong political pressure to urbanize and industrialize the area, the process of transforming the Areais da Ribanceira into an SDR has been halted. The situation was further aggravated in 2010 by a court order to expel the farmers from a major part of the area. However, the community and their partners continue to put up fierce resistance to this order, and still aim to find a solution that includes the establishment of the SDR. It should, however, be noted that the desired outcome of the process of empowerment (third degree of empowerment) (i.e. the transformation of imbalances in power relationships for the farming community associated with the Areais da Ribanceira) has not (yet) been achieved.

Agency

The term 'agency' is based on Paulo Freire's (1973) 'agents of change', in which people take greater control over their lives through conscious action. In a context of changing imbalances in power, agency is about what Bartlett (2008) calls lasting changes in perceptions and relationships, thus an intrinsic rather than instrumental form of empowerment. Long (2001, pp112–113) emphasizes that collective rather than individual agency results in empowerment, he states that 'agency is only manifested, and can only become effective, when individuals interact . . . agency entails a complex set of social relationships . . . that include individuals, organizations, relevant technologies, financial and material resources . . . How they are cemented together is what counts in the end.' When agency reaches consolidation, and perceptions and relationships are transformed, empowerment is achieved (i.e. when self-determination emerges, the outcomes of the transformation processes are determined by the collective and purposeful actions of farmers or farming communities, rather than by development professionals).

We use the example of the relationship between Mekelle University and a group of farmers in Habes, Ethiopia, to understand aspects of agency. During the initial collaboration, the farmer research group contributed to the participatory barley breeding programme by hosting and contributing to PVS trials. We consider this to be participation, since the initiative was undertaken by the breeders and the farmers had to wait for the results of their contribution until the variety could be formally released. However, the interactions between the farmers' group and the university changed with the LSB project. The farmers manage their own crop and variety portfolio in their SPC and if they wish to include a variety that is released by the university, because of their relationship with the university, they can access basic seed to further engage themselves in seed business. This relationship has been transformed through purposeful action, and in this way empowerment is achieved.

Another example could be where a farming community with a community seed bank similar to the one illustrated in Jeypore is faced with changes in rainfall patterns and seeks support from the gene bank to either restore or introduce early maturing materials in their collection. Consequently, the community develops the capacity to approach a formal institution (gene bank), and through this relationship and its conscious action they increase their ability to cope with changing conditions, which we can understand as an expression of agency.

Case study 5: Community seed banks and the CBM trust fund in Barra, Nepal

One of the practices of the agrobiodiversity project in Barra, Nepal (described by Subedi *et al.*, Chapter 1.2), is the establishment of a community seed bank (CSB). The local facilitating NGO, Local Initiatives for Biodiversity, Research and Development (LI-BIRD), together with the local CBM committee, explored ways to sustain the community seed bank without the support of a project or external organization. A solution was found in linking the CSB with a CBM trust fund, which is described in more detail by Shrestha *et al.* (Chapter 2.9). The fund was set up with the aim of providing microcredit to poor farming households; these same households also maintain local varieties. An innovative element of the fund is that it requires farming households who access the fund to commit themselves to multiplying the less popular varieties of the CSB. This allows the continuous regeneration of even the most unpopular materials of the CSB, while the microcredit contributes to social inclusion through value addition. LI-BIRD, together with government and NGO partners, are currently scaling-up this modality in many districts of the country.

With regards to the case study in Nepal, in the context of empowerment, did the CBM process contribute to the development of institutional mechanisms for sustaining the management of agrobiodiversity? This mechanism contributes to the empowerment of a community in organizing itself (agency), and assists it in providing access to financial services to poorer farming households. This illustrates the importance of agency for contributing to empowerment in a CBM process (second and third assumptions in Figure 7.2.1).

Structure

Nevertheless, the transformation of imbalances in power relationships through agency brings us to another aspect of transformation that lies beyond the control of rural people or farming communities and to a large degree beyond those of conservation and development professionals. Alsop and Heinsohn (2005) define 'structure' in the context of empowerment as the formal and informal context within which actors operate, covering rules, social forces or institutions (such as social class, religion, gender, ethnicity and customs). Structure defines the rules of the game that limit or influence the opportunities that determine actions of individuals and groups of individuals. The presence and operation of formal and informal laws, regulations, norms, and customs determine whether individuals and groups have access to psychological, informational, organizational, material, social, financial, but also human and biological assets, and whether these individuals and groups can use those assets to achieve the desired outcomes. These assets are not independent, several of them interact with one another, and coping with that poses many challenges. When describing structure, the discussion on assets links empowerment to the strategies of livelihood development and community resilience, for which assets and their interactions are vital, as further described in the final chapter of this book.

In the context of informal seed supply, several formal and informal rules emerge concerning the practical management of PGR. In many countries, plant variety protection and seed laws define the room for manoeuvre for farmers, as can be seen in the case studies from Ethiopia and France. These laws specify whether or not farmers are allowed to exchange or sell seed of local or improved varieties, which has direct implications on informal seed systems and farmers' management of agrobiodiversity (Louwaars *et al.*, Chapter 6.1; Santilli, Chapter 6.7). If, for example, the SPC in Habes wants to market the seed of local varieties, it must adhere to those limitations within the formal regulations; that is, they must keep to formal boundaries (i.e. structure). However, empowerment processes can also be robust, with social norms and traditions, as well as expressions of structure, guiding the SPC in Habes towards bartering instead of selling seed. In this way, local traditions create a solution for by-passing the formal seed laws, or even overruling formal regulations, as can be seen in rural areas or in the markets of Tigray for example. Lipper *et al.* (2010) describe similar cases for the functioning and organization of traditional seed markets in several countries of Africa.

In the ISSD Programme in Ethiopia, Mekelle University and other universities are documenting these aspects of seed business and are facilitating the implementation of innovative mechanisms at all levels, including that of policy development. They aim to contribute to processes of structural transformation so that farmers are allowed to produce and market the seed of local varieties under specific conditions. In this way, the university is contributing to empowerment through transformation at the level of structure, which is linked to transformation at the level of agency described above.

AgroBio Périgord in France has been able to develop a mechanism for facilitating the exchange of seed of local varieties among farmers (agency). The Réseau Semences Paysannes, in which AgroBio Périgord participates, supports farmers in coping with the limitations that variety and seed laws impose on small-scale farmers (Kastler and Moÿ, Chapter 6.6; Kastler, Chapter 6.8). In the case studies from both Ethiopia and France, structure poses limitations on the farmers' groups for achieving their aims, but through innovative arrangements or transformations as expressions of agency they are able find their own solutions, which in the end may contribute to transformation at a more structural level.

Case study 6: Farmers' empowerment through access and benefit-sharing in Brazil

In Brazil, access and benefit-sharing (ABS) concerning genetic resources is embedded in a legal structure with regulations, as described by Albuquerque *et al.* (Chapter 6.5). One of the objectives of the ABS Law is to empower farming communities to control access to genetic resources and associated traditional knowledge. It thereby contributes to a process of transformation of imbalances in power relationships, as expressed in the third objective of the CBD and the ITPGRFA, which is the fair and equitable sharing of the benefits arising from the use of genetic resources and associated traditional knowledge. The ABS Law in Brazil contributes to the implementation of those international agreements. It has created mechanisms for 'prior informed consent' (PIC), in which scientists are required to obtain permission from traditional farming communities to access genetic resources and/or associated traditional knowledge, when applying for a permit from the responsible government agency. The question remains as to whether traditional farming communities understand that they have the choice to accept or refuse a PIC initiated by scientists. Furthermore, are the communities capable of making that choice; and if so, in the case of signing a PIC, do they have the capability to assess whether or not their choice has resulted in the desired outcomes? Do the traditional farming communities have access to the resources and information necessary to consider the options and make the best choice within a PIC procedure? In ABS, empowerment is not that much an issue of structure; a key challenge to be addressed is agency, or, to be more specific, collective agency, i.e. the capabilities of traditional farming communities to exercise their 'rights' in a collective and purposeful manner. The example of ABS is basically applicable to any country where an ABS regime has been put in place. Moreover, it is an illustration from our world of PGR of the reciprocal relationship between agency and structure in achieving empowerment.

CBM and empowerment: lessons learned

Faced with the challenge to capture the meaning of empowerment in the context of CBM, we matched pre-established concepts on empowerment with experiences in applying the methodology, its practices, related research, and relevant advocacy and policy processes. In Figure 7.2.2, we illustrate the relationship between CBM and the various types of empowerment, which then helps us to identify a series of lessons learned.

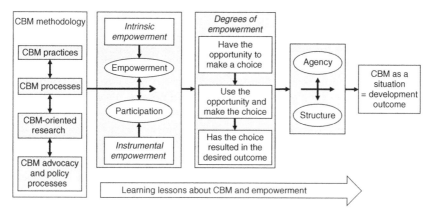

Figure 7.2.2 Relationship between CBM and the various concepts of empowerment.

Source: Authors based on Alsop and Heinsohn (2005).

A first lesson we learned in distinguishing between intrinsic and instrumental empowerment is that we are now better able to differentiate between participation and empowerment. Many of the actions that are described as CBM practices are of a participatory nature, but we now realize that they do not necessary contribute to empowerment. We recognize that those practices relevant for achieving *in situ* conservation result in participation, not in the transformation of imbalances in power relationships required for achieving empowerment. But at the same time, many professionals and their organizations are engaged in addressing farmers' rights, encouraging custodianship by traditional farming communities and enhancing the autonomy of farming communities in agrobiodiversity management. Those professionals and organizations take an approach that is geared towards intrinsic empowerment.

The degrees of empowerment illustrate that empowerment is a process in which a community, first, should have opportunities to choose; second, should use the opportunities to choose; and third, should be able to reflect on whether the choice has resulted in the desired outcome in terms of transformations of imbalances in power relationships. In understanding the degrees of empowerment, we realize that this cannot be achieved by single CBM practices; moreover, a CBM process should involve several steps for reaching the desired outcome, resulting in the required transformation. In understanding the degrees of empowerment, we are no longer obliged to look at CBM in purely technical or practical terms of conservation, but also in social, economic, cultural and political terms of emancipation of farming communities.

It can easily be concluded that with many of the CBM practices and processes shared in this book, we primarily focus on strengthening the capabilities of local institutions or community-based organizations. Should we then conclude that we solely address agency and perhaps limit ourselves by not adequately addressing structure? Are we therefore conducting a partial job in the process of empowerment, which will restrict our efforts to make significant changes in imbalances in power relationships to make CBM work? To answer those questions requires some nuances. A number of cases in the book show that many conservation and development organizations combine their activities at grassroots level with advocacy work influencing policy and institutional discussions. As shared in several chapters in the book, NGOs such as the M.S. Swaminathan Research Foundation (MSSRF, India), Local Initiatives for Biodiversity, Research and Development (LI-BIRD, Nepal), South-East Asia Regional Initiatives for Community Empowerment (SEARICE; South-East Asia), Semiarid Network in Paraíba (ASA/PB, Brazil) and Réseau Semences Paysannes (RSP, France) link grassroots action in CBM processes with advocacy, contributing to policy processes including up- and out-scaling. In this way, these NGOs show that they consciously balance agency and structure in their operations in their efforts to contribute to empowerment. For cases in which universities are engaged in CBM, their research is often associated with participatory and local action, providing both a grassroots and scientific basis for informing or influencing policy processes. They contribute to aspects of agency at local level through their participatory action research, and take on a facilitation role in processes of transformation and learning, which places them in a good position for contributing to empowerment both at the level of policy and decision-makers, and at the level of traditional norms and values.

A good example of this can be seen in the role played by Mekelle University in Ethiopia in facilitating innovation both at the level of farmers' research groups and seed cooperatives, and at the level of seed policy and regulatory frameworks. In the case of the Areais da Ribanceira in Brazil, the Federal University of Santa Catarina is engaged in supporting the community organization by providing technical and legal advice, and addressing both the modalities required to make the SDR work at the community level, and the overall modalities and design. The strategic role played by NGOs and universities in contributing to the right balance in agency and structure is illustrated in Figure 7.2.2, in which CBM-oriented research and CBM advocacy and policy processes are shown as contributing to the CBM methodology.

With our increased understanding of agency, we realize that by contributing to community-based structures we have taken a collective rather than individual position in terms of agency. Our original reasoning was from a conservation perspective; our argument was that by strengthening collective structures we reach CBM as a more sustainable situation, which is a prerequisite for *in situ* conservation. This reasoning was reinforced by the lesson learned that in order for agency to be effective in transformation, what matters are interactions and social relationships (Long, 2001). We can thus conclude that our reasoning has a foundation from both a conservation and empowerment perspective.

When exploring the relationship between agency and structure, many of the case studies in the book appear to begin with agency before addressing structure, although we realize that this may be biased by our selection of authors. However, the Brazilian ABS Law is an example of the reverse (i.e. it begins with structure before addressing agency); the major aim of ABS is the empowerment of farming communities. For its achievement, ABS systems depend greatly on the capabilities of farming communities to express and assume their rights (i.e. to make decisions in a collective and purposeful manner, in terms of providing conscious access to their genetic resources and associated traditional knowledge).

For the final lessons learned in this chapter, we return to Figure 7.2.1, which places empowerment in a context of reaching a situation where farming communities manage their agrobiodiversity in a collective and purposeful manner, thereby contributing to *in situ* conservation. Revisiting the definition of Robert Chambers that 'empowerment means that people, especially poorer people, are enabled to take more control over their lives', we need to understand that we as professionals, together with farming communities, engage ourselves in processes of transformation, with the aim that farming communities take more control over their livelihoods through their agrobiodiversity management. Bartlett (2008) emphasizes the voice of the author in relation to empowerment. When translating his self-reflection into our context of CBM, we realize that we continue to distinguish between professionals and farming communities. We as professionals are always talking about the empowerment of others (i.e. the empowerment of farming communities). If we take the matter of agency seriously in the relationship between CBM and empowerment, then this distinction no longer holds; we as professionals become subject to the development or empowerment processes ourselves. We cannot promote transformations in relationships of rural people, between rich and poor, between community-based organizations and NGOs, universities or governments, or between farming communities and professionals, without also being open to our own relationships with the people we are motivating to take control over their own development and actions in the conservation and use of agrobiodiversity. As professionals we cannot escape from this transformation if we, and our organizations, claim to contribute to the empowerment of communities in their agrobiodiversity management.

7.3 Community biodiversity management
Promoting resilience

Walter Simon de Boef, Marja Thijssen, Nivaldo Peroni and Abishkar Subedi

Poverty and agrobiodiversity

The state of vulnerability and powerlessness in making meaningful choices about livelihood development defines poverty; this state is complemented by restricted tangible assets for livelihood development and goods for consumption (Fisher *et al.*, 2008). Herewith, we meet what the World Bank (2001) refers to as three dimension of poverty: (i) powerlessness, (ii) vulnerability, and (iii) lack of assets. In the current chapter we explore the three dimensions of poverty as a means to disentangle the relationship between community biodiversity management (CBM) and community resilience.

Agrobiodiversity – expressed as the diversity at genetic, species, agro-ecosystem and landscape levels – is one of the options for poor farming households and communities to meet their livelihood needs in times of seasonal shortage and in times of crises, such as drought, crop or even market failure (Fisher *et al.*, 2008; Jarvis *et al.*, 2011). As such, farming households or communities integrate the conservation and use of agrobiodiversity in their livelihood development pathways.

In the 1990s and early 2000s, conservation and development organizations used participatory approaches to address both poverty alleviation and (agro) biodiversity conservation. Where initially many interactions between these organizations and farming communities were of a more consultative nature, they gradually developed into the kind of relationships that fostered learning and action among the poor. The organizations began to address awareness, knowledge and action crucial for motivating (poor) rural communities to assume and take control over their options in biodiversity and natural resource management (Borrini-Feyerabend *et al.*, 2007; Fisher *et al.*, 2008; Wilmsen *et al.*, 2008). CBM, by linking conservation and development in the field of agrobiodiversity management, thus builds upon this body of expertise.

CBM: addressing powerlessness and vulnerability

Empowerment is understood as being vital for the CBM methodology to result in CBM as a situation. Empowerment may also be considered a perquisite for CBM to contribute to the realization of on-farm management of agrobiodiversity. The diversity of experiences, such as community seed banks, value addition, community-based seed entrepreneurship and the establishment of conservation units that target sustainable development, illustrate how the CBM methodology and its practices contribute to community empowerment in terms of both agency and structure (De Boef *et al.*, Chapter 7.2). It can be concluded that the CBM methodology has the potential to address powerlessness (i.e. the first dimension of poverty).

Vulnerability, the second dimension of poverty, can be defined as the opposite of resilience. Wilson (2012) describes vulnerability as the exposure and sensitivity of an individual or a group of individuals who are unable to cope with risks, hazards or catastrophic changes. We assume that efforts to promote resilience through the CBM methodology and its practices focus on enhancing the capabilities of communities to engage in and follow 'good' livelihood pathways, in a collective and purposeful manner. Of course, with this assumption, we realize that 'good' is quite subjective and can be perceived differently by individuals or stakeholder groups.

Below, we refer to a number of examples shared in the book where CBM processes and practices contribute to reducing such vulnerability. The process of using diversity kits in Guaraciaba, Brazil, is an illustration of where farming households and communities regain security at an individual and community level in terms of food, crops and varieties (Canci *et al.*, Chapter 1.5). In Central America, development organizations have been supporting farming communities in the establishment of community seed reserves, which have gradually made farmers and communities less vulnerable to losing their varieties during frequent natural disasters. The farmers regain access to seed and varieties through community institutions instead of being dependent on relief or seed markets in which they are unable to access seed of their preferred varieties (Fuentes and Alonzo, Chapter 2.6). Shrestha *et al.* (Chapter 2.8) show how community seed banks throughout the world and in very different contexts are drivers of the CBM process, promoting agrodiversity conservation and use while building community structures. Several methods of participatory crop improvement enhance farmers' access to a wider diversity of varieties, reducing farmers' vulnerability to natural and anthropogenic disturbances (De Boef *et al.*, Chapter 5.1). A striking example is where the introduction of exotic germplasm of taro in Samoa has been vital to sustaining the cropping system (Iosefa *et al.*, Chapter 5.10). The case studies on marketing and value addition of local wheat varieties from the mountainous region of the Lubéron, France (Kastler and Moÿ, Chapter 4.6), is an illustration of where a local livelihood-based food market system reconnects farmers with consumers, while valuing diversity and supporting the establishment of community institutions. These CBM practices not only empower farming communities in their agrobiodiversity management, they also decrease vulnerability.

We assume that efforts aimed at promoting resilience through the CBM methodology and its practices focus on strengthening community capabilities for using 'good' livelihood pathways. Through the collective and purposeful management of agrobiodiversity, community resilience is an emerging property of the CBM processes and practices.

If we want to explore the ways in which CBM contributes to community resilience, we must do so within the larger context of poverty alleviation, livelihood development and empowerment. Here we address ways in which resilience is defined in ecological and social systems, and then analyse whether the CBM methodology and its practices are instrumental in promoting community resilience.

CBM: defining community resilience

In the field of ecology, resilience is defined as the ability of an ecosystem to reorganize itself following perturbations. Resilience is measured by the extent of disturbance a system can tolerate while maintaining its structures and functions. Research on the resilience of ecosystems has shown that when seeking resilience, most disturbances are often linked to changes in social systems (Berkes *et al.*, 1998; Gunderson *et al.*, 1995). While promoting resilience, both natural and anthropogenic disturbances and their interactions need to be addressed. Table 7.3.1 illustrates the diversity of natural and anthropogenic disturbances.

An increasing amount of attention is being placed on understanding how human systems respond to disturbances (i.e. social or community resilience). Wilson (2012, p17) defines community resilience as the ability of a social system to respond to, recover from and reorganize itself following disturbances; he includes those inherent conditions that allow a social system to absorb impacts and cope with an event, as well as a post-event. He includes adaptive processes that facilitate the ability of a social system to reorganize and learn in response to a change. The notion of learning defines community resilience more as a dynamic process that is continuously adapting to changes or disturbances, rather than returning to a previous stage of equilibrium. As such, dynamism and learning are incorporated as ways in which we define but also approach community resilience. Resilience is thus approached both as an outcome and a process that is linked to the capacity of a community to learn, and its willingness to seek out and take responsibility for 'good' livelihood pathways. When

Table 7.3.1 Examples of anthropogenic and natural disturbances affecting community resilience

Natural disturbances		
Weather-related disasters	*Geological disasters*	*Natural disasters resulting from human mismanagement*
Cold events	Landslides	Global climate change
Hurricanes	Volcanic eruptions	Pollution
Drought/desertification	Earthquakes	Human-induced desertification
Floods	Meteorite impacts	Pest and disease outbreak affecting agriculture
	Tsunamis	Biodiversity loss

Anthropogenic disturbances			
Socio-political disturbances	*Disturbances linked to energy availability*	*Economic disturbances*	*Disturbances linked to globalization*
War	Peak oil	Shift in global trade	Loss of socio-cultural values
Revolution	Energy shortages	Recession	Modernization
Power/ governance shifts		Food prizes fluctuation	Technological change Migration

Source: Adapted by the authors from Wilson (2012).

communities assume their responsibilities in those pathways, in a collective and purposeful manner, and engage in processes of social learning, the CBM methodology and its practices emerge as potential strategy for fostering community resilience.

Spatial aspects and community resilience

In order to define community resilience, we need to first establish clearly what we consider constitutes a community. In this context, community is approached as an open social system with inflows and outflows of people, goods, knowledge, information, skills and resources (Cumming *et al.*, 2006). As illustrated in Figure 7.3.1, we recognize that communities are not homogenous entities, and are comprised of individuals, households and stakeholder groups, each of which has its own resilience pathway. What may be good for a community as a whole may not be beneficial or result in resilience at the level of individuals, households or specific stakeholders groups. Similarly, it can be easily reasoned that what may be good for one household may not necessarily benefit other households, or the community itself (e.g. the accumulation of wealth through creating some exclusive access rights to seed of a certain local variety).

With communities as open systems, community resilience is influenced by pathways that are defined at higher scales, such as the outcomes of policies and decisions. Wilson (2012) argues that larger-scale processes or actions that influence decision-making (e.g. national policies, and/or global drivers of change) are ultimately mediated by the household within its community. For example, national genetic resource policies result in actions that have tangible effects in a given locality (e.g. they create options for a farming household or group of households in a community to claim certain exclusive rights in the production and marketing of seed of specific local variety). The accumulated effect of such individual actions can be seen at local community level, where farmers or groups of farmers actually exercise their rights, or engage themselves in a collective effort, such as establishing a community seed bank, and ensuring that farming households in the community and beyond can access the seed

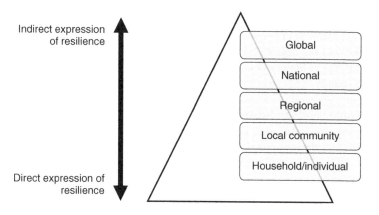

Figure 7.3.1 The human scales of resilience.

Source: Adapted by the authors from Wilson (2012).

of local varieties in a more equalitarian manner. Various genetic resource policies bring global and national policies to the level of communities and farming households as drivers of change (Louwaars *et al.*, Chapter 6.1). However, for delineating and up-scaling resilience, actions are defined by the degree to which individuals or groups of individuals become engaged in those actions that are addressed by such global and national genetic resource policies.

Resilience, assets and capital endowments

In the early 2000s, the focus in rural development moved from participatory approaches to livelihood development. Components of improved livelihoods usually include good health, higher income, resilience, empowerment, food security, and a more sustainable use of the natural resource base (Carney, 1998; DFID, 1999). Assets, which are also identified as the third dimension of poverty, are considered to be the basis for livelihood development; they include physical, natural and biological assets; human and social assets; and economic and financial assets. When we build upon the three pillars of sustainability, these assets can be structured into economic, social and environmental capital (Cutter *et al.*, 2008; Ostrom, 2009). Instead of indi-vidual livelihood assets, we refer to these assets as capital endowments for defining community resilience.

Economic capital is defined as monetary income and financial assets. Factors such as the levels of community or household income, the development of community infrastructure and the availability of funding are associated with strong economic capital, which many would argue is overemphasized as an indicator of community resilience. Social capital is generally described as the social, political and cultural networks and relationships in a community (Bordieu, 1987). What is relevant to community resilience is the description of social capital as being the ability and willingness of community members to participate in actions directed to community objectives and to processes of engagement (Magis, 2010). Wilson (2012) highlights functions such as bonding (group cohesion), bridging (ties between groups), and linking (vertical relationships). Environmental capital has been immersed in debates on sustainability since the 1980s; today, it is self-evident that a community relies on a healthy environment for its survival. With the complex and strong linkages between social and ecological systems described above, the importance of environ-mental capital has generally become more accepted as one of three pillars when defining community resilience. Environmental capital addresses physical, natural and biological (agrobiodiversity) assets, which are often highly interdependent on social and economic assets. Environmental capital is influenced by individual and collective actions, but also presents opportunities and constraints to communities for their livelihood development and for promoting community resilience (Magis, 2010).

Based on Wilson (2012), community resilience and vulnerability can be concep-tualized as a triangle of economic, social and environmental capital (Figure 7.3.2). The strongest resilience is achieved when all three capitals are well developed; such situations can be found when there are dynamic interactions between strong eco-nomic, social and environmental capital, as shown in the overlap of the three circles

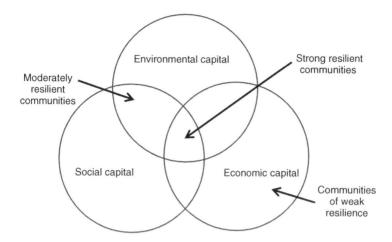

Figure 7.3.2 Community resilience, vulnerability, and economic, social and environmental capital.

Source: Wilson (2010, 2012).

in Figure 7.3.2. Communities with only two capitals are characterized as moderately resilient or moderately vulnerable; while the communities with only one well developed capital can be characterized as of weak resilience and high vulnerability. Community resilience can be seen as the capacity of the community to balance their economic, social and environmental needs in their livelihood pathways in response to disturbances. The relationships between the three capitals are of more relevance than the capitals themselves are individually, thus for reaching community resilience an integrated approach is required.

Many of the contributing authors of this book share experiences with CBM processes and practices that address the three capitals. The common and global reference in the development of the CBM methodology is Local Initiatives of Biodiversity, Research and Development (LI-BIRD) and partner organizations in Nepal (Subedi *et al.*, Chapter 1.2; Shrestha *et al.*, Chapter 1.3). Experiences with more than nine CBM practices implemented by LI-BIRD are shared in this book. For example, the community seed bank and community biodiversity register constitute local institutions for strengthening agrobiodiversity management (natural and social capital). Above all, these practices motivate communities to engage themselves in collective and purposeful action, which results in what Wilson (2012) describes as bonding, bridging and linking, as features of social capital. The community seed bank, coupled with the CBM trust fund, enables poor farming households to obtain access to credit (social and economic capital), and get involved in value addition and the marketing of specific crops and varieties (natural and economic capital), while at the same time providing services in the conservation of unpopular varieties (natural capital) (Shrestha *et al.*, Chapter 2.9). A diversity of participatory crop improvement methods, as used by LI-BIRD, contributes to the availability of local and improved varieties (natural capital). The methods include the participatory genetic enhancement of local varieties of a major food crop, such as rice (Silwal *et al.*, Chapter 5.5) and grassroots breeding, which targets neglected and underutilized fruit and vegetable species in particular (Sthapit *et al.*, Chapter 5.2).

The other global reference for CBM processes is the M.S. Swaminathan Research Foundation (MSSRF) in India. It has taken an integrated approach in supporting tribal communities in the Kolli Hills, through raising awareness and enhancing the capabilities of farming communities to maintain and use a diversity of small grain species. In using this approach, the foundation aimed to ensure food and nutritional security in harsh climatic conditions (natural and social capital), and address changes in food and agricultural systems. MSSRF has promoted the establishment of community seed banks (natural and social capital) and supports value addition processes (economic capital) in promoting small grain products in rural and urban markets (King *et al.*, Chapter 4.3).

Similarly to the examples from India and Nepal, the Seeds of Passion network, facilitated by the Semiarid Network in Paraíba, Brazil (ASA/PB), enhances the use of farmer-produced seed of local varieties (natural capital) (Dias, Chapter 2.7). In terms of social capital, the Seeds of Passion Network bonds farmers, fostering cohesion within communities; builds bridges between communities, facilitating varietal and seed flow; and establishes linkages in terms of vertical relationships with, for example, policy bodies. With regards to economic capital, farmers' autonomy increases so they do not have to access seed and varieties from governmental distribution programmes. Through the vertical integration, even the use of their assets in natural and social capital can be fostered for engaging in income generation as an expression of their economic capital. As such, the Seeds of Passion network does not only contribute to on-farm management of agrobiodiversity, it also enhances the autonomy of farmers and communities in semi-arid areas in seed, varietal and food security. By addressing the three capitals in an integrated approach, the network promotes community resilience. The example is powerful since it clearly expresses how several scales of resilience are interrelated.

Similar considerations relating to CBM processes and practices, and capital endowments, can be made for many other experiences shared in this book. Several chapters address araucaria landscape management in Brazil (Assis *et al.*, Chapter 3.2; Peroni *et al.*, Chapter 3.4; Reis *et al.*, Chapter 3.5), in which traditional peoples' rights are fostered, management systems are supported, and market access and value addition is promoted. Several authors share experiences in promoting farmers' involvement in crop improvement coupled with community-based strategies of seed production. Experiences are shared from Ethiopia (Mohammed *et al.*, Chapter 5.9), Thailand and Vietnam (Doctor, Chapter 1.6; Thomas and Anh, Chapter 4.5) and Central America (Alonzo *et al.*, Chapter 5.8). Plant genetic resource programmes support the establishment of agrobiodiversity-oriented cooperatives in Ecuador (Tapia and Carrera, Chapter 2.3) and Ethiopia (Feyissa *et al.*, Chapter 1.4). In France, a community seed bank and various methods of participatory crop improvement support the farmers' dynamic management of local varieties (Kendall and Gras, Chapter 1.7; Lassaigne and Kendall, Chapter 5.4). Basically, most experiences link livelihood development and agrobiodiversity management in one way or another.

Following the argumentation built upon the cases from Nepal, India and Brazil, and complemented by several others, the CBM methodology motivates conservation and development organizations to take an integrated approach. With the focus on agrobiodiversity, the organizations strengthen the capacities of communities to manage their biological assets in a more purposeful manner as part of their environmental capital. They target both individual and collective capabilities, and strengthen community institutions and structures, i.e. enhance human and social assets as part of social capital. Many experiences shared in the book promote livelihood-based food market systems (De Boef *et al.*, Chapter 5.1). They show how communities are positioned in the overlap between the three circles of capital endowments, as illustrated in Figure 7.3.2. We can therefore conclude that through the diversity of approaches that constitute the CBM methodology and practices, these experiences contribute to community resilience.

Community resilience, the CBM methodology and its practices

In the conclusion of this chapter and the book, the experiences and emerging insights on poverty alleviation, livelihood development, empowerment and conservation of agrobiodiversity shape CBM as a methodology that promotes community resilience. CBM is introduced by Shrestha *et al.* (Chapter 1.3) as a methodology for contributing to the conservation and sustainable utilization of biodiversity at local level, with an emphasis on agrobiodiversity or plant genetic resources, where it focuses on enhancing the capacity of communities to make collective decisions and purposefully implement the practices that contribute to securing access to and control over their agrobiodiversity for sustainable livelihood development. Moreover, De Boef and Thijssen (Chapter 1.8) describe CBM as a situation in which on-farm management of agrobiodiversity is achieved.

Here, in this chapter, we can formulate a similar argumentation where CBM as a situation promotes community resilience. Again, CBM can be approached as a methodology and a situation that describes what people or communities do. CBM is thus a process but also an end-point, where CBM results in a resilient community. The practices that are elements of the CBM methodology can be single actions that are geared towards raising awareness of conservation and diversity, enhancing understanding, building capabilities within community-based organizations, and encouraging communities to make informed decisions over their agrobiodiversity in times of change. CBM incorporates such multi-year processes for value addition and market development, and participatory crop improvement. Following this logic, new features can be incorporated in Figure 7.2.1 that illustrate the relationship between CBM as a methodology and a situation with empowerment and *in situ* conservation. In Figure 7.3.3 we incorporate community resilience in this conceptual framework.

A critical assumption remains that for CBM as a methodology to reach CBM as a situation it should target empowerment. However, when such a situation is reached, community resilience and on-farm management emerge as properties of that situation. Linking these outcomes with the discussion on genetic resource policies and rights (Louwaars *et al.*, Chapter 6.1), farmers' rights also emerge as a property of

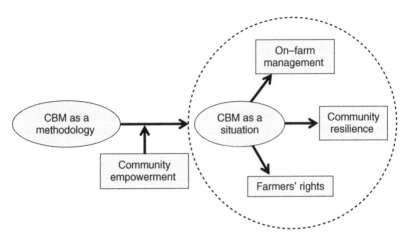

Figure 7.3.3 Visualization of the assumptions underlying the relationship between community biodiversity management (CBM), empowerment, on-farm management, community resilience and farmers' rights.

CBM as a situation. In this way, we meet the three objectives of both the Convention on Biological Diversity (CBD) and the International Treaty on Plant Genetic Resources for Food and Agriculture (ITPGRFA).

When we first decided to compile this book with case studies and insights, our objective was that of assessing to what degree we, as conservation and development professionals and our organizations, have advanced in achieving those three objectives related to the conservation and use of agrobiodiversity, and the fair and equitable sharing of benefits arising from the use of agrobiodiversity. By incorporating farmers' rights in this conceptual framework, and linking them to a situation where community resilience and on-farm management are promoted, we have been able to link the three objectives of the CBD and the ITPGRFA to the current challenges of climate and social–environmental changes.

A comparative advantage of adding community resilience to this conceptual framework is that we can benefit from advances that have been made over the past two decades in the field of agrobiodiversity management and livelihood development in support of farming communities. As we have learned from contributing to on-farm management and promoting farmers' rights, we do not need to start from scratch when approaching the subject of community resilience. We can build upon a body of knowledge and experiences, as well as insights, for strengthening farming communities in their livelihood pathways to be more resilient to the unpredictable challenges they, and we as a global community, face.

References

Abay, F. and Bjørnstad, A. (2008) 'Specific adaptation of barley varieties in different locations in Ethiopia', *Euphytica*, vol 167, no 2, pp181–195

Abay, A., Waters-Bayer, A. and Bjørnstad, Å. (2008) 'Farmers' seed management and innovation in varietal selection: implications for barley breeding in Tigray, Northern Ethiopia', *AMBIO*, vol 37, no 4, pp312–320

Abay, F., De Boef, W.S. and Bjørnstad, A. (2011) 'Network analysis of barley seed flows in Tigray, Ethiopia: supporting the design of strategies that contribute to on-farm management of plant genetic resources', *Plant Genetic Resources*, vol 9, no 4, pp495–505

Abrasem (2008) 'Semente: inovação tecnológica', *Anuário 2008*, Associação Brasileira de Sementes e Mudas, Brasília, DF, Brazil

Adams, W.M., Aveling, R., Brockington, D., Dickson, B., Elliott, J., Hutton, J., Roe, D., Vira, B. and Wolmer, W. (2004) 'Biodiversity conservation and the eradication of poverty', *Science*, vol 306, pp1146–1149

African Union (2011) *Communiqué on Integrated Seed Sector Development*. African Union, Addis Ababa, Ethiopia

AgroBio Périgord (2010) *L'Aquitaine cultive la biodiversité: expérimentations semences biologiques de populations en Aquitaine, edition 2009*. AgroBio Périgord, Périgueux, France

Aguilar-Espinoza, E. (2007) 'Evaluating a participatory plant breeding approach in a local seed system of northern Nicaragua: challenges for upscaling technologies and institutions', MSc thesis, University of Life Sciences, Ås, Norway

Albuquerque, U.P. (2010) 'Etnobotânica aplicada à conservação da biodiversidade', in U.P. Albuquerque, R.F.P. Lucena and L.V.F.C.C. Cunha (eds) *Métodos e técnicas na pesquisa etnobiológica e etnoecológica*, NUPEEA, Recife-PE, Brazil, pp351–364

Albuquerque, U.P. and Andrade, L.H.C. (2002) 'Uso dos recursos vegetais da caatinga: o caso do agreste do Estado de Pernambuco (Nordeste do Brasil)', *Interciencia*, vol 27, no 7, pp336–345

Albuquerque, U.P., Nascimento, L.G.S., Vieira, F.J., Almeida, C.M.A. and Silva, A.C.O. (2010) '"Return" and extension actions after ethnobotanical research: the perceptions and expectations of a rural community in semi-arid Northeastern Brazil', *Journal of Agricultural and Environmental Ethics*, vol 25, no 1, pp19–32

Alcorn, J. (1995) 'The scope and aims of ethnobotany in a developing world', in R.E. Schultes and S.V. Reis (eds) *Ethnobotany*, Dioscorides Press, Portland, OR, USA, pp23–39

Alexiades, M.N. (2003) 'Ethnobotany in the third millennium: expectations and unresolved issues', *Delpinoa*, vol 45, pp15–28

Allaire, G. (2002) 'L'économie de la qualité, en ses secteurs, ses territoires et ses mythes', *Géographie, Économie, Société*, vol 4, no 2, pp155–180

Almeida, P. and Cordeiro, A. (2002) *Semente da paixão: estratégia comunitária de conservação de variedades locais no semi-árido*. AS-PTA, Rio de Janeiro-RJ, Brazil

Almekinders, C.J.M. and De Boef, W.S. (2000) *Encouraging Diversity: conservation and development of plant genetic resources.* Intermediate Technology, London

Almekinders, C.J.M. and Elings, E. (2001) 'Collaboration of farmers and breeders: participatory crop improvement in perspective', *Euphytica,* vol 122, pp425–438

Almekinders, C.J.M. and Louwaars, N.P. (1999) *Farmers' Seed Production: new approaches and practices.* Intermediate Technology, London

Almekinders, C., Hardon, J., Christinck, A., Humphries, S., Pelegrina, D., Sthapit, B., Vernooy, R., Visser, B. and Weltzien, E. (eds) (2006) *Bringing Farmers Back into breeding: experiences with participatory plant breeding and challenges for institutionalization, AgroSpecial 5.* Agromisa, Wageningen, the Netherlands, http://web.idrc.ca/uploads/user-S/11798591651Bringing_farmers_back.pdf, accessed 10 May 2012

Andersen, R. (2005a) 'Results from an international stakeholder survey on farmers' rights', *FNI Report 9/2005,* the Fridtjof Nansen Institute, Lysaker, Norway, http://www.fni.no/doc&pdf/FNI-R0905.pdf, accessed 10 May 2012

Andersen, R. (2005b) 'The history of farmers' rights: a guide to central documents and literature', *FNI-Report 8/2005,* the Fridtjof Nansen Institute, Lysaker, Norway, http://www.fni.no/doc&pdf/FNI-R0805.pdf, accessed 10 May 2012

Andersen, R. (2006) 'Realising farmers' rights under the International Treaty on Plant Genetic Resources for Food and Agriculture, summary of findings from the farmers' rights project (phase 1)', *FNI Report 11/2006,* the Fridtjof Nansen Institute, Lysaker, Norway, http://www.fni.no/doc&pdf/FNI-R1106.pdf, accessed 10 May 2012

Andersen, R. (2008) *Governing Agrobiodiversity: Plant Genetics and Developing Countries.* Ashgate, Aldershot, UK

Andersen, R. (2011) 'Plantemangfold i jordbruket og bonders rettigheter i Norge' ('Farmers' rights in Norway – a case study'), *FNI Report 11/2011,* the Fridtjof Nansen Institute, Lysaker, Norway, http://www.fni.no/doc&pdf/FNI-R1111.pdf, accessed 9 May 2012

Andersen, R. and Winge T. (2008) 'Success stories from the realization of farmers' rights related to plant genetic resources for food and agriculture', *FNI Report 4/2008,* the Fridtjof Nansen Institute, Lysaker, Norway, http://www.fni.no/doc&pdf/FNI-R0408.pdf, accessed 9 May 2012

Andersen, R. and Winge T. (2011) 'Global consultations on farmers' rights in 2010', *FNI Report 1/2011,* the Fridtjof Nansen Institute, Lysaker, Norway, http://www.farmersrights.org/pdf/FNI%20Report%201–2011%20Farmers%20Rights.pdf, accessed 9 May 2012

Arunachalam, V. (2000) 'Participatory conservation: a means of encouraging community biodiversity', *Plant Genetic Resources Newsletter,* vol 122, pp1–6

Arunachalam, V. (ed.) (2007) 'Participatory plant breeding and knowledge management for strengthening rural livelihoods', *Papers presented in an international symposium held at M.S. Swaminathan Research Foundation, Chennai, TM, India,* http://www.mssrf.org/bd/bd-pub/PPB-book.pdf, accessed 6 February 2012

Arunachalam, V., Chaudhury, S.S., Sarangi, S.K., Ray, T., Mohanty, B.P., Nambi, V.A. and Mishra, S. (2006) *Rising on Rice. The story of Jeypore,* M.S. Swaminathan Research Foundation, Chennai, TM, India, http://www.mssrf.org/bd/bd-pub/rising%200n%20rice_booklet.pdf, accessed 6 February 2012

Assis, A.L., Devkota, R., Ramesh, V., Subedi, A., Swain, S. and Zank, S. (2010a) *Global study on CBM and empowerment – Brazil exchange report,* Wageningen University and Research Centre/Centre for Development Innovation, Wageningen, the Netherlands, http://www.cdi.wur.nl/NR/rdonlyres/8A95C897-C42C-4A57-9DDB-5EECFF725066/156964/CBM_Exchange_Brazildef.pdf, accessed 30 May 2012

Assis, A.L., Hanazaki, N., Reis, M.S., Mattos, A.G. and Peroni, N. (2010b) 'Espécie-chave cultural: indicadores e aplicabilidade em etnoecologia', in A.G.C. Alves, F.J.B. Souto and N. Peroni (eds) *Etnoecologia em perspectiva: natureza, cultura e conservação,* NUPEEA, Recife, PE, Brazil, pp145–153

Bala Ravi, S. and Parida, A. (2007) 'Recognition and reward to the tribal and farming communities for conservation of agro-biodiversity', *Current Science,* vol 92, no 5, pp581–584

Balcha, G. and Tanto, T. (2008) 'Conservation of genetic diversity and supporting informal seed supply in Ethiopia', in M.H. Thijssen, Z. Bishaw, A. Beshir and W.S. de Boef (eds) *Farmers' Seeds and Varieties: supporting informal seed supply in Ethiopia.* Wageningen International, Wageningen, the Netherlands, pp141–149, http://edepot.wur.nl/18448, accessed 6 February 2012

Balée, W. (1989) 'The culture of Amazonian forests', *Advances in Economic Botany,* vol 7, pp63–71

Balée, W. (2006) 'The research program of historical ecology', *Annual Review of Anthropology,* vol 35, no 1, pp75–98

Bänziger, M. and De Meyer, J. (2002) 'Collaborative maize variety development for stress-prone environments in southern Africa', in D.A. Cleveland and D. Soleri (eds) *Farmers, Scientists and Plant Breeding: integrating knowledge and practice.* CABI, Wallingford, UK, pp269–296

Barbieri, R. and Tempel, A.C. (2012) 'Description for Tavares, Brazil', in M.H. Thijssen, W.S. de Boef and G. Borman (eds) *Community Biodiversity Management and Empowerment: documentation of the sites included in a global study.* Wageningen UR/CDI, Wageningen, the Netherlands, pp19–21, http://www.cdi.wur.nl/NR/rdonlyres/8A95C897-C42C-4A57–9DDB-5EECFF725066/157080/Tavares.pdf, accessed 17 February 2012

Bartlett, A. (2008) 'No more adoption rates! looking for empowerment in agricultural development programmes', *Development in Practice,* vol 18, nos 4–5, pp524–538

Bellon, M.R. (2006) 'Crop research to benefit poor farmers in marginal areas of the developing world: a review of technical challenges and tools', *CAB Review: Perspectives in Agriculture, Veterinary Science, Nutrition and Natural Resources,* vol 2006, no 1:70, http://www.bioversityinternational.org/fileadmin/bioversity/news/documents/Bellon.pdf, accessed 5 February 2012

Berkes, F., Folke, C. and Gadgill, M. (1995) 'Traditional ecological knowledge, biodiversity, resilience and sustainability', in C. Perrings, K.G. Mäler, C. Folke, C.S. Holling and B.O. Jansson (eds) *Bioversity Conservation: policy issues and options.* Kluwer, Dordrecht, the Netherlands, pp281–299

Berkes, F., Folke, C. and Colding, J. (eds) (1998) *Linking Social and Ecological Systems: management practices and social mechanisms.* Cambridge University Press, Cambridge, UK

Bezabih, M. (2008) 'Agrobiodiversity conservation under an imperfect seed system: the role of community seed banking schemes', *Agricultural Economics,* vol 38, pp77–87

Bhag Mal, Padulosi, S. and Bala Ravi, S. (eds) (2010) *Minor millets in South Asia: learnings from IFAD-NUS project in India and Nepal.* Bioversity International, Maccarese, Rome, Italy; and the M.S. Swaminathan Research Foundation, Chennai, India, http://www.bioversityinternational.org/fileadmin/bioversity/publications/pdfs/1407_Minor%20millets%20in%20South%20Asia%20learnings%20from%20IFAD-NUS%20project%20in%20India%20and%20Nepal.pdf, accessed 14 February 2012

Bio d'Aquitaine/AgroBio Périgord (2011) *L'Aquitaine cultive la biodiversité 2001–2011: 10 ans d'expériences et d'expérimentation sur les variétés paysannes de maïs et de tournesol.* AgroBio Périgord, Périgueux, France, http://www.agrobioperigord.fr/upload/biodiv/rapport10ans-semences.pdf, accessed 10 May 2012

Bishaw, Z., Sahlu, Y. and Simane, B. (2008) 'The status of the Ethiopian seed industry' in M.H. Thijssen, Z. Bishaw, A. Beshir and W.S. de Boef (eds) *Farmers, Seeds and Varieties: supporting informal seed supply in Ethiopia,* Wageningen International, Wageningen, the Netherlands, pp23–33, http://edepot.wur.nl/18448, accessed 7 February 2012

Bitencourt, A.L.V. and Krauspenhar, P.M. (2006) 'Possible prehistoric anthropogenic effect on *Araucaria angustifolia* (Bert.) Kuntze expansion during the late Holocene', *Revista Brasileira de Paleontologia,* vol 9, no 1, pp109–116

Bonneuil, C. and Thomas, F. (2009) *Gènes, pouvoir et profits: recherche publique et régimes de production des savoirs de Mendel aux OGM,* Éditions Quae, Versailles, France

Bontempo, P., Mita, L., Miceli, M., Doto, A., Nebbioso, A., De Bellis, F., Conte, M.,

Minichiello, A., Manzo, F., Carafa, V., Basile, A., Rigano, D., Sorbo, S., Castaldo Cobianchi, R., Schiavone, E. M., Ferrara, F., De Simone, M., Vietri, M., Cioffi, M., Sica, V., Bresciani, F., De Lera, A.R., Altucci, L. and Molinari, A.M. (2007) '*Feijoa sellowiana* derived natural Flavone exerts anti-cancer action displaying HDAC inhibitory activities', *International Journal of Biochemistry and Cell Biology*, vol 39, no 10, pp1902–1910

Bordieu, P. (1987) 'What makes a social class? On the theoretical and practical existence of groups', *Berkeley Journal of Sociology*, vol 59, pp20–28

Borrini-Feyerabend, G., Pimbert, M., Taghi Farvar, M. Kothari A. and Renard, Y. (2007) *Sharing Power: a global guide to collaborative management of natural resources*. Earthscan, London

Brazil (1997) *Lei de proteção de cultivares*, Presidência da República, Casa Civil, Brasília, DF, Brazil, http://www.planalto.gov.br/ccivil_03/leis/L9456.htm, accessed 19 February 2012

Brazil (2003) *Lei de sementes*, Presidência da República, Casa Civil, Brasília, DF, Brazil, http://www.planalto.gov.br/ccivil_03/leis/2003/L10.711.htm, accessed 19 February 2012

Brazil (2007) *Decreto No 6040; Política Nacional de Desenvolvimento Sustentável dos Povos e Comunidades Tradicionais, Presidência da República*, Casa Civil, Brasília, DF, Brazil, http://www.planalto.gov.br/ccivil_03/_at02007–2010/2007/decreto/d6040.htm, accessed 22 February 2012

Brush, S.B. (2004) *Farmers' Bounty: the survival of crop diversity in the modern world*. Yale University Press, New Haven, CT, USA

Byerlee, D., De Janvry, A., Sadoulet, E., Townsend R. and Klytchnikova, I. (2007) *World Development Report, 2008: agriculture for development*. World Bank, Washington, DC, http://siteresources.worldbank.org/INTWDR2008/Resources/WDR_00_book.pdf, accessed 23 February 2012

Caillon, S., Quero-Garcia, J., Lescure, J.P. and Lebot, V. (2006) 'Nature of taro (*Colocasia esculenta* (L.) Schott) genetic diversity prevalent in a Pacific Ocean Island, Vanua Lava, Vanuatu', *Genetic Resources and Crop Evolution*, vol 53, no 6, pp1273–1289

Camus, P. and Lebot, V. (2010) 'On–farm assessment of clonal introduction of root crops diversity in Vanuatu, Melanesia', *Experimental Agriculture,* vol 46, pp541–559

Canci, A., Alves, A.C. and Guadagnin, C.A. (eds) (2010) *Kit diversidade: estratégias para a segurança alimentar e valorização das sementes locais*. Instituto de Agrobiodiversidade e Desenvolvimento Socioambiental, Guaraciaba, SC, Brazil

Canci, I.J. (2006) 'Relações dos Sistemas Informais de Conhecimento no Manejo da Agrobiodiversidade no Oeste de Santa Catarina', Masters thesis, Federal University of Santa Catarina, Florianópolis, SC, Brazil

Carney, D. (ed,) (1998) *Sustainable Rural Livelihoods: What contribution can we make?* Department for International Development, London

Carraro, I. (2005) *A empresa de sementes no ambiente de proteção de cultivares no Brasil*, PhD thesis, Federal University of Pelotas, Pelotas, RS, Brazil, http://www.ufpel.edu.br/tede/tde_arquivos/2/TDE-2005–06–14T06:50:42Z-18/Publico/tese_ivo_marcos_carraro.pdf, accessed 23 February 2012

Cavalcanti, N.B., Lima, J.B., Resende, G.M. and Brito, L.T.L. (2000) 'Ciclo reprodutivo do umbuzeiro (Spondias tuberosa Arruda) no semiárido do nordeste brasileiro', *Revista Ceres*, vol 47, no 272, pp421–439

CBD (Convention on Biological Diversity) (1992a) *The Convention on Biological Diversity*. http://www.cbd.int/convention/text/, accessed 16 February 2012

CBD (1992b) 'Article 8(j) – Traditional knowledge, innovations and practices', Secretariat of the Convention on Biological Diversity, Montreal, Canada, http://www.cbd.int/traditional/, accessed 11 January 2012

Ceccarelli, S. (1994) 'Specific adaptation and breeding for marginal conditions', *Euphytica*, vol 77, pp205–219

Ceccarelli, S., Erskine, W., Hamblin, J. and Grando, S. (1994) 'Genotype by environment interaction and international breeding programs', *Experimental Agriculture*, vol 30, pp177–187

Ceccarelli, S., Guimarães, E.P. and Weltzien, E. (eds) (2009) *Plant Breeding and Farmer Participation*. FAO, Rome, Italy, http://www.fao.org/docrep/012/i1070e/i1070e00.htm, accessed 5 February 2012

Chambers, R. (1993) *Challenging the Professions; frontiers for rural development*. Intermediate Technology, London

Chambers, R., Pacey, A. and Thrupp, L.A. (1989) *Farmer First: farmer innovation and agricultural research*. IT Publications, London

Chaudhary, P., Gauchan, D., Rana, R.B., Sthapit, B.R. and Jarvis, D. (2004) 'Potential loss of rice landraces from a Terai community in Nepal: a case study from Kachorwa, Bara', *Plant Genetic Resources Newsletter*, vol 137, pp14–21. FAO-Bioversity, Rome, Italy, http://www2. bioversityinternational.org/publications/pgrnewsletter/article.asp?id_article=3&id_ issue=137, accessed 17 June 2012

Chaudhury, S.S., Ray, T., Sarangi, S.K., Mohanty, B.P., Nambi, V.A. and Arunachalam, V. (2007) *Jeypore Kalajeera: Orissa's gift to quality rice lovers*. M.S. Swaminathan Research Foundation, Chennai, TN, India, http://www.mssrf.org/bd/bd-pub/Kalajeera%20Booklet.pdf, accessed 14 February 2012

Chevalier, J.M. and Buckles, D.J. (2011) *A Handbook for Participatory Action Research, Planning and Evaluation*. SAS2 Dialogue, Ottawa, Canada, http://www.participatoryactionresearch. net/sites/default/files/sites/all/files/manager/pdf/sas2_module4_sept11_red_en.pdf, accessed 11 January 2012

Cifuentes Jara, M. (2010) 'ABC of climate change in Mesoamerica', Technical report no. 383. Climate Change Programme, CATIE, Turrialba, Costa Rica

CIP and FEDECH (2006) *Catálogo de variedades de papa nativa de Huancavelica-Perú*. Centro Internacional de la Papa and la Federación Departamental de Comunidades Campesinas, Lima, Peru, http://cipotato.org/publications/pdf/003524.pdf, accessed 13 February 2012

Clement, C. (1999) '1492 and the loss of Amazonian crop genetic resources: I The relation between domestication and human population decline', *Economic Botany*, vol 53, no 2, pp188–202

Clement, C.R., De Cristo-Araújo, M.D., Eeckenbrugge, G.C., Alves Pereira, A. and Picanço-Rodrigues, D. (2010) 'Origin and domestication of native Amazonian crops', *Diversity*, vol 2010, no 2, pp72–106

Coelho-de-Souza, G. (2010) 'Modernização da agricultura e o agravamento da insegurança alimentar no Brasil: o papel das populações locais e sua agrobiodiversidade', in V.A. da Silva, A.S de Almeida and U.P. de Albuquerque (eds) *Etnobiologia e etnoecologia: pessoas e natureza na América Latina*, NUPEEA, Recife-PE, Brazil, pp65–85

Comissão Pró-Indio de São Paulo (2011) *Comunidades Quilombolas no Brasil*. www.cpisp.org.br/, accessed 21 November 2011

Cooper, D., Vellvé, R. and Hobbelink, H. (1992) *Growing Diversity: genetic resources and local food security*. Intermediate Technology, London

Cordeiro, A. (2007) *Documentação participativa do PAA: Intervenções governamentais em segurança alimentar e nutricional com geração de renda e valorização da biodiversidade–resultados do PAA–Programa de Aquisição de Alimentos da agricultura familiar*. CONAB, Brasília-DF, Brazil

Cordeiro, A., Alves, A.C. and Ogliari, J.B. (2008) 'Challenges for co-existence in small-scale farming: the case of maize in Brazil', in B. Breckling, H. Reuter and R. Verhoeven (eds) *Implications of GM-Crop Cultivation at Large Spatial Scales*, Peter Lang, Frankfurt, Germany, pp134–140

Correa, C. and von der Weid, J.M. (2007) 'Variedades crioulas na lei de sementes: avanços e impasses', *Agriculturas*, vol 3, no 1, pp11–14

Cumming, G.S., Cumming, D.H. and Redman, C.L. (2006) 'Scale mismatches in social-ecological systems: causes, consequences, and solutions', *Ecology and Society*, vol 11, no 1. http:// www.ecologyandsociety.org/vol11/iss1/art14/, accessed 23 March 2012

Cunha, M. C. (1999) 'Populações tradicionais e a conservação da biodiversidade', *Estudos*

Avançados, vol 13, no 36, http://www.scielo.br/scielo.php?pid=S0103–4014199900020000
8&script=sci_arttext#back, accessed 19 February 2012

Cunha, M.C. (2009) *Cultura com aspas e outros ensaios*. Cosac Naify, São Paulo-SP, Brazil

Cunningham, A.B. (2001) *Applied Ethnobotany: people, wild plant use and conservation*. Earthscan, London

Cutter, S.L., Barnes, L., Berry, M., Burton, C., Evans, E., Tate, E. and Webb, J. (2008) 'A place-based model for understanding community resilience to natural disasters', *Global Environmental Change*, vol 18, pp598–606

Dao The Anh, Vu Trong Binh and Lê Duc Thinh (2003) 'Changes in food production', in P. Moustier, D.T. Anh and M. Figuié (eds) *Food Markets and Agricultural Development in Vietnam*. The Gioi Publishers, Hanoi, Vietnam, pp48–67

De Boef, W.S. (2000) 'Tales of the Unpredictable: learning about institutional frameworks that support farmer management of agro-biodiversity', PhD thesis, Wageningen University, Wageningen, the Netherlands

De Boef, W.S. and Ogliari, J.B. (2007) 'Seleção de variedades e melhoramento genético participativo', in W.S. de Boef, M.H. Thijssen, J.B. Ogliari and B.R. Sthapit (eds) *Biodiversidade e agricultures: fortalecendo o manejo comunitario*. L&PM, Porto Alegre, RS, Brazil

De Boef, W.S. and Thijssen, M.H. (2007) *Participatory Tools Working with Crops, Varieties and Seeds*. Wageningen International, Wageningen, the Netherlands, http://edepot.wur.nl/194067, accessed 17 February 2012

De Boef, W.S., Dempewolf, H., Byakweli, J.M. and Engels, J.M.M. (2010) 'Integrating genetic resource conservation and sustainable development into strategies to increase the robustness of seed systems', *Journal of Sustainable Agriculture*, vol 34, pp504–531

De Boef, W.S., Thijssen M.H., Shrestha, P., Subedi, A., Feyissa, R., Gezu, G., Canci, C., De Fonseca Ferreira, M.A.J., Dias, T., Swain, S. and Sthapit, B.R. (2012) 'Moving beyond the dilemma: practices that contribute to the on-farm management of agrobiodiversity', *Journal of Sustainable Agriculture*, vol 36, pp788–809

De Jonge, B. (2011) 'What is fair equitable benefit-sharing?' *Journal of Agricultural and Environmental Ethics*, vol 24, pp127–146

Demeulenaere, E. (2008) 'De la rehabilitation des variétés anciennes à la pratique collective d'une sélection paysanne', in H. Zaharia (ed.) *Voyage autour des blés paysans*, Réseau Semences Paysannes, Brens, France, pp4–13

Demeulenaere, E., Bonneuil, C. Balfourier, F., Basson, A., Berthellot, J.-F., Chesneau, V., Ferté, H., Galic, N., Kastler, G., Koenig, J., Mercier, F. Payement, J. Pommart, A., Ronnot, B., Rossele, Y., Supiot, N., Zahariaria, H. and Goldringer, I. (2008) Étude des complémentarités entre gestion dynamique à la ferme et gestion statique en collection: cas de la variété de blé Rouge de Bordeaux, *Actes du colloque Bureau de Ressources Génétiques*, vol 7, pp 117–138, http://hal.archives-ouvertes.fr/docs/00/45/95/72/PDF/Demeulenaere-et-al-ActesBRG2008.pdf, accessed 14 February 2012

DFID (UK Department for International Development) (1999) Sustainable livelihood sheets, Eldis Department Store, Institute of Development Studies, Brighton, UK, http://www.eldis.org/go/topics/dossiers/livelihoods-connect&id=41731&type=Document, accessed 16 August 2012

Dias, T.A.B., Zarur, S.B.B., Alves, R.B.N., Costa, I.R.S. and Bustamante, P.G. (2007) 'Etnobiologia e conservação de recursos genéticos, o caso do povo Craô, Brasil', in L.L. Nass (ed.) *Recursos genéticos vegetais*. Embrapa CENARGEN, Brasília, DF, Brazil, pp651–681

Diegues, A.C. (1999) *Biodiversidade e comunidades tradicionais no Brasil*, NUPAUB-USP, São Paulo-SP, Brazil

Diegues, A.C. (2000) 'Saberes tradicionais e etnoconservação' in A.C Diegues and V.M. Viana (eds) *Comunidades tradicionais e manejo dos recursos naturais da Mata Atlântica*, Hucitec, São Paulo, SP, Brazil, pp9–22

Douglas, J. (1980) *Successful Seed Programs: a planning and management guide*. Westview Press, Boulder, CO, USA

Ducroquet, J.P.H.J., Hickel, E.R. and Nodari, R.O. (2000) 'Goiabeira serrana (*Feijoa sellowiana*)', *Série Frutas Nativas*, No 5, FUNEP, Jaboticabal, SP, Brazil

EC (2008) *Council Directive 2008/62/EC*, Commission of the European Union, Brussels, Belgium, http://eur-lex.europa.eu/LexUriServ/LexUriServ.do?uri=OJ:L:2008:162:0013:0019:EN: PDF, accessed 21 February 2012

EC (2009) *Council Directive 2009/145/EC*, Commission of the European Union, Brussels, Belgium, http://eur-lex.europa.eu/LexUriServ/LexUriServ.do?uri=OJ:L:2009:312:0044:0054:EN: PDF, accessed 21 February 2012

Edwards, S. (1991) 'Crops with wild relatives found in Ethiopia', in J.M.M. Engels, G. Hawkes and M. Worede (eds) *Plant Genetic Resources of Ethiopia*, Cambridge University Press, New York, pp43–72

Engels, J. and Hawkes, J.G. (1991) 'The Ethiopian gene centre and its genetic diversity', in J.M. Engels, G. Hawkes and M. Worede (eds) *Plant Genetic Resources of Ethiopia*. Cambridge University Press, New York, pp21–37

Engels, J.M.M. and Visser, L. (2003) 'A guide to effective management of germplasm collections', *IPGRI Handbooks for Genebanks No. 6*. IPGRI, Rome, Italy

Engels, J.M.M., Dempewolf, H. and Henson-Apollonio, V. (2011) 'Ethical considerations in agro-biodiversity research, collecting and use', *Journal of Agricultural and Environmental Ethics*, vol 24, pp107–126

Engels, J.M.M., Polreich, S. and Dulloo, M.E. (2008) 'Role of community gene/seedbanks in the conservation and use of crop genetic resources in Ethiopia' in M.H. Thijssen, Z. Bishaw, A. Beshir and W.S. de Boef (eds) *Farmers' Seeds and Varieties: supporting informal seed supply in Ethiopia*. Wageningen International, Wageningen, The Netherlands, pp149–160, http:// edepot.wur.nl/18448, accessed 6 February 2012

Ertug Firat, A. and Tan, A. (1997) '*In situ* conservation of genetic diversity in Turkey' in N. Maxted, B.V. Ford-Lloyd and J.G. Hawkes (eds) *Plant Genetic Conservation. The* in situ *approach*. Chapman and Hall, London, pp254–262

EU (1998) *Council Directive 98/95/EC of 14 December 1998*, Commission of the European Union, Brussels, Belgium, http://faolex.fao.org/cgi-bin/faolex.exe?rec_id=014387&database=f aolex&search_type=link&table=result&lang=eng&format_name=@ERALL, accessed 21 February 2012

FAO (Food and Agriculture Organization) (1983) *International Undertaking on Plant Genetic Resources for Food and Agriculture*, http://www.fao.org/ag//CGRFA/iu.htm, accessed 16 February 2012

FAO (1989) *Farmers' Rights, Resolution 5/89, Report of the Conference of FAO, 25th Session, Rome 11–29 November, 1989*, FAO Doc. No.C89/REP. FAO, Rome, Italy

FAO (1996a) *Report on the State of the World's Plant Genetic Resources for Food and Agriculture*. Prepared for the International Technical Conference on Plant Genetic Resources Leipzig, Germany, 17–23 June, 1996, FAO, Rome, Italy, http://apps3.fao.org/wiews/docs/swrfull.pdf, accessed 20 March 2012

FAO (1996b) *Global Plan of Action for the Conservation and Sustainable Utilization of Plant Genetic Resources for Food and Agriculture*. FAO, Rome, Italy, http://typo03.fao.org/fileadmin/templates/agphome/documents/PGR/GPA/gpaeng.pdf, accessed 21 February 2012

FAO (2001) *The International Treaty on Plant Genetic Resources for Food and Agriculture*. FAO, Rome, Italy, http://www.fao.org/legal/treaties/033t-e.htm, accessed 16 February 2012

FAO (2010) *The Second Report on the State of the World's Plant Genetic Resources for Food and Agriculture*. FAO, Rome, Italy, http://www.fao.org/docrep/013/i1500e/i1500e00.htm, accessed 12 January 2012

Fernandes-Pinto, É. (2010) 'Unidades de conservação e populações tradicionais: possiblidades de contribuição da etnobotânica', in *Anais do 61° Congresso Nacional de Botânica, INPA,* Manaus, AM, Brazil, pp166–171

Feyissa, R. (2000) 'Community seed banks and seed exchange in Ethiopia: a farmer-led approach', in E. Friss-Hansen and B. Sthapit (eds) *Participatory Approaches to the Conservation and Use of Plant Genetic Resources.* International Plant Genetic Resources Institute, Rome, Italy, pp142–148, ftp://ftp.cgiar.org/ipgri/Publications/pdf/603.pdf, accessed 10 May 2012

Fisher, R.J. (1995) *Collaborative Management of Forests for Conservation and Development,* IUCN, Gland, Switzerland

Fisher, R.J., Maginnis, S., Jackson, W., Barrow, E. and Jeanrenaud, S. (eds) (2008) *Linking Conservation and Poverty Reduction: landscapes, people and power.* Earthscan, London

Fonseca Ferreira, M.A.J. and De Azevedo, S.G. (2012) 'Description of Porteirinha, Brazil', in M.H. Thijssen, W.S. de Boef and G. Borman (eds) *Community Biodiversity Management and Empowerment: documentation of the sites included in a global study.* Wageningen UR/CDI, Wageningen, the Netherlands, pp15–17, http://www.cdi.wur.nl/NR/rdonlyres/8A95C897-C42C-4A57-9DDB-5EECFF725066/157079/Porteirinha.pdf, accessed 27 February 2012

Frankel, O.H. and Soulé, M.E. (1981) *Conservation and Evolution.* Cambridge University Press, Cambridge, UK

Fraser, J.A., Junqueira, A.B., Kawa, N.C., Moraes, C.P. and Clement, C.R. (2011) 'Crop diversity on anthropogenic dark earths in central Amazonia', *Human Ecology,* vol 39, pp395–406

Freeman, W.H. and Barwale, B.R. (2010) *Seeds of Change-Growth of the Indian Seed Industry, 1961 and Beyond.* Barwale Foundation, Hyderabad, AP, India

Freire, P. (1973) *Education: the practice of freedom.* Writers and Readers Publishing Cooperative, London

Frison, C., López, F. and Esquinas-Alcázar, J.T. (eds) (2011) *Plant Genetic Resources and Food Security: stakeholder perspectives on the International Treaty on Plant Genetic Resources for Food and Agriculture.* Earthscan, London

Fukuoka, S., Suu, D.T., Ebana, K., Trinh, L.N., Nagamine, T. and Okuno, K. (2006) 'Diversity in phenotypic profiles in landrace populations of Vietnamese rice: a case study of agronomic characters for conserving crop genetic diversity on farm', *Genetic Resources and Crop Evolution,* vol 53, pp753–761

Garibaldi, A.E and Turner, N.J. (2004) 'Cultural keystone species: implications for ecological conservation and restoration', *Ecology and Society,* vol 9, no 3 (1), http://www.ecologyandsociety.org/v0l9/iss3/art1/, accessed 21 January 2012

Gauchan, D., Subedi, A., Vaidya, S.N., Upadhyay, M.P., Baniya, B.K., Rijal, D.K. and Chaudhary, P. (1999) 'National policy and its implication for agrobiodiversity conservation and utilization in Nepal', in B.R. Sthapit, M. Upadhyay and A. Subedi (eds) *A Scientific Basis of in-situ Conservation of Agrobiodiversity On-farm: Nepal's contribution to global project, Nepal working paper No. 1/99,* NARC/LI-BIRD/IPGRI, Kathmandu, Nepal, pp111–128

Gaudin, P., Zaharia, H., Delmond, F., Kastler, G., Moulène, S., Chable, V. and Kendall, J. (2009) *Variétés paysannes de maïs et tournesol pour une agriculture écologique et économe, Cahier technique.* Reseau Semences Paysanne, AgroBio Périgord and Bio d'Anquitaine, Cazalan, France

Gezu, G. and Seboka, H. (2012) 'Site description for Chefe Donsa, Ethiopia', in M.H. Thijssen, W.S. de Boef and G. Borman (eds) *Community Biodiversity Management and Empowerment: documentation of sites in a global study.* Wageningen UR/CDI, Wageningen, the Netherlands, pp23–27, http://www.cdi.wur.nl/NR/rdonlyres/8A95C897-C42C-4A57-9DDB-5EECFF725066/157081/ChefeDonsa.pdf, accessed 6 February 2012

Glowka, L., Burhenne-Guilmin, F. and Synge, H. (1996) 'A guide to the Convention on Biological Diversity', *IUCN Environmental Policy and Law Paper No 30.* Environmental Law Centre, Gland, Switzerland; and Cambridge, UK, http://data.iucn.org/dbtw-wpd/edocs/EPLP-n0.030.pdf, accessed 16 February 2012

Govindaswami, S., Krishnamurty, A. and Shastry, N.S. (1966) 'The role of introgression in the varietal variability in rice in the Jeypore tract of Orissa', *Oryza*, vol 3, no 1, pp74–85

Grisa, C.A., Job Schmitt, C., Mattei, L.F, Maluf, R.S. and Leite, S.P. (2011) 'Brazil's PAA: policy driven food systems', *Farming Matters*, vol 27, no 3, pp34–36

Gruère, G., Nagarajan, L. and King, E.D.I.O. (2009) 'The role of collective action in the marketing of underutilized plant species: lessons from a case study on minor millets in South India', *Food Policy*, vol 34, pp39–45

Guerra, M.P., Silveira, V., Reis, M.S. and Schneider, L. (2002) 'Exploração, manejo e conservação da araucária (*Araucaria angustifolia*)', in L.L. Simões and C.F. Lino (eds) *Sustentável Mata Atlântica: a exploração de seus recursos florestais*, Editora SENAC, São Paulo, SP, Brazil

Gunderson, L.H., Holling, C.S. and Light, S.S. (eds) (1995) *Barriers and Bridges to Renewal of Ecosystems and Institutions*, Columbia University Press, New York

Gyawali, S., Sthapit, B.R., Bhandari, B., Bajracharya, J., Shrestha, P.K., Upadhyay, M.P. and Jarvis, D.I. (2010) 'Participatory crop improvement and formal release of Jethobudho rice landrace in Nepal', *Euphytica*, vol 176, pp 59–78

Halewood, M. and Nnadozie, K. (2008) 'Giving priority to the commons: The International Treaty on Plant Genetic Resources for Food and Agriculture (ITPGRFA)', in G. Tansey and T. Rajotte (eds) *The Future Control of Food: a guide to international negotiations and rules on intellectual property, biodiversity and food security*. Earthscan, London, pp115–140

Hanazaki, N. (2003) 'Comunidades, conservação e manejo: o papel do conhecimento ecológico local', *Biotemas*, vol 16, pp23–47

Hanazaki, N., Gandolfo, E.S., Bender, M.G., Giraldi, M., Moura, E.A., Souza, C.S., Printes, R., Denardi, M. and Kubo, R.R. (2010) 'Conservação biológica e valorização sócio-cultural: explorando conexões entre a biodiversidade e a sociodiversidade', in A.G. Chaves Alves, F.J. Bezerra Souto and N. Peroni (eds) *Etnoecologia em perspetica natureza, cultura e conservação*, NUPEEA, Recife-PE, Brazil, pp89–102

Hardon, J.J. and De Boef, W.S. (1993) 'Linking farmers and plant breeders in local crop development', in W.S. de Boef, K. Amanor, K. Wellard and A. Bebbington (eds) *Cultivating Knowledge: genetic diversity, farmers experimentation and crop research*. Intermediate Technology, London, pp64–171

Haverkort, B., van 't Hooft, K. and W. Hiemstra (eds) (2002) *Ancient Roots, New Shoots: endogenous development in practice*. Zed Books, London

Hawkes, J.G., Maxted, N. and Ford-Lloyd, B.V. (2000) *The Ex Situ Conservation of Plant Genetic Resources*. Kluwer, Dordrecht, the Netherlands

Hegde, N.R. and Vasudeva, R. (2010) 'Ecological cost of drying uppage (*Garcinia gummi- gutta*) fruits in the Western Ghats', in R. Vasudeva, B.S. Janagoudar, B.M.C. Reddy, B.R. Sthapit and H.P. Singh (eds) *Garcinia Genetic Resources: linking diversity, livelihood and management*, College of Forestry, Sirsi, KA, India, pp1–188

Heiden, G., Barbieri, R.L. and Neitzke, R.S. (2007) 'Chave para identificação das espécies de abóboras (Cucurbita, Cucurbitaceae) cultivadas no Brasil', Documentos, 197, Embrapa Clima Temperado, Pelotas, RS, Brazil, http://www.cpact.embrapa.br/publicacoes/download/documentos/documento_197.pdf, accessed 30 January 2012

Hocdé, H., Rosas, J.C. and Araya, R. (2010) *Co-desarrollo de variedades entre agricultores, científicos y profesionales, biodiversidad y otras cosas: Enseñanzas de un programa centroamericano de gestión local de la biodiversidad y de fitomejoramiento participativo*, ISDA, Montpellier, France, 12pp

Holley, C., Gunningham, N. and Shearing, C. (2012) *The New Environmental Governance*. Earthscan, London

Hunter, D. and Heywood, D. (eds) (2011) *Crop Wild Relatives: a manual of in situ conservation*. Earthscan, London

Hunter, D.G., Iosefa, T., Delp, C.J. and Fonoti, P. (2001) 'Beyond taro leaf blight: a participatory approach for plant breeding and selection for taro improvement in Samoa', in

Proceedings of the International Symposium on Participatory Plant Breeding and Participatory Plant Genetic Resource Enhancement, CGIAR System-wide Program on Participatory Research and Gender Analysis for Technology Development and Institutional Innovation, CIAT, Cali, Colombia, pp219–227, http://www.spc.int/tarogen/Documents/Misc_Publications/PPB.pdf, accessed 7 February 2012

IBPGR (International Board for Plant Genetic Resources) (1991) *Dictionary of Plant Genetic Resources.* Elsevier, Amsterdam

ICEPA (2005) *Síntese anual da agricultura de Santa Catarina,* ICEPA, Florianópolis, SC, Brazil

ISE (International Society of Ethnobiology) (2011) *Declaration of Belém.* International Society of Ethnobiology, Bristol, VT, USA, http://ethnobiology.net/global-coalition/declaration-of-belem, accessed 21 November 2011

Ishizawa, J. (2004) 'Cosmovisions and environmental governance: the case of *in situ* conservation of native cultivated plants and their wild relatives in Peru', in *Millennium Ecosystem Assessment. Bridging Scales and Epistemologies; linking local knowledge with global science in multi-scale assessments,* Conference at the Bibliotheca Alexandrina, Egypt, 17–20 March, 2004, http://www.maweb.org/documents/bridging/papers/ishizawa.jorge.pdf, accessed 12 February 2012

Jansen, T. (2002) *Hidden Taro, Hidden Talents: a study of on-farm conservation of* Colocasia esculenta *(taro) in Solomon Islands,* Solomon Islands Planting Materials Network and Kastom Garden Association, Honiara, Solomon Islands, http://www.spc.int/tarogen/Documents/Misc_Publications/Hidden%20taro%20hidden%20talents%20PMN%20on%20farm%20May%202002.pdf, accessed 7 February 2012

Jarvis, D.I., Brown, A.H.D., Cuong, P.H., Collado-Panduro, L., Latournerie-Moreno, L., Gyawali, S., Tanto, T., Sawadogo, M., Mar, I., Sadiki, M., Hue, N.T.N., Arias-Reyes, L., Balma, D., Bajracharya, J., Castillo, F., Rijal, D., Belqadi, L., Rana, R., Saidi, S., Ouedraogo, J., Zangre, R., Rhrib, K., Chavez, J.L., Schoen, D., Sthapit, B., De Santis, P., Fadda, C., and Hodgkin, T. (2008) 'A global perspective of the richness and evenness of traditional crop variety diversity maintained by farming communities', *Proceedings of the National Academy of Sciences of the USA,* vol 108, pp 5326–5331

Jarvis, D.I., Hodgkin, T., Sthapit, B.R., Fadda, C. and Lopez-Noriega, I. (2011) 'An heuristic framework for identifying multiple ways of supporting the conservation and use of traditional crop varieties within the agricultural production system', *Critical Reviews in Plant Science,* vol 30, no 1–2, pp125–176

Johnson, L.M. and Hunn, E.S. (2010) 'Landscape ethnoecology', in L.M. Johnson and E.S. Hunn (eds) *Landscape Ethnoecology: concepts of biotic and physical space,* Berghahn Books, New York, pp1–11

Karkkainen, B. (2004) 'Reply, "new governance" in legal thought and in the world: some splitting as antidote to overzealous lumping', *Minnesota Journal of Law, Science and Technology,* vol 7, no 1, pp59–78

Kellert, S,R., Mehta, J.N., Ebbin, S.A. and Lichtenfeld, L.L. (2000) 'Community natural resource management: promise, rhetoric, and reality', *Society and Natural Resources,* vol 13, pp705–715

King, E.D.I.O., Nambi, V.N. and Nagarajan, L. (2009) 'Integrated approaches in small millet conservation. A case from Kolli Hills, India', in H. Jaenike (ed.) *Proceedings of the International Symposium on underutilized plants for food security, nutrition, income and sustainable development: Arusha, Tanzania, March 3–6, 2008, ISHS Acta Horticulturae,* vol 806, no 1, pp79–84

Kiros, H. (2011) 'Impact of participatory varietal selection in varietal diversification and seed dissemination in highlands of Tigray', MSc thesis, Mekelle University, Mekelle, Ethiopia

Kist, V. (2006) 'Seleção recorrente de famílias de meio-irmãos em população composta de milho (*Zea mays* L.) procedente de Anchieta–SC', Master thesis, Federal University of Santa Catarina, Florianópolis, SC, Brazil

Kist, V., Ogliari, J.B., Alves, A.C. and Miranda Filho, J.B. (2010) 'Genetic potential analysis of a maize population from Southern Brazil by modified convergent-divergent selection scheme', *Euphytica*, vol 176, pp25–36

Korten, D.C. (1980) 'Community organization and rural development: a learning process approach', *Public Administration Review*, vol 40, pp480–510

Korten, D.C. and Klauss, R. (1984) *People-centered Development*. Kumarian Press, West Hartford, CT, USA

Kuhnen, S., Lemos, P.M.M., Campestrani, L.H., Ogliari, J.B., Dias, P.F. and Maraschin, M. (2011) 'Carotenoid and anthocyanin contents of grains of Brazilian maize landraces', *Journal of the Science of Food and Agriculture*, vol 91, pp1548–1553

Labouisse, J.P., Bellachew, B., Kotecha, S. and Bertrand, B. (2008) 'Current status of coffee (*Coffea arabica* L.) genetic resources in Ethiopia: implications for conservation', *Genetic Resources and Crop Evolution*, vol 55, pp1079–1093

Leakey, R.R.B. and Akinnifesi, F.K. (2008) 'Towards a domestication strategy for indigenous fruit trees in the tropics', in F.K. Akinnifesi, R.R.B. Leakey, O.C. Ajayi, G. Sileshi, Z. Tchoundjeu, P. Matakala and F.R. Kwesiga (eds) *Indigenous Fruit Trees in the Tropics: domestication, utilization and commercialization*, CABI, Wallingford, UK, pp28–49

Lebot, V., Ivancic, A. and Abraham, K. (2005) 'The geographical distribution of allelic diversity, a practical means of preserving and using minor root crop genetic resources', *Experimental Agriculture*, vol 41, no 4, pp475–489

LI-BIRD Seed Unit (2011) *Annual report*, LI-BIRD, Pokhara, Nepal

Lincoln, Y.S., Lynham, S.A. and Guba, E.G. (2011) 'Paradigmatic controversies, contradictions, and emerging confluences, revisited', in N.K. Denzin and Y.S. Lincoln (eds) *The Sage Handbook of Qualitative Research*. Sage Thousand Oaks, CA, USA, pp97–128

Lins Neto, E.M.F., Peroni, N. and Albuquerque, U.P. (2010) 'Traditional knowledge and management of Umbu (*Spondias tuberosa*, *Anacardiaceae*): an endemic species from the semi–arid region of Northeastern Brazil', *Economic Botany*, vol 64, no 1, pp11–21

Lipper, L. Anderson, C.L. and Dalton, T. (2010) *Seed Trade in Rural Markets: implications for crop diversity and agricultural development*. Earthscan, London

Lockwood, M., Davidson, J., Curtis, A., Stratford, E. and Griffith, R. (2010) 'Governance principles for natural resource management', *Society and natural resources*, vol 23, no 10, pp986–1001

Londres, F. and Almeida, P. (2009) *Impacto do controle corporativo no setor de sementes sobre agricultores familiares e sistemas alternativos de distribuição: estudo de caso do Brasil*. AS-PTA, Rio de Janeiro-RJ, Brazil, http://aspta.org.br/wp-content/uploads/2011/05/Estudo-Sementes-ASPTA-WoW-AA-2009-FINAL.pdf, accessed 15 January 2012

Long, N. (2001) *Development Sociology: actor perspectives*. Routledge, London

Louwaars, N. (2006) 'Ethics watch: controls over plant genetic resources – a double-edged sword', *Nature Reviews Genetics*, vol 7, p241

Louwaars, N. (2007) 'Seeds of confusion: the impact of policies on seed systems', PhD thesis, Wageningen University, Wageningen, the Netherlands, edepot.wur.nl/121915, accessed 12 February 2012

Louwaars, N.P. and De Boef, W.S. (2012) 'Integrated seed sector development in Africa: a conceptual framework for creating coherence between practices, programs, and policies', *Journal of Crop Improvement*, vol 26, pp39–59

Louwaars, N., Dons, H., van Overwalle, G., Raven, H., Arundel, A., Eaton, D. and Nelis, A. (2009) 'Breeding business: the future of plant breeding in the light of developments in patent rights and plant breeder's rights', *CGN-Report 2009–14*, Centre for Genetic Resources, Wageningen, the Netherlands, http://edepot.wur.nl/141258, accessed 12 February 2012

Louwaars, N., Edeme, J. and De Boef, W.S. (2013) 'Integrated seed sector development: a basis for seed policy and law', *Journal of Crop Improvement* (accepted)

Löwen Sahr, C.L. and Cunha, L.A.G.O. (2005) 'Significado social e ecológico dos faxinais: reflexões acerca de uma política agrária sustentável para a região da mata com Araucária no Paraná', *Emancipação*, vol 5, no 1, pp 89–104

Mace, E.S., Mathur, P.N., Godwin, I.D., Hunter, D., Taylor, M.B., Singh, D., De Lacy, I.H. and Jackson, G.V.H. (2010) 'Development of a regional core collection (Oceania) for taro, *Colocasia esculenta* (L.) Schott, based on molecular and phenotypic characterization', in R.V. Rao, P.J. Matthews, P.B. Eyzaguirre and D. Hunter (eds) *The Global Diversity of Taro: ethnobotany and conservation*, Bioversity International, Rome, Italy, pp185–201

Maffi, L. and Woodley, E. (eds) (2010) *Biocultural Diversity Conservation: a global source book*. Earthscan, London

Magis, K. (2010) 'Community resilience: an indicator of social sustainability', *Society of Natural Resources*, vol 23, pp401–416

McMannis, C. (ed.) (2007) *Biodiversity and the Law: intellectual property, biotechnology and traditional knowledge*. Earthscan, London

MMA (Ministerio de medio ambiente) (2001) *Provisional Act no 2,186–16*, Ministerio de medio ambiente, Brasília, DF, Brazil http://www.mma.gov.br/estruturas/sbf_dpg/_arquivos/mp2186i.pdf, accessed 19 February 2012

MMA (2002a) *Lei n. 9.985, de 18 de Julho de 2000, 4ª Edição*, MMA/SBF, Brasília, DF, Brazil, http://www.planalto.gov.br/ccivil_03/leis/L9985.htm, accessed 22 February 2012

MMA (2002b) *Decreto n. 4.340, de 22 de Agosto de 2002, 4ª Edição*, MMA/SBF, Brasília, DF, Brazil, http://www.planalto.gov.br/ccivil_03/decreto/2002/d4340.htm, accessed 22 February 2012

Morris, M.L. and Bellon, M.R. (2004) 'Participatory plant breeding research: opportunities and challenges for the international crop improvement system', *Euphytica*, vol 136, pp21–35

Moustier, P., Tam, P.T.G., Anh, D.T., Binh, V.T. and Loc, N.T.T. (2010) 'The role of farmer organizations in supplying supermarkets with quality food in Vietnam', *Food Policy*, vol 35, no 1, pp69–78

MSSRF (M.S. Swaminathan Research Foundation) (2009) *'Kalajeera' a local variety to strengthen the economy of tribal communities*, M.S. Swaminathan Research Foundation, Chennai, TN, India

MSSRF (2011a) *Annual report 2010–2011*, M.S. Swaminathan Research Foundation, Chennai, TN, India

MSSRF (2011b) 'Livelihood options for tribal communities through value chain approach for rice', in *Scouting and Upscaling Livelihoods Innovations in Odisha*, ACCESS Development Services, New Delhi, India, pp70–77, http://www.accessdev.org/downloads/case_study_book.pdf, accessed 4 February 2012

Musungu, S.F. and Dutfield, G. (2003) Multilateral agreements and TRIPS-plus world: the World Intellectual Property Organisation, *TRIPS Issues Papers No 3*, The Quaker United Nations Office, Geneva, http://www.geneva.quno.info/pdf/WIPO(A4)final0304.pdf, accessed 21 November 2011

Neate, P.J.H. and Guei, R.G. (2011) *Promoting the Growth and Development of Smallholder Seed Enterprises For Food Security Crops*, FAO, Rome, Italy, http://typ03.fao.org/fileadmin/templates/agphome/documents/PGR/PubSeeds/seedpolicyguide6.pdf, accessed 7 February 2012

Ogliari, J.B. and Alves, A.C. (2007) 'Manejo e uso de variedades de milho como estratégia de conservação em Anchieta', in W.S. de Boef, M.H. Thijssen, J.B. Ogliari and B.R. Sthapit (eds) *Biodiversidade e agricultores: fortalecendo o manejo comunitário*, L&PM, Porto Alegre, RS, Brazil, pp226–234

Oliveira, F.C., Albuquerque, U.P.A., Fonseca-Kruel, V.S. and Hanazaki, N. (2009) 'Avanços nas pesquisas etnobotânicas no Brasil', *Acta Botanica Brasilica*, vol 23, no 2, pp590–605

Ostrom, E. (1990) *Governing the Commons; the evolution of institutions for collective action*, Cambridge University Press, New York

Ostrom, E. (2009), 'A general framework for analyzing sustainability of social-ecological systems', *Science,* vol 325, pp419–422

Padulosi, S., Hoeschle-Zeledon, I. and Bordoni, P. (2008) 'Minor crops and underutilized species: lessons and prospects', in N. Maxted, B.V. Ford-Lloyd, S.P. Kell, J.M. Iriondo, M.E. Dulloo and J. Turok (eds) *Crop Wild Relative Conservation and Use,* CAB International, Wallingford, UK, pp605–625

Padulosi, S., Bhag Mal, Bala Ravi, S., Gowda, J., Gowda, K.T.K., Shanthakumar, G., Yenagi, N. and Dutta, M. (2009) 'Food security and climate change: role of plant genetic resources of minor millets', *Indian Journal of Plant Genetic Resources,* vol 22, no 1, pp1–16

Park, Y.J., Dixit, A., Ma, K.-H., Kang, J.-H., Rao, V.R. and Cho, E.-G. (2005) 'On-farm conservation strategy to ensure crop genetic diversity in changing agro-ecosystems in the Republic of Korea', *Journal of Agronomy and Crop Science,* vol 191, no 6, pp401–410

Paudel, B., Shrestha, P., Tamang, B.B. and Subedi, A. (2010) 'Implementing access and benefit sharing regime in Nepal through community based biodiversity management framework', *Journal of Agriculture and Environment,* vol 11, pp148–157

Peroni, N. and Hanazaki, N. (2002) 'Current and lost diversity of cultivated varieties, especially cassava, under swidden cultivation systems in the Brazilian Atlantic Forest', *Agriculture, Ecosystems and Environment,* vol 92, pp171–183

Peroni, N. and Martins, P.S. (2000) 'Influência da dinâmica agrícola itinerante na geração de diversidade de etnovariedades cultivadas vegetativamente', *Interciencia,* vol 25, no 1, pp22–29

Pilgrim, S. and J. Pretty (2010) *Nature and Culture: rebuilding lost connection.* Earthscan, London

Pimbert, M.P. and Pretty, J.N. (1997) 'Parks, people and professionals: putting 'participation' into protected area management', in K.B. Ghimire and M.P. Pimbert (eds) *Social change and conservation; environmental politics and impacts of national parks and protected areas.* Earthscan, London, pp297–330

Pistorius, R. (1997) *Scientists, Plants and Politics: the history of the Plant Genetic Resources Movement.* IPGRI, Rome, Italy

Platten, S. and Henfrey, T. (2009) 'The cultural keystone concept: insights from ecological anthropology', *Human Ecology,* vol 37, no 4, pp491–500

Poudel, D. and Johnsen, F.H. (2009) 'Valuation of crop genetic resources in Kaski, Nepal: farmers' willingness to pay for rice landraces conservation', *Journal of Environmental Management,* vol 90, no 1, pp483–491

Poudel, D., Sthapit, B. and Shrestha, P. (2009) *Application of Social Network Analysis to Understand On Farm Agro Biodiversity Conservation: case study from Nepal.* Paper presented at the 6th Conference on Applications of Social Network Analysis 2009, University of Zurich, 27–28 August 2009, University of Zurich, Zurich, Switzerland

PPVFR Act (2001) *The Protection of Plant Variety and Farmers' Rights Act, Ministry of Agriculture, Government of India.* New Delhi, India, http://plantauthority.gov.in/pdf/PPV&FRAct2001.pdf, accessed 15 February 2012

Pretty, J.N. (1994) 'Alternative systems of inquiry for sustainable agriculture', *IDS Bulletin,* vol 25, no 2, pp37–48

Pretty, J.N. (1995) *Regenerating Agriculture: policies and practice for sustainability and self-reliance.* Earthscan, London

Pretty, J. and Smith, D. (2004) 'Social capital in biodiversity conservation and management', *Conservation Biology,* vol 18, no 3, pp631–638

Priori, D., Barbieri, R.L., Neitzke, R.S., Vasconcelos, C.S., Oliveira, C.S., Mistura, C.C. and Costa, F.A. (2010) *Acervo do banco ativo de germoplasma de Cucurbitáceas da Embrapa Clima Temperado.* Embrapa Clima Temperado, Pelotas, RS, Brazil

Rao, R.V., Matthews, P.J., Eyzaguirre, P.B. and Hunter, D. (eds) (2010) *The Global Diversity of Taro: ethnobotany and conservation,* Bioversity International, Rome, Italy, http://www.cropwildrelatives.org/fileadmin/bioversity/publications/pdfs/1402_The%20global%20diversity

%200f%20taro%20Ethnobotany%20and%20conservation%20.pdf, accessed 7 February 2012

Reijntjes, C., Haverkort, B. and Waters-Bayer, A. (1992) *Farming for the Future: an introduction to low-external-input and sustainable agriculture*. Macmillan Press, London

Reis, M.S., Peroni, N., Mariot, A., Steenbock, W., Filippon, S., Vieira da Silva, C. and Mantovani, A. (2010) 'Uso sustentável e domesticação de espécies da Floresta Ombrófila Mista', in L.C. Ming, M.C.M. Amorozo and C.W. Kffuri (eds) *Agrobiodiversidade no Brasil: experiências e caminhos da pesquisa*, NUPEEA, Recife, PE, Brazil, pp183–214

Rengalakshmi, R., Dhanapal, D., King, E.D.I.O. and Boopathy, P. (2003) 'Institutionalizing traditional seed exchange networks through community seed banks in Kolli Hills, India', in *Conservation and Sustainable Use of Agricultural Biodiversity: a source book*, CIP-UPWARD, in collaboration with GTZ, IDRC, IPGRI and SEARICE, Los Banõs, the Philippines, pp302–308

Rijal, D.K., Rana, R.B., Subedi, A. and Sthapit, B.R. (2000) 'Adding value to landraces: community-based approaches for *in-situ* conservation of plant genetic resources in Nepal', in E. Friis-Hansen and B.R. Sthapit (eds) *Participatory Approaches to the Conservation and Use of Plant Genetic Resources*. IPGRI, Rome, Italy, pp166–172

Rijal, D.K., Subedi, A., Upadhyay, M.P., Rana, R.B., Chaudhary, P., Tiwari, R.K., Sthapit, B.R. and Gauchan, D. (2003) 'Community biodiversity register: developing community based database for genetic resources and local knowledge of Nepal', in B.R. Sthapit, M.P. Upadhyay, B.K. Baniya, A. Subedi and B.K. Joshi (eds) *On farm Management of Agricultural Biodiversity in Nepal, Proceedings of a National Workshop, 24–26 April 2001, Lumle, Nepal*. IPGRI, LI-BIRD and NARC, Kathmandu, Nepal, pp28–40, http://idl-bnc.idrc.ca/dspace/bitstream/10625/27838/1/122788.pdf, accessed 11 January 2012

Rosas, J.C. (2001) 'Aplicación de metodologías participativas para el mejoramiento genético de frijol en Honduras', *Agronomía Mesoamericana*, vol 12, no 2, pp219–228

Rosas, J.C., Gallardo, O.O. and Jiménez, J. (2003) 'Mejoramiento genético del frijol común mediante enfoques participativos en Honduras', *Agronomía Mesoamericana*, vol 14, no 1, pp1–9

Rosas, J.C., Gallardo, O.O. and Jiménez, J. (2006) 'Mejoramiento de maíces criollos de Honduras mediante la aplicación de metodologías de fitomejoramiento participativo', *Agronomía Mesoamericana*, vol 17, no 3, pp375–383

Ruiz, M. and Vernooy, R. (eds) (2012) *The Custodians of Biodiversity: sharing access and benefits to genetic resources*. Earthscan, London

Salazar, R. Louwaars, N. and Visser, B. (2007) 'Protecting farmers' new varieties: new approaches to rights on collective innovations in plant genetic resources', *World Development*, vol 35, no 9, pp1015–1528

Santilli, J. (2010) 'Human-inhabited protected areas and the law: integration of local communities and protected areas in Brazilian law', *Journal of Sustainable Forestry*, vol 29, no 2–4, pp390–402

Santilli, J. (2012) *Agrobiodiversity and the Law: regulating genetic resources, food security and cultural diversity*. Earthscan, London

Santos, A.J., Corso, N.M., Martins, G. and Bittencourt, E. (2002) 'Aspectos produtivos e comerciais do pinhão no Estado do Paraná', *Revista Floresta*, vol 32, no 2, pp63–169

Santos, K.L. (2005) 'Bases genéticas de características de importância agronômica em goiabeira-serrana (*Acca sellowiana*)', Masters thesis, Federal University of Santa Catarina, Florianópolis, SC, Brazil

Santos, K.L. (2009) 'Diversidade cultural, genética e fenotípica da goiabeira-serrana (Acca sellowiana): implicações para a domesticação da espécie', PhD thesis, Federal University of Santa Catarina, Florianópolis, SC, Brazil

Santos, K.L., Peroni, N., Guries, R.P., Nodari, R.O. (2009) 'Traditional knowledge and management of Feijoa (*Acca sellowiana*) in southern Brazil', *Economic Botany*, vol 63, no 2, pp204–214

Schmidt, C.J. and Guimarães, L.A. (2008) 'O mercado institucional como instrumento para o fortalecimento da agricultura familiar de base ecológica', *Agriculturas*, Vol 5, No 2, pp7–13

Secretariat of the Convention on Biological Diversity (2011) *Nagoya Protocol on Access to Genetic Resources and the Fair and Equitable Sharing of Benefits*, Secretariat of the Convention on Biological Diversity, Montreal, Canada, Quebec, Canada, http://www.cbd.int/abs/doc/protocol/nagoya-protocol-en.pdf, accessed 16 February 2012

Seixas, C.S and Berkes, F. (2005) 'Mudanças socioecológicas nas pesca da lagoa de Ibiraquera, Brasil', in P.F. Vieira, F. Berkes and C.S. Seixas (eds) *Gestão integrada e participativa de recursos naturais – conceitos, métodos e experiências*, Secco/APED, Florianólis, SC, Brazil, pp113–146

Senbeta, F. (2006) 'Biodiversity and ecology of Afromontane rainforests with wild *Coffea arabica* L. populations in Ethiopia', *Ecology and Development Series*, vol 38, Cuvillier Verlag, Göttingen, Germany

Senbeta, F., Tesfaye, K. and Woldemariam, T. (2007) 'Matching the traditional wild coffee management systems and biosphere reserve approach for biodiversity conservation and sustainable livelihood of the local community', in E. Kelbessa and C. De Stoop (eds) *Participatory Forest Management, Biodiversity and Livelihoods in Africa, Proceedings of the International Conference, March 19–21, 2007, Addis Ababa, Ethiopia*, Government of Ethiopia, Addis Ababa, Ethiopia, pp93–102, http://www.pfmp-farmsos.org/Docs/pfm%20conference_proceeding.pdf, accessed 23 January 2012

Shrestha P., Subedi, A., Paudel, B. and Bhandari, B. (2011) *Community Seed Bank Resource Book*. LI-BIRD, Pokhara, Nepal

Shrestha, P., Subedi, A., Rijal, D.K., Singh, D., Sthapit, B.R. and Upadhyay, M.P. (2005) 'Enhancing local seed security and on-farm conservation through a community seed bank in Bara district of Nepal', in B.R. Sthapit, M.P. Upadhyay, P.K. Shrestha and D.I. Jarvis (eds) *On-farm Conservation of Agricultural Biodiversity in Nepal Vol II: Managing diversity and promoting its benefits, Proceedings of the Second National Workshop, 25–27 August 2004, Nagarkot, Nepal.* IPGRI, Rome, Italy, pp70–77

Shrestha, P., Subedi, A., Sthapit, S., Rijal, D., Gupta, S.K. and Sthapit, B.R. (2006) 'Community seed bank: reliable and effective option for agricultural biodiversity conservation', in B.R. Sthapit, P.K. Shrestha and M.P. Upadhyay (eds) *Good Practices: on-farm management of agricultural biodiversity in Nepal.* NARC, LI-BIRD, IPGRI and IDRC, Kathmandu, Nepal

Shrestha, P., Sthapit, B.R., Shrestha, P.K., Upadhyay, M.P. and Yadav M. (2008) 'Community seed banks: experiences from Nepal', in M.H. Thijssen, Z. Bishaw, A. Beshir and W.S. de Boef (eds) *Farmers, Seeds and Varieties Supporting Informal Seed Supply in Ethiopia.* Wageningen International, Wageningen, the Netherlands, pp103–108, http://edepot.wur.nl/18448, accessed 06 February 2012

Singh, D., Subedi, A. and Shrestha, P. (2006) 'Enhancing local seed security and on-farm conservation through a community seed bank in Bara district of Nepal', in R. Vernooy (ed.) *Social and Gender Analysis in Natural Resource Management: learning studies and lessons from Asia.* Sage, New Delhi, India, pp99–128

Singh, D., Hunter, D., Iosefa, T., Okpul, T., Fonoti, P. and Delp, C. (2010) 'Improving taro production in the South Pacific through breeding and selection', in R.V. Rao, P.J. Matthews, P.B. Eyzaguirre and D. Hunter (eds) *The Global Diversity of Taro: ethnobotany and conservation,* Bioversity International, Rome, Italy, pp168–184

Singh, R.P., Jayaprakash, G.K. and Sarkaria, K.K. (1995) 'Hydroxy citric acid from *Garcinia cambogia*', *Biological Memoirs*, vol 21, pp27–33

Snapp, S., Kanyama-Phiri, G., Kamanga, B., Gilbert, R. and Wellard, K. (2002) 'Farmer and researcher partnerships in Malawi: developing soil fertility technologies for the near-term and far-term', *Experimental Agriculture*, vol 38, pp411–431

Solórzano-Filho, J.A. (2001) 'Demografia, fenologia e ecologia da dispersão de sementes de

Araucaria angustifolia (Bert.) O. Kuntze (Araucariaceae), numa população relictual em Campos do Jordão', Masters thesis, University of São Paulo, São Paulo, Brazil

Song, Y. and Vernooy, R. (eds) (2010) *Seeds and Synergies: innovating rural development in China*. Practical Action Publishing, Burton on Dunsmore, UK, and IDRC, Ottawa, Canada

Song, Y., Li, J. and Vernooy, R. (2012) 'China: designing policies and laws to ensure fair and equitable benefit sharing of genetic resources and participatory plant breeding products', in M. Ruiz and R. Vernooy (eds) *The Custodians of Biodiversity: sharing access to and benefits of genetic resources*. Earthscan, London, pp94–120

Soutullo, A. (2010) 'Extent of the global network of terrestrial protected areas', *Conservation Biology*, vol 24, no 2, pp362–363

Souza, R.M.S. (2007) 'Mapeamento social dos Faxinais no Paraná', in A.W.B. Almeida and R.M.S. Souza (eds) *Terras de Faxinais*, Editora UEA, Manaus-AM, Brazil, pp29–88

Sperling, L. and Loevinsohn, M. (1996) *Using Diversity: enhancing and maintaining genetic resources on-farm, Proceedings of a Workshop held on 19–21 June 1995, New Delhi, India*. IDRC, New Delhi, India

Sperling, L., and McGuire, S. (2010) 'Understanding and strengthening informal seed markets', *Experimental Agriculture*, vol 46, no 2, pp119–136

Sperling, L., Cooper, H.D. and Remmington, T. (2008) 'Moving towards more effective seed aid', *Journal of Development Studies*, vol 44, no 4, pp586–612

Sthapit, B.R. and Jarvis, D. (2005) 'Good practices of community based on-farm management of agricultural biodiversity in Nepal: lessons learned', in B.R. Sthapit, M.P. Upadhyay, P.K. Shrestha and D.I. Jarvis (eds) *On-farm Conservation of Agricultural Biodiversity in Nepal Vol II: Managing diversity and promoting its benefits, Proceedings of the Second National Workshop, 25–27 August 2004, Nagarkot, Nepal*. IPGRI, Rome, Italy, pp91–112

Sthapit, B.R. and Rao, V.R. (2009) 'Consolidating community's role in local crop development by promoting farmer innovation to maximise the use of local crop diversity for the well being of people', *Acta Horticulturae*, vol 806, no 2, pp669–676

Sthapit, B.R., Joshi, K.D. and Witcombe, J.R. (1996) 'Farmer participatory crop improvement III: participatory plant breeding, a case study for rice in Nepal', *Experimental Agriculture*, vol 32, pp479–496

Sthapit, B.R., Shrestha, P., and Upadhyay, M.P. (eds) (2006) *Good Practices: on-farm management of agricultural biodiversity in Nepal*. NARC, LI-BIRD, IPGRI and IDC, Kathmandu, Nepal

Sthapit, B.R., Subedi, A., Shrestha, P., Chaudhary, P., Shrestha, P.K. and Upadhyay, M. (2008a) 'Practices supporting community management of farmers' varieties', in M.H. Thijssen, Z. Bishaw, A. Beshir and W.S. de Boef (eds) *Farmers, Seeds and Varieties: supporting informal seed supply in Ethiopia*. Wageningen International, Wageningen, the Netherlands, pp166–171

Sthapit, B.R., Shrestha, P.K., Subedi, A., Shrestha, P., Upadhyay M.P. and Eyzaguirre, P. (2008b) 'Mobilizing and empowering communities in biodiversity management', in M.H. Thijssen, Z. Bishaw, A. Beshir and W.S. de Boef (eds) *Farmers, Seeds and Varieties: supporting informal seed supply in Ethiopia*. Wageningen International, Wageningen, the Netherlands, pp160–165, http://portals.wi.wur.nl/files/docs/Sthapit%20et%20al,%202008b%20-%20Practices%20s upporting%20community%20management(2).pdf, accessed 15 January 2012

Subedi, A., Shrestha, P., Sthapit, B.R., Rijal, D.K., Rana, R.B., Upadhyay, M.P. and Shrestha, P.K. (2005a) 'Community biodiversity management (CBM): lessons learned from the *in situ* conservation project', in B.R. Sthapit, M.P. Upadhyay, P.K. Shrestha and D.I. Jarvis (eds) *On-farm Conservation of Agricultural Biodiversity in Nepal Vol II: Managing diversity and promoting its benefits, Proceedings of the Second National Workshop, 25–27 August 2004, Nagarkot, Nepal*. IPGRI, Rome, Italy, pp56–70

Subedi, A., Poudel, I., Regmi, B., Baral, K., Suwal, R., Rijal, D., Sthapit, B.R. and Shrestha P.K. (2005b) 'Strengthening the community biodiversity registers for agriculture, forest and wetland biodiversity management: users and livelihood perspectives', in A. Subedi, B.R. Sthapit,

M.P. Upadhay and D. Gauchan (eds) *Learning from Community Biodiversity Register in Nepal, Proceedings of National Workshop, 27–28 October 2005, Kathmandu, Nepal*. LI-BIRD, Kathmandu, Nepal, pp41–54, http://idl-bnc.idrc.ca/dspace/bitstream/10625/27838/1/122788.pdf, accessed 12 May 2012

Subedi, A., Sthapit, B.R., Shrestha, P., Gauchan, D. and Upadhyay, M.P. (2005c) 'Emerging methodology of community biodiversity registers: a synthesis', in A. Subedi, B.R. Sthapit, M.P. Upadhyay and D. Gauchan (eds) *Learning from Community Biodiversity Register in Nepal. Proceedings of the National Workshop, 27–28 Oct 2005, Khumaltar, Nepal*. NARC, Kathmandu, Nepal, pp75–83, http://idl-bnc.idrc.ca/dspace/bitstream/10625/27838/1/122788.pdf, accessed 11 January 2012

Sundar, N. (2000) 'Unpacking the "joint" in joint forest management', *Development and Change*, vol 31, pp255–279

Swaminathan, M.S. (2003) 'Towards an evergreen revolution in agriculture: technology, planning and management', *Rites Journal*, vol 7, pp1–14

Tanto, T. and Balcha, G. (2003) *Terminal Report on 'A Dynamic Farmer-based Approach to the Conservation of Ethiopia's Plant Genetic Resources Project*. IBC and UNDP, Addis Ababa, Ethiopia

Tapia, M.E. and Rosa, A. (1993) 'Seed fairs in the Andes: a strategy for local conservation of plant genetic resources', in W.S. de Boef, K. Amanor, K. Wellard and A. Bebbington (eds) *Cultivating Knowledge: genetic diversity, farmer experimentation and crop research*. Intermediate Technology, London, UK, pp111–118

Taylor, M., Hunter, D., Rao, V.R., Jackson, G.V.H., Sivan, P. and Guarino, L. (2010), 'Taro collecting and conservation in the Pacific region', in R.V. Rao, P.J. Matthews, P.B. Eyzaguirre and D. Hunter (eds) *The Global Diversity of Taro: ethnobotany and conservation*, Bioversity International, Rome, Italy, pp150–167

Tedla, B. (2012) 'LSB–local entrepreneurship', *LSB Newsletter*, vol 10, pp5–6

Teixeira, C. (2005) 'O desenvolvimento sustentável em unidade de conservação: a "naturalização" do social' *Revista Brasileira de Ciências Sociais*, vol 20, no 59, pp51–66

Teketay, D. and Tigneh, A. (1994) 'A study on landraces of Harer coffee in Eastern Ethiopia', in J.H. Seyani and A.C. Chikuni (eds) *Proceedings of the 13th Plenary Meeting of AETFAT, 2–11 April 1991, Zomba, Malawi*, Malawi National Herbarium and Botanic Gardens, Zomba, Malawi, pp161–169

Teklu, Y. and Hammer, K. (2006) 'Farmers' perception and genetic erosion of tetraploid wheat landraces in Ethiopia', *Genetic Resources and Crop Evolution*, vol 53, pp1099–1113

Tesfaye, K. (2006) 'Genetic Diversity of wild *Coffea arabica* populations in Ethiopia as a contribution to conservation and use planning', *Ecology and Development Series*, vol 44, University of Bonn, Bonn, Germany

Teshome, A., Fahrig, L., Torrance, J.K., Lambert, J.D., Arnason, T.J. and Baum, B.R. (1999) 'Maintenance of sorghum (*Sorghum bicolor, Poaceae*) landrace diversity by farmers' selection in Ethiopia', *Economic Botany*, vol 53, no 1, pp79–88

Tin, H.Q., Cuc, N.H., Be, T.T., Ignacio, N. and Berg, T. (2011) 'Impacts of seed clubs in ensuring local seed systems in the Mekong Delta, Vietnam', *Journal of Sustainable Agriculture*, vol 35, no 8, pp840–854

Tripathy, S., Gurung, P. and Sharma, S.D. (2005) 'Intellectual property contributions with regard to rice genetic resources by tribes of south Orissa, India', *Plant Genetic Resources Newsletter*, vol 141, pp70–73

Tripp, R. (2001) *Seed Provision and Agricultural Development: The Institutions of Rural Change*, James Currey, Oxford, UK

Tripp, R. and Louwaars, N. (1998) 'Seed regulation: choices on the road to reform', *Food Policy*, vol 22, no 5, pp433–446

Tsegaye, B. (2005) 'Incentives for on-farm conservation in a center of diversity: the case study of durum wheat (*Triticum turgidum* L.) landraces from East Shewa, Central Ethiopia', PhD dissertation, Norwegian University of Life Sciences, As, Norway

Tuxill, J. and Nabhan, G. P. (2001) *People, Plants and Protected Areas: a guide to in situ management.* Earthscan, London

UPOV (1978) *The International Convention for the Protection of New Varieties of Plants*, of December 2, 1961, as revised at Geneva on November 10, 1972, and on October 23, 1978, http://www.upov.int/upovlex/en/conventions/1978/act1978.html, accessed 21 February 2012

UPOV (1991) *The International Convention for the Protection of New Varieties of Plants*, of December 2, 1961, as revised at Geneva on November 10, 1972, on October 23, 1978, and on March 19, 1991, http://www.upov.int/upovlex/en/conventions/1991/act1991.html, accessed 20 February 2012

Vasudeva, R., Raghu, H.B., Dasappa, Uma Shaanker, R. and Ganeshaiah, K.N. (2006) 'Population structure, reproductive biology and conservation of *Semecarpus kathalekanensis:* a critically endangered swamp tree species of the Western Ghats', in R. Uma Shaanker, K.N. Ganeshaiah and K.S. Bawa (eds) *Forest Genetic Resources: status, threats and conservation strategies,* Oxford & IBH Publishing. New Delhi, India, pp211–223

Vasudeva, R. and Hombe Gowda, H.C. (2009) 'Domestication of *Garcinia gummi-gutta* (L.) Robson: An important non-timber forest product yielding species of the Central Western Ghats', in R. Uma Shaanker, A.J. Hiremath, G.C. Joseph and N.D. Rai (eds) *Non-timber Forest Products Conservation, Management and Policy in the Tropics,* Ashoka Trust for Research in Ecology and the Environment, Bangalore, KA, India, pp65–77

Vasudeva, R., Janagoudar, B.S., Reddy, B.M.C., Sthapit, B.R. and Singh, H.P. (2010) 'Garcinia genetic resources: linking diversity, livelihood and management', in R. Vasudeva, B.S. Janagoudar, B.M.C. Reddy, B.R. Sthapit and H.P. Singh (eds) *Garcinia Genetic Resources: linking diversity, livelihood and management,* College of Forestry, Sirsi, KA, India, pp1–18

Vianna, L.P. (2008) *De invisíveis a protagonistas: populações tradicionais e unidades de conservação,* Annablume Editora, São Paulo-SP, Brazil

Vicente, N.R., Fantini, A.C., Alves, A.C. and Canci, A. (2010) 'Avaliação da efetividade do kit de diversidade', in A. Canci, A.C. Alves and C.A. Guadagnin (eds) *Kit diversidade: estratégias para a segurança alimentar e valorização das sementes locais.* Instituto de Agrobiodiversidade e Desenvolvimento Socioambiental, Guaraciaba, SC, Brazil, pp171–185

Vieira da Silva, C. and Reis, M.S. (2009) 'Produção de pinhão na região de Caçador, SC: aspectos da obtenção e sua Importância para comunidades locais', *Ciência Florestal,* vol 19, no 4, pp363–374

Vieira, A.R.R, Suertegaray, C.E.O., Heldwein, A.B., Maraschin, M. and Silva, A.L. (2003) 'Influência do microclima de um sistema agroflorestal na cultura da erva-mate (*Ilex paraguariensis* St. Hil)', *Revista Brasileira de Agrometeorologia,* vol 11, no 1, pp91–97

Visser, B., Pistorius, R. van Raalte, R., Eaton, D. and Louwaars, N. (2005) 'Options for non-monetary benefit sharing, an inventory', *Background Study Paper* no 30, Commission on Genetic Resources for Food and Agriculture, FAO, Rome, Italy, ftp://ftp.fao.org/docrep/fao/meeting/014/j6639e.pdf, accessed 12 February 2012

Vogel, J .H. (2005) 'Sovereignty as a Trojan horse: how the convention on biological diversity morphs biopiracy into biofraud' in B. Hocking (ed.) *Unfinished Constitutional Business? Rethinking indigenous self-determination.* Aboriginal Studies Press, Canberra, Australia, pp228–247

Vogt, G.A. (2005) 'A dinâmica do uso e manejo de variedades locais de milho em propriedades agrícolas familiares', Masters thesis, Federal University of Santa Catarina, Florianópolis, SC, Brazil

Vu Trong Binh and Dao Huan Duc (2007) *Geographical Indication and Appellation of Origin in Vietnam: reality, policy, and perspective,* Agricultural Publishing House, Hanoi, Vietnam

Vuotto, M.L., Basile, A., Moscatiello, V., De Sole, P., Castaldo-Cobianchi, R., Laghi, E. and Ielpo, M.T.L. (2000) 'Antimicrobial and antioxidant activities of *Feijoa sellowiana* fruit', *International Journal of Antimicrobial Agents,* vol 13, pp197–201

Wakjira, D.T. (2007) 'Forest cover change and socioeconomic drivers in southwest Ethiopia', MSc thesis, Technische Universität München, München, Germany

Walker, N. (2006) 'EU constitutionalism and new governance', in G. De Burca and J. Scott (eds) *Law and New Governance in the UE and the US*, Hart Publishing, Oxford, UK, pp 868–872

Warren, D.M., Slikkeveer, L.J. and Brokensha, D. (eds) (1995) *The Cultural Dimension of Development: indigenous knowledge systems*. Intermediate Technology, London

Wilmsen, C. (2008) 'Negotiating community, participation, knowledge and power in participatory research', in C. Wilmsen, C., W, Elmendorf, L. Fisher, J. Ross, B. Sarathy and G. Wells (eds) *Partnerships for Empowerment: participatory research for community based natural resource management*. Earthscan, London, pp1–22

Wilmsen, C., Elmendorf, W., Fisher, L., Ross, J, Sarathy, B. and Wells, G. (eds) (2008) *Partnerships for Empowerment: participatory research for community-based natural resource management*. Earthscan, London

Wilson, G.A. (2010) 'Multifunctional "quality" and rural community resilience', *Transactions of the Institute of British Geographers*, vol 35, no 3, pp364–381

Wilson, G.A. (2012) *Community Resilience and Environmental Transitions*. Earthscan, London

Witcombe, J.R., Joshi, A., Joshi, K.D. and Sthapit, B.R. (1996) 'Farmer participatory crop improvement I: varietal selection and breeding methods and their impact on biodiversity', *Experimental Agriculture*, vol 32, pp445–460

Witcombe, J.R., Joshi, K.D., Rana R.B. and Virk, D.S. (2001) 'Broadening genetic diversity in high potential production systems by participatory varietal selection', *Euphytica*, vol 122, pp575–588

Worede, M. and Mekbib, H. (1993) 'Linking genetic resource conservation to farmers in Ethiopia', in W.S. de Boef, K. Amanor, K. Wellard and A. Bebbington (eds) *Cultivating Knowledge. Genetic diversity, farmers' experimentation and crop research*. Intermediate Technology, London, pp78–84

Worede, M., Tesemma, T. and Feyissa, R. (1999) 'Keeping diversity alive: an Ethiopian perspective', in Brush, S.B. (ed.) *Genes in the Field: on-farm conservation of crop diversity*. Lewis, Boca Raton, FL, USA, pp143–163

World Bank (2001) *World Development Report 2001/2002: Attacking poverty*. Oxford University Press, New York

WTO (World Trade Organization) (1994) *Agreement on Trade-related Aspects of Intellectual Property Rights*, http://www.wto.org/english/tratop_e/trips_e/t_agm0_e.htm, accessed 16 February 2012

Zaharia, Z. (2008) 'De la relance du blé Meunier d'Apt à la création d'un réseau local valorisant les variétés anciennes', in H. Zaharia (ed.) *Voyage autour des blés paysans*, Réseau Semences Paysannes, Brens, France, pp91–98

Zaharia, H. (2009) 'Expériences paysannes', in P. Gaudin, H. Zaharia, F. Delmond, G. Kastler, S. Mouène, V. Chable and J. Kendall (eds) *Variétés paysannes de maïs et tournesol pour une agriculture écologique et économe, Cahier technique*. Réseau Semences Paysannes, Brens, France, pp80–82

Zank, Z. and Hanazaki, N. (2012) 'Exploring the links between ethnobotany, local therapeutic practices, and protected areas in Santa Catarina coastline, Brazil', *Evidence-Based Complementary and Alternative Medicine*, vol 2012, Article ID 563570, http://www.hindawi.com/journals/ecam/2012/563570/, accessed 28 January 2012

Index

.